DISCOVERING
ASTRONOMY

DISCOVERING ASTRONOMY

ROBERT D. CHAPMAN

NASA-Goddard Space Flight Center, Greenbelt, Maryland

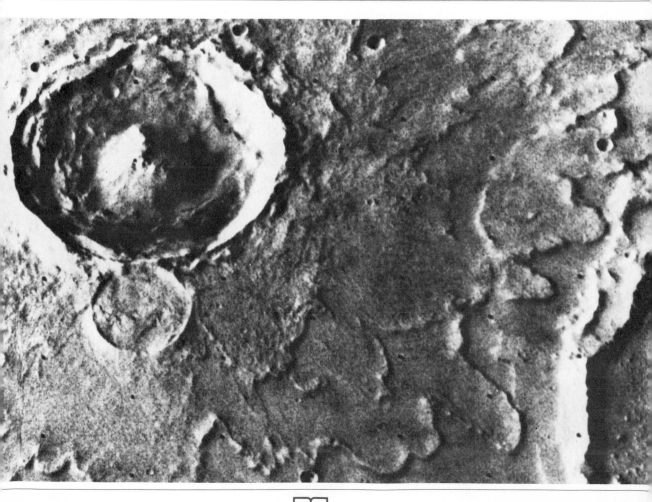

W. H. Freeman and Company
San Francisco

Indication of the author's affiliation with the National Aeronautics and Space Administration on the title page of this book does not constitute endorsement of any of the material contained herein by NASA. The author is solely responsible for any views expressed.

Library of Congress Cataloging in Publication Data

Chapman, Robert DeWitt, 1937–
    Discovering astronomy.

    Includes index.
    1. Astronomy. I. Title.
QB45.C48   520   77-16024
ISBN 0-7176-0034-4
ISBN 0-7167-0033-6 pbk.

Printed in the United States of America

9 8 7 6 5 4 3 2 1

To Susan, Jennifer, and Christopher

CONTENTS

Preface   xi

1   **OUR PLACE IN THE UNIVERSE**   1
What Is Astronomy?   3
Why Study Astronomy?   3
How Do We Study Astronomy?   4
What Will We Study?   7

2   **THE SKY**   9
Risings and Settings   10
The Constellations   15
The Annual Motion of the Sun   20
Wanderers: Motions of the Planets   22
Time and the Calendar   25
Review Questions   29

3   **THE HISTORICAL DEVELOPMENT
OF GRAVITATIONAL ASTRONOMY**   31
The Motions of the Planets   32
The Beginning of Modern Science   46
Review Questions   58

4   **LIGHT AND THE TELESCOPE**   59
The Nature of Light   60
Telescopes   70
Detecting Light   86

Invisible Radiation 88
The Inverse-Square Law 93
Review Questions 94

5   THE EARTH AS A PLANET   95
The Revolution of the Earth 96
The Rotation of the Earth 102
Dimensions of the Earth 108
The Interior of the Earth 114
Review Questions 120

6   THE MOON   121
The Phases of the Moon 123
Distance, Size, and Motion of the Moon 128
Eclipses of the Moon 132
The Surface of the Moon 135
The Origin of the Lunar Surface Features 149
Review Questions 155

7   ATOMS AND LIGHT   157
The Spectrum 158
The Continuous Spectrum 163
Atoms and Light 169
The Doppler Effect 177
Review Questions 179

8   THE SUN   181
The Visible Sun 182
Solar Rotation 189
Solar Power Output 191
Solar Structure 194
The Photosphere 199
Sunspots and the Sunspot Cycle 201
Monochromatic Solar Images 205
The Chromosphere 206
The Corona 211
The Sun from Space 216
Flares 218
The Theory of Solar Activity 220
Review Questions 221

9   THE PLANETS   223
Planetary Dimensions 224
Mercury 226

Venus   231
Mars   238
Jupiter   252
Saturn   257
Uranus and Neptune   261
Pluto   262
Review Questions   264

10   **THE SOLAR SYSTEM**   265
The Form of the Solar System   266
Asteroids   268
Meteoroids, Meteors, and Meteorites   273
Comets   281
The Origin of the Solar System   292
Review Questions   295

11   **THE SUN AND THE EARTH**   297
Solar Photons   299
The Earth's Atmosphere   301
Solar Wind   307
The Magnetosphere   308
Solar Flares and the Earth   314
The Maunder Minimum   315
Solar–Terrestrial Relations   316
Review Questions   317

12   **INTRODUCING THE STARS**   319
Descriptions   320
Stellar Distances   326
Stellar Motions   328
Stellar Distances Again   333
Absolute Magnitude   333
Stellar Spectra   334
The Hertzsprung–Russell Diagram   341
Stellar Sizes   343
Binary Stars and Stellar Masses   345
Mass-Luminosity Relation   351
Common Stars   352
Cepheid Variable Stars: Stellar Mileposts   353
Review Questions   356

13   **DISCOVERING OUR GALAXY**   359
Star Clusters   360
Interstellar Matter   366

The Galaxy 378
Stellar Populations 388
The Center of the Galaxy 389
Review Questions 390

14   **LIFE HISTORIES OF STARS**   **391**
The Internal Structure of a Star 393
Stellar Evolution 398
Late Stages of Stellar Evolution 403
Supernovae, Pulsars, and Neutron Stars 414
The Birth of a Star 424
Review Questions 428

15   **THE UNIVERSE**   **431**
Galaxies 432
The Distance Scale 446
The Expanding Universe 447
Radio Galaxies 449
Cosmology 458
Review Questions 465

16   **LIFE IN THE UNIVERSE**   **467**
How Many Planets Are There? 468
The Origin of Life 469
How Many Civilizations Are There? 476
Interstellar Communication 477
Interstellar Travel 481
Review Questions 482

Appendix A   The Skinny Triangle   483
Appendix B   Powers of Ten   487
Appendix C   Temperature Conversion   489
Appendix D   The Elements   490
Appendix E   The Greek Alphabet   493
Appendix F   Star Charts   494

Glossary   501

Index   511

PREFACE

Astronomy is a vital, exciting field that continually inspires scientists to push outward the limits of our understanding of the physical universe. New developments have been occurring at an ever-increasing pace in recent times. As this preface is being written, the Viking Spacecraft are sitting on the surface of Mars, sending us a steady stream of information about that planet. In the coming years, our understanding of Mars will surely change, as it has in the last few years. Topic after topic is in a state of rapid development. This state of affairs makes astronomy one of the most interesting fields to be working in, but at the same time makes writing a textbook frustrating. It is inevitable that some parts of this (or any) book will be out of date by the time the ink has dried.

My aim in writing this book has been to produce a text that can be used for a one-semester, college-level course in astronomy for students who are not physical science majors. Like any technical subject, astronomy has its specialized vocabulary, terms that are essential for the unambiguous communication of scientific concepts. These terms have been defined and used where appropriate in the text. Otherwise, I have tried to keep the language as simple as possible; jargon for its own sake has been avoided.

Many teachers have experimented with an outside-in approach to astronomy. They begin the subject by looking at the overall structure of the universe, then focus on its parts, working from the large to the small, until they finally reach the Earth. This approach can be developed into a logically consistent story. I like the historical approach to teaching science. Astronomy did not develop outside-in. People first became aware of their immediate surroundings—the Earth and the sky. Among the first attempts to quantify our knowledge of the universe were theories of the motions of the planets: we became aware of the solar system at an early stage. Our present picture of our Galaxy began to develop in the eighteenth century with William Herschel. However, it was not until the twentieth century that everything jelled into our current concept of the Galaxy and its place in the universe. Historically, we started with the Earth and worked our way out to the universe. I have chosen this traditional approach in presenting the science of astronomy here.

I owe a debt of gratitude to many colleagues for help with the preparation of the text and illustrations. For illustrations, I would like to offer hearty thanks to F. J. Betts, R. Davis, Jr., R. B. Dunn, R. M. Goldstein, A. S. Krieger, J. C. Lopresto, G. A. Newkirk, H. H. Nininger, G. Westerhout, and J. P. Wild.

Several colleagues have read all or part of the manuscript and have made valuable suggestions for its improvement. I would particularly like to thank Larry Toy and Joseph Miller, who read the entire book with great care. Their suggestions have been extremely useful to me. E. D. Ortell and R. N. Hartman have each read sections of the manuscript and have made many helpful suggestions. The main burden of the typing of the text was ably carried out by Bonnie Noble. She deserves a medal for managing to read my handwriting.

Last, but far from least, I am happy to record my thanks to my family for moral support and understanding while I was locked up in my workroom so many weekends and evenings. My wife Susan has provided assistance with typing, proofreading, grammatical usage, and all manner of other details of putting the whole thing together.

September 1977   Robert D. Chapman

DISCOVERING
ASTRONOMY

1

OUR PLACE
IN THE UNIVERSE

To the Egyptians the universe was a box. A bowl-shaped Earth, with Egypt at its center, formed the floor of the box, and the sky its roof. This seemingly naive concept of the universe makes an assumption that is common in the mythologies of early civilizations: that the abode of mankind is the center of the universe. If we go outside on a clear, moonless night, far from the pollution and lights of our cities, we begin to understand the origin of those early ideas: the sky does appear to be a star-studded overturned bowl with us standing at its center. Our senses lead us to believe that we are the center of all we see.

The first serious doubt of our early concept of our place in the universe came in the Renaissance, when it was suggested that the Sun, not the Earth, was the center of the universe. Earlier thinkers had put forward a similar idea, but it was not until the rebirth of intellectualism in Europe that the thought gathered momentum. The change came about because of our scientific study of nature. We no longer merely go outside and sense the beauty of the night-time sky. We satisfy our innate human curiosity by studying the Sun, the Moon, the planets, and the stars in more detail. This close scrutiny reveals that their beauty is more than skin deep. Nature remains fascinating and lovely as we probe more and more deeply to discover her secrets.

Today, we believe that Earth occupies an insignificant corner of the universe. Even more importantly, we have become aware that we may be only one of many intelligent creatures inhabiting the cosmos. In fact, there may be life forms out there that are as far advanced beyond us as we are beyond the manlike creatures that walked the Earth a million years ago. Our aim in this volume is to describe the universe as we understand it today, and to trace the course of scientific discoveries that led to this understanding.

WHAT IS ASTRONOMY?

What is astronomy? It is the study of the nature of the universe. Speaking loosely, we can think of astronomy as physics applied to problems outside the Earth. The principles that are applied are the same as those discovered by physicists in their terrestrial laboratories. We will see, as this book unfolds, that there has been considerable cross-fertilization between physics and astronomy during recent centuries. The thinking that led to the modern science of mechanics, for instance, began in part with attempts to predict the observed motions of the planets.

What this discussion boils down to is, simply, that we must understand some of the fundamentals of physics before we can understand astronomy. There are places in this book where we must talk about physics. In each case, the digressions into physics are essential for our understanding of the astronomical topics.

Of course, we must have something to which we can apply the physical principles. That something is the body of observed facts about the universe. That body of facts is immense. However, only a specialist need wade into it in its full detail. Our aim in this book is to present a selected sample of facts covering a selected sample of astronomical topics. The selection is designed to give the flavor of the entire field, to describe some of the big problems that astronomers are trying to solve today, and to illustrate how science in general is done.

WHY STUDY ASTRONOMY?

Why study astronomy, or physics, or geology? The increase of knowledge for its own sake will not buy so much as a cup of coffee. Why do poets write their verses? Why do conservationists work so hard to save whales? Mankind can exist without poems. Whale oil no longer lights our lamps. The answer to these questions is simple. When we increase our knowledge of nature or compose poetry or save a species from extinction, we enrich our daily lives. We contribute to our cultural heritage. In the long run we make progress. No civilization has thrived for long without seeking knowledge for its own sake.

Most astronomers are intensely curious men and women. They are motivated first and foremost by this curiosity. The sum total of the information in this book has been learned because of the studies of these inquiring minds. I hope, as you read it, that you, too, will become curious about nature, and will continue to delve into her secrets beyond these few brief pages.

HOW DO WE STUDY ASTRONOMY?

There is a myth that there exists a profound thing called the scientific method, which is a technique for studying nature that leads its practitioner successfully from great discovery to great discovery. This myth is amplified if one reads scientific papers published in professional journals. The majority of these papers are beautifully logical, step-by-step descriptions of a problem and its solution.

In reality, the scientist at work has much in common with the detective. Both the scientist and the detective must continually be on the alert for important clues. Both must work very hard to find the needed clues, and must expect to follow many incorrect leads, before they can unravel a mystery. Both the scientist and the detective must keep open minds. The detective cannot afford to believe everything he is told by suspects in the case. The scientist doubts all traditional explanations, and seeks new and novel ways to look at phenomena. Typically, the great scientific theories have flown in the face of tradition. We will see examples of this in the work of Copernicus, Galileo, and Einstein in physics and astronomy. Other scientific fields have had similar thinkers. Men like Darwin and Pasteur revolutionized biological science with very unorthodox theories for their day. We must bear in mind that, for every Darwin with a theory that changes the fabric of scientific thought, there are thousands of brilliant scientists who lay the foundations of human thought.

Einstein himself stated the most important tool of the scientist when he said that discovery favors the prepared mind. The greatest progress is made by individuals who thoroughly understand their subject.

To sum up, the so-called scientific method is keeping alert, keeping an open mind, and keeping aware of progress in scientific research and in technological developments that might provide new research tools.

When Sherlock Holmes summed up his cases with, "It's elementary, my dear Watson," he always gave a brilliantly logical description of the major steps that led to his solution of the case. He didn't review all the detailed work that led to the solutions. Published

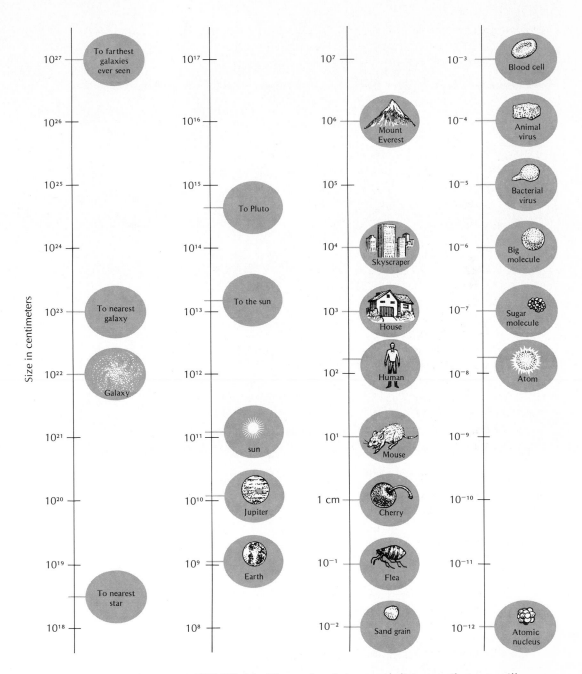

FIGURE 1.1 The scale of sizes and distances that we will encounter throughout our studies of the universe. (From *New Horizons in Astronomy* by J. C. Brandt and S. P. Maran. W. H. Freeman and Company. Copyright © 1972.)

FIGURE 1.2   The Whirlpool Galaxy in the constellation Canes
Venatici. This gigantic celestial wonder is composed of stars, gas,
and dust. (Lick Observatory photograph.)

scientific papers are like that summation: *It's elementary, my dear colleagues . . .*

We must hasten to add that the chief reason why current research papers are written in this summary way is that so many papers are now being published. If the authors were permitted to ramble on about their research efforts, the libraries could not hold the volume of published material. In fact, papers published a century ago tended to be more informal, with the author describing his entire thought process, including unsuccessful thoughts. If you are near a large library with scientific journals from the nineteenth century, you might find it amusing to read some of the older papers.

WHAT WILL WE STUDY?

The universe exists on a number of scales that cover an immense range (Figure 1.1). When one talks about things as small as the nuclei of atoms or as large as galaxies, the discussion must take on an abstract quality. We can and will define our size scales very precisely. But it is not easy to close our eyes and imagine the scales of these objects at the extremes of the range of sizes.

Light travels 300,000 kilometers each second. At that speed, one could circumnavigate the Earth seven times in a second. Yet at that speed a light beam would take 100,000 years to cross our galaxy. The size of a galaxy is so far beyond our everyday experience that it is virtually unimaginable (Figure 1.2). Planets and their moons; comets, meteoroids, and asteroids; stars and star groups; and great clouds of glowing gas all lie between atoms and galaxies in scale. We will examine all these objects in enough detail to discover the importance of the niche that each occupies in the cosmos.

Size is not the only quantity that takes on unimaginable extremes. We will talk about the interiors of stars, where temperatures reach tens or even hundreds of millions of degrees. We will talk about the beginning of our Sun, our Earth, and the universe itself, and encounter time-scales of billions of years. We will talk about objects made of material that is extremely dense, and with immense gravitational pulls at their surfaces, where a human being would be crushed under his or her own weight.

Let's get on with it. We will start at home and work our way to the very edge of the universe.

2

THE SKY

It is very difficult to find words to describe the beauty of the clear night sky. And, unfortunately, it is becoming increasingly hard to experience this beauty firsthand. However, on a clear, moonless night one can go into the desert or to a high mountain meadow, far from the smoke and lights of civilization, and look up at the sky. One's first impression is that the sky is filled with an uncountable multitude of stars. If it is summer, the Milky Way can be seen stretching from horizon to horizon. Perhaps the brightest points of light are not stars at all, but are one or more of the bright planets. When one comes to be on intimate terms with the sky, it is easy to decide whether a point of light is a planet or a star. Even on the clearest of nights, there are fewer than two thousand stars visible at any instant, and they fall into easily recognized patterns called **constellations**. It can be a rewarding experience to become familiar with these constellations.

For humans who lived millennia ago, intimate knowledge of the sky was more than a rewarding personal experience; it was a matter of survival. The sky was the calendar, and people timed their wanderings and, in later ages, their agriculture by the positions of the stars. Three thousand years before Christ, long before Homer wrote his great epics, much was already known about the sky. And you or I can easily rediscover this knowledge by observing the sky night after night. What are the basic facts? In this chapter we will answer that question. Do go out and observe the sky yourself, and see how many of the ideas you can verify independently.

RISINGS AND SETTINGS

Certainly one of the most obvious phenomena of the celestial sphere is the rising and setting of the Sun. Nearly as obvious is the

FIGURE 2.1 Polar star trails. This photograph was made by pointing a tripod-mounted camera at the north celestial pole and opening the shutter for about eight hours. As the Earth rotated under the sky, the apparent motion of the stars trailed their images. You can easily make a similar photograph. Start with a ten- or fifteen-minute exposure, unless you live far from city lights and your sky is very dark. (Lick Observatory photograph.)

fact that most other celestial objects rise and set as well. The **apparent motion** of the celestial sphere, which causes these risings and settings, we can call its daily or **diurnal** motion.

The early Greeks regarded the sky as a crystalline sphere—the celestial sphere—to which the stars are attached. The diurnal motion of the sky was thought to be a simple rotation of this gigantic globe, which carried the attached stars across the sky from east to west. We can easily discover that the point in the northern-hemisphere sky about which the celestial sphere appears to pivot (Figure 2.1) is near the so-called north star, Polaris, which is situated at the end of the handle of the constellation known as the Little Dipper. During the course of a night, the Little Dipper can be observed to swing about the end of its handle like the hour hand of a clock (Figure 2.2). The two points about which the sky rotates are called the celestial poles—and the celestial pole near Polaris is the **north celestial pole**, for it lies directly above the north pole of the Earth (Figure 2.3).

If you were standing shivering at the north pole looking at the sky, the north celestial pole would appear directly overhead. You would then see all the stars move around the sky in circles centered at the **zenith**, the point directly overhead (Figure 2.4). Therefore you would see no stars rise or set. Each star would swing around the sky, maintaining a constant distance from the horizon. On the other

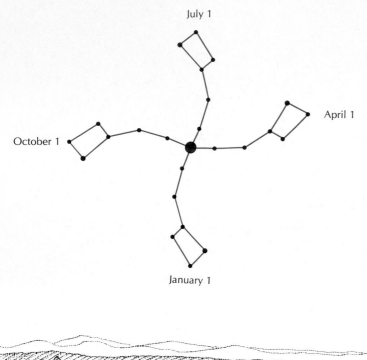

July 1

April 1

October 1

January 1

FIGURE 2.2   The Little Dipper at four different seasons. The figure shows the orientation of the Little Dipper at 8:00 in the evening on the dates indicated.

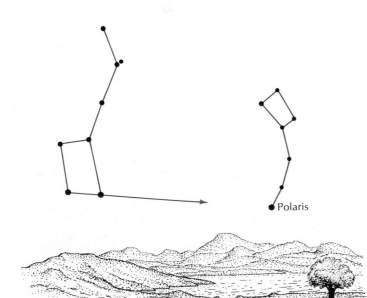

Polaris

FIGURE 2.3   The end two stars in the bowl of the Big Dipper point to the pole star.

FIGURE 2.4 The apparent rotation of the sky as seen by an observer at the North Pole. The stars appear to move around the sky, keeping the same distance from the horizon.

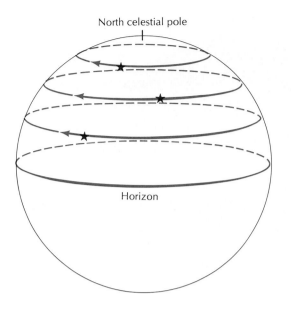

North celestial pole

Horizon

hand, if you were in the midwestern United States (Figure 2.5), you would see the north celestial pole roughly halfway between the horizon and the zenith. (We will be more precise on this point shortly.)

Some stars, like the stars of the Little Dipper, never rise or set. As the sky rotates, sometimes the Little Dipper is higher than the north celestial pole, sometimes lower—but it is always above the horizon. We call the stars that do not rise or set, but are always in the sky, **circumpolar** stars. All the stars that you could see in the sky when you were standing at the north pole of the Earth were circumpolar.

What happens if you travel south through the U.S., Mexico, and Central America, say, on the Pan-American Highway, observing the sky as you go? You would find that the north celestial pole became nearer to the horizon as you traveled south, and fewer and fewer stars were circumpolar (Figure 2.6). As you crossed the equator near Quito, Ecuador, the celestial pole would be on the horizon and no stars would be circumpolar. All the while that you traveled, you had to look toward the north to see the celestial pole. Now as you travel further south, beyond the equator and toward the southern tip of South America, the south celestial pole rises higher and higher in the sky, and you must look toward the south to see it. As

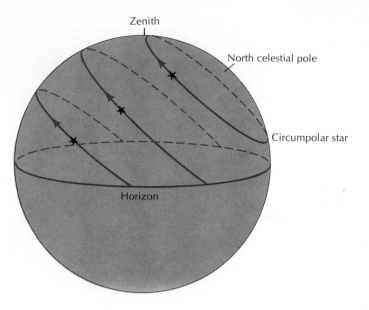

Zenith

North celestial pole

Circumpolar star

Horizon

FIGURE 2.5 The apparent rotation of the sky as seen from the latitude of Washington, D.C. (39° N).

you move further south, more stars become circumpolar. These are not the stars you saw in the northern hemisphere, however. Instead, these are the south circumpolar stars. In contrast to the north, no bright star marks the location of the south celestial pole, and the pole is therefore more difficult to find. From Tierra del Fuego, you can cross to the forbidding Antarctic to reach the Earth's south pole and see the south celestial pole at the zenith.

All the foregoing discussion points out a fact well-known to mariners, and well-used by them, before the day of modern electronic navigation beacons. The angular distance that the celestial pole stands above the horizon—its **altitude**—is equal to the observer's latitude. At the poles your latitude is 90° north or south, and the north or south celestial pole is directly overhead, 90° from the horizon. At the equator, your latitude is 0°, and the celestial poles are right on the horizon. In the Midwest, say, near St. Louis, your latitude is roughly 38°, and the north celestial pole is 38° above the horizon—and any star that is less than 38° away from the pole is circumpolar.

In this section we have talked about two facts: first, that celestial objects seem to rise and set because of the daily motion of the celestial sphere; second, that the celestial sphere rotates about poles,

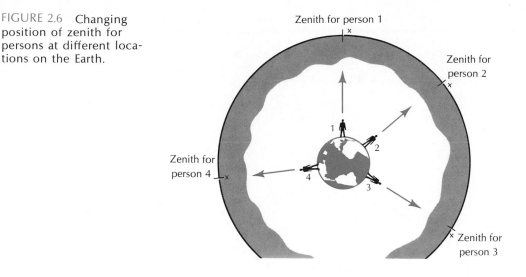

whose position in the sky depends on the observer's latitude. By the fourth century B.C., when Aristotle wrote his many works on natural philosophy, the Greeks had dropped the notion that the celestial sphere was made of crystal. Aristotle considered it to be an imaginary sphere. However, even today, when one wants to discuss apparent positions and motions in the sky, as we have in this section, the celestial sphere remains a useful concept.

THE CONSTELLATIONS

When people looked at the sky ages ago, they saw their legends embodied in the patterns formed by the stars. Today, we recognize 88 figures, the **constellations**, on the celestial sphere. These are simply chance groupings of stars that human imagination has pictured as a bull, a unicorn, a beautiful princess, or one of innumerable other creatures (Figure 2.7).

The myths associated with the constellations make interesting stories, and can sometimes help one remember the figures that are grouped together in an area of the sky. For instance, there is a myth which ties together the constellations Perseus, Pegasus, Andromeda, Cepheus, Cassiopeia, and Cetus. These constellations are

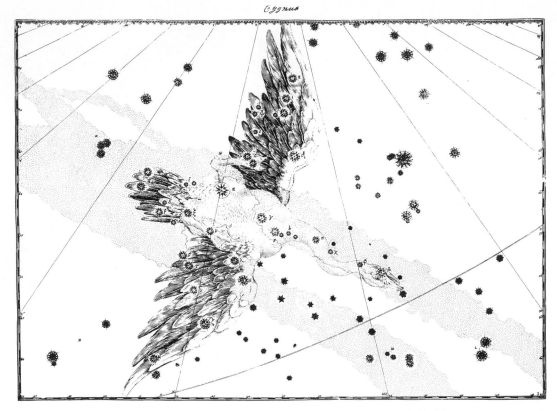

Cygnus

FIGURE 2.7 The constellation Cygnus as portrayed in the *Uranometria*, a celestial atlas published by J. Bayer in 1603. (Yerkes Observatory.)

grouped together and are conspicuous in the sky for northern observers, in the late fall, around 9:00 P.M. (Figure 2.8).

Perseus slew Medusa, the terrible monster, who was so frightening that a mortal looking at her would be turned to stone. Her hair, the story tells, was a mass of writhing, hissing snakes. To execute the sword stroke which severed Medusa's head, Perseus looked at her reflection in his highly polished shield and was not petrified. Legend says that as Medusa's blood soaked into the ground, Pegasus, the winged horse, sprang forth. Perseus mounted Pegasus and went flying off home, carrying the terrible head in one hand.

Meanwhile, another drama was unfolding in a nearby ancient kingdom. Cassiopeia, the queen, who was an overly vain woman, dared to openly compare her beauty to that of the Nereids (sea nymphs). Neptune, the father of the Nereids, became angry and

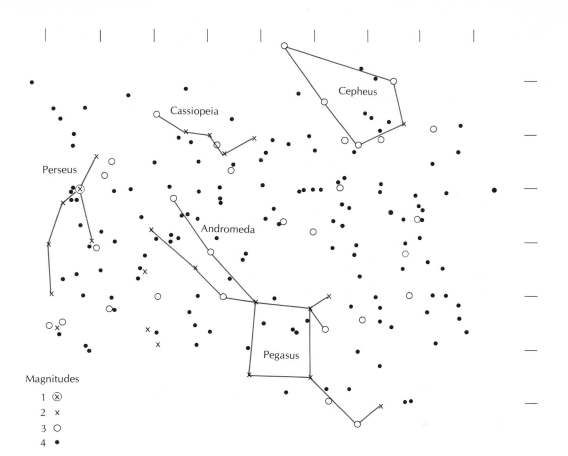

Magnitudes

1 ⊗

2 ×

3 ○

4 •

FIGURE 2.8 The portion of the night sky which includes the constellations Perseus, Andromeda, Cassiopeia, Pegasus, and Cepheus. All the stars portrayed are easily visible to the naked eye under clear, dark-sky conditions.

sent Cetus, a sea monster, to wreak havoc on the seacoast to repay Cassiopeia for her arrogance. Cepheus, the king, learned in a dream that if he and Cassiopeia sacrificed their daughter Andromeda to Cetus, the gods would be appeased. So poor Andromeda was chained to a rock at the edge of the sea, to await the horrible monster. Just as Cetus crawled from the depths to devour her, however, Perseus came winging on the scene. He slew Cetus, unchained Andromeda, and lifted her onto Pegasus, and they rode off into the sunset to live happily ever after. And that, briefly, is one of the many myths.*

---

* *The Age of Fable,* by Thomas Bulfinch (1796–1867), gives a readable version of many other classic myths related to the constellations.

Earlier we mentioned that the group of constellations in our story could be seen in late fall around 9:00 P.M. Why do we have to be specific about time and date? There are two reasons, the first of which we discussed in the last section. That is, the celestial sphere rotates once daily, causing some stars to set and others to rise, and in the process the constellations seem to move westward across the sky. An observer at the latitude of, say, St. Louis, would see Andromeda approximately at his zenith at 9:00 P.M. in mid-November. If he were to look at the sky again roughly six hours later, at 3:00 A.M., Andromeda would just be setting. The constellation Gemini would be near the zenith.

The second reason why we must be specific about time and date has to do with an annual effect. Each day a given star rises four minutes *earlier* than it did the day before, reaches its highest point in the sky four minutes *earlier,* and sets four minutes *earlier.* Thus at 9:00 P.M. tonight the sky has the same orientation it had at 9:04 P.M. last night; that is, at a time four minutes later last night. One month ago the same orientation of the sky as seen at 9:00 P.M. tonight was seen $4 \times 30 = 120$ minutes or two hours later, that is, at 11:00 P.M. We will return to the reasons for this effect in the next section.

To become truly familiar with the sky, then, one must learn the shape of each constellation and its position with respect to neighboring constellations. If, some winter night, you note that Orion can be seen over a particular building at 9:00 P.M., you can be sure that it will not be there at midnight, or at 9:00 P.M. in the spring. Therefore learning the constellations relative to the landscape is sure to fail. So what is the best way to learn the constellations?

There is no single answer to this question. But the Big Dipper is a good place to start. First, locate a dark field or other area that is away from bright lights and that has an unobstructed view of the northern horizon. Remember, the celestial pole is at an angular height above the horizon equal to your latitude. In the fall, the Big Dipper will be near the northern horizon, directly below the pole. This is the time of year when the Dipper is most difficult to find. In January, the Dipper stands on its handle, to the right of the pole as you face north; by April the Dipper is directly above the pole; and in July the Dipper stands handle upward to the left of the pole as you face northward (see Figure 2.9).

Once you have located the Big Dipper, you can easily find the Little Dipper—if the sky is clear and dark. The two end stars in the bowl of the Big Dipper (opposite the side where the handle attaches) point to the north star (see Figure 2.3). The Little Dipper

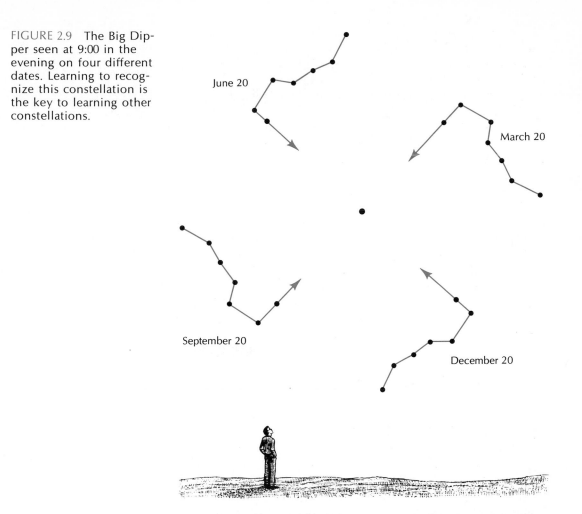

June 20

March 20

September 20

December 20

roughly parallels the Big Dipper, with the handle pointing opposite to the Big Dipper's handle. If the sky is the least bit overcast or bright, the Little Dipper will be difficult to find. You might see only Polaris and the two extreme opposite bowl stars. The stars in between are somewhat fainter and are hard to see under adverse conditions.

Having found the Dippers, you might next try for Cassiopeia, which is a w-shaped constellation, roughly on the opposite side of the Little Dipper from the Big Dipper and about as far away. With these landmarks learned, you can take one of the star charts in this

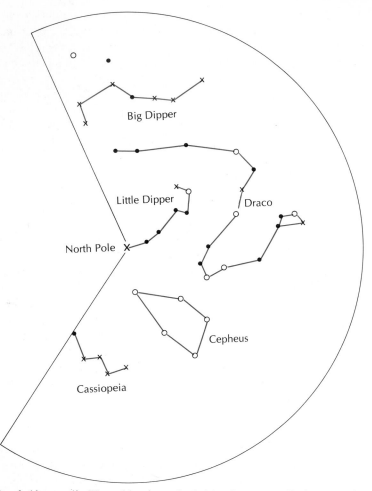

book (Appendix F) and begin to find the other constellations one by one (Figure 2.10). We must stress again, learn them as you would a map, that is, in their relationships with one another. In that way, no matter what the season or the time, you can look at the sky on a clear night and recognize the stars.

THE ANNUAL MOTION OF THE SUN

In the last section, we described the fact that each star rises four minutes earlier on successive nights, causing the constellations to slowly shift their positions westward when viewed at a fixed time of night. We pointed out that, in one month, this shift amounts to the distance the sky rotates in two hours. If we continue this argument, we arrive at the conclusion that in six months the shift amounts to the distance the sky rotates in twelve hours. Since the sky rotates completely in twenty-four hours, it rotates half a turn in twelve hours. The stars we see in the sky at midnight tonight will be in the sky at noon six months from now—and were in the sky at noon six months ago. This means, of course, that six months from now the Sun will be in the part of the celestial sphere that is seen at nighttime today.

The slow westward shift of the sky night after night is due to the **annual motion** of the Sun. The Sun moves roughly 1° eastward among the stars each day. (Actually, the Sun moves 360° in 365¼ days—but this is close enough to 1° per day for our purposes in this section.) Furthermore, since the sky rotates 360° in 24 hours because of its diurnal motion, it rotates 15° in one hour or 1° in 4 minutes. Later in this chapter we will describe how time is determined by the position of the Sun in the sky. The day is the length of time it takes the celestial sphere to rotate once, taking the Sun completely around the sky. In that day, however, the Sun will move 1° eastward among the stars, that is, in the direction opposite to the westward diurnal rotation of the celestial sphere. Therefore, the time it takes a particular star to rotate completely around the sky once is 23 hours and 56 minutes; that is, 4 minutes less than the Sun takes; so the star will rise 4 minutes earlier tonight than it did last night.

It is by just this type of reasoning that the Greek philosophers arrived at the annual motion of the Sun among the stars. But then they went a step further. They noted the position of the Sun in the sky at some hour, and asked themselves the question: what constellation was in that position twelve hours later and six months ago? The answer to their question told them which constellation the Sun was situated in at the time when they noted its position. In such a manner the Greeks discovered the path of the Sun's annual motion among the stars.

We call the Sun's path around the sky the **ecliptic**. The ecliptic is the centerline of a band around the celestial sphere known as the **zodiac**. The zodiac is divided into twelve equal parts, each represented by a constellation. The twelve constellations are Aries, Taurus, Gemini, Cancer, Leo, Virgo, Libra, Scorpius, Sagittarius, Capricornus, Aquarius, and Pisces. The **sign** of the zodiac for a given date is the constellation in which the Sun is found on that date.

In this section we have talked about the annual motion of the Sun around the sky and described the path it follows. We have shown how this information can be inferred from a study of how the constellations change position with the passage of time.

We have been glibly talking about the daily rotation of the sky and the annual motion of the Sun so far. From our modern vantage point, we know that these motions are only apparent, caused by real motions of the Earth. However, it took many centuries and much human suffering to bring us to this state of knowledge. In the next chapter we will trace the history of this important intellectual step, after laying further groundwork in the following sections.

WANDERERS: MOTIONS OF THE PLANETS

Even during hundreds of years, the shapes of the constellations remain unchanged. The stars do not appear to move relative to one another—they appear to be fixed to the celestial sphere. The ancients recognized seven exceptions, objects they called **planets** or "wanderers." These are the Sun, the Moon, Mercury, Venus, Mars, Jupiter, and Saturn. These planets move around the sky relative to the stars and constellations, remaining near the ecliptic, within the constellations of the zodiac.*

The Moon is the most rapidly moving celestial body as seen from the Earth. In fact, in one hour it appears to move its own diameter eastward along the ecliptic, moving 13° in one day, and completing a circuit around the entire sky in about a month. As a result of this rapid motion, the Moon rises about 50 minutes later each night than it did the night before. Of course, we also can easily observe that the Moon changes its apparent shape from night to night: it exhibits phases, which it cycles through in a month.

The motions of the five planets Mercury, Venus, Mars, Jupiter, and Saturn are more complex than the motion of the Sun and the Moon. Furthermore, Mercury and Venus move differently from Mars, Jupiter, and Saturn. We will discuss the motions of these two groupings separately.

---

*Incidentally, notice how three of our modern names for the days of the week come from this list of the names of planets: *Sun*day, *Mon*day, . . . *Satur*day. The names for the other four days come from Teutonic gods: Tiev, Woden, Thor, and Freya. Of course, our modern usage of the term planet does not encompass the Sun and the Moon. Since the Greek word *planetos* simply meant "wanderer," it was used to contrast the seven bodies which appeared to move against the background of the apparently unmoving, "fixed" stars.

Mercury and Venus never move far away from the Sun. They can be seen in the morning sky, rising before the Sun, or in the evening sky, setting after the Sun. Depending on whether it is seen in the morning or evening, Venus is frequently called the **morning star** or **evening star**. The Greeks called the morning star Phosphorus, the evening star Hesperus, and at first thought them to be two separate objects.

In a period of 584 days, Venus swings from one side of the Sun to the other and back, in the process passing close to the Sun and becoming difficult to view for a short period. Mercury makes the same swing back and forth in 116 days.

Venus is never more than 47° from the Sun, which means that, when the Sun is just setting, Venus is *at most* 47° above the horizon; that is, at most halfway between the horizon and the zenith. However, the ecliptic (as we will see later) is typically not perpendicular to the horizon, but at an angle. As a result, the altitude of Venus, even when it is at maximum distance from the Sun, is therefore usually less than 47°. These same comments apply also to Mercury, except that Mercury never gets more than 28° away from the Sun—and is therefore usually less than 28° away from the horizon. Mercury is frequently difficult to see: it is often lost in the glare of the setting Sun, or in haze near the horizon. Of course, the two planets spend only a small part of the time at maximum distance from the Sun. The remainder of the time is spent moving from the extreme position on one side of the Sun to the extreme on the other side.

Venus is the brightest object in the sky, after the Sun and Moon. It is so bright that it can be seen in broad daylight—if you know where to look. When Venus is at maximum brightness, it is about ten times brighter than Mercury is at maximum brightness. A beautiful sight to behold is Venus and the crescent Moon near one another in the west just at sunset on a crisply clear night. On rare occasions one can see Venus, Mercury, and the Moon clustered together in the darkening sky—a spectacular sight indeed.

The motion of Mars, Jupiter, and Saturn is considerably different from that of Mercury and Venus. In the first place, these three planets can lie anywhere along the ecliptic. They are not constrained to keep within a certain distance of the Sun. In fact, the situation when one of these planets is on the opposite side of the celestial sphere from the Sun is given a name: it is called **opposition**. Most of the time, Mars, Jupiter, and Saturn move like the Sun and Moon; that is, they move slowly eastward along a path that is close to the ecliptic. However, periodically they do a little "dance" known as **retrograde motion**. During a period of retrograde

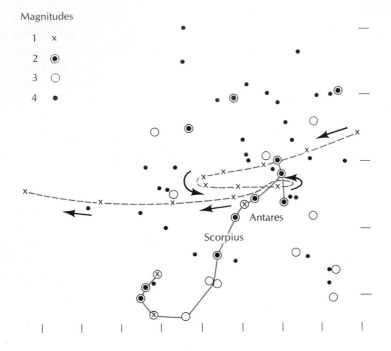

Magnitudes

1 x
2 ◉
3 ○
4 •

FIGURE 2.11 The motion
of Mars from February to
October 1969. The plot-
ted points indicated by
x mark the position of
the planet at 20-day in-
tervals. When Mars is
near the star Antares, the
two bodies are almost
undistinguishable in
brightness and color. An-
tares is Greek for "rival
of Mars."

motion, the planet stops its slow eastward motion and begins to
move slowly westward; that is, its motion is backward. While it
moves westward, it also moves away from the ecliptic. Finally the
westward motion ceases, and the direct, normal motion begins
again. Meanwhile the planet has also moved back near the ecliptic.
If one plots the planet's motion on a star map, the retrograde
"dance" looks like a small loop in the otherwise nearly straight path
(Figure 2.11). The retrograde motion occurs when each planet is
near opposition. The period of time between retrograde loops is
780 days for Mars, 399 days for Jupiter, and 378 days for Saturn.

In this section, we have described the motions of the planets, and
have contrasted the motions of Mercury and Venus, on the one
hand, with the motions of Mars, Jupiter, and Saturn, on the other
hand. As was mentioned at the end of the last section, we are only
discussing appearances: things any astute observer could discover.
We will defer explanations until the next chapter.

Having discussed many motions so far, we now come to the most
practical application of a knowledge of the heavens: keeping time.

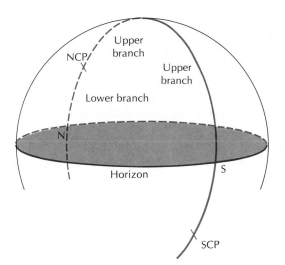

FIGURE 2.12 Upper and lower branches of the meridian.

The diagram is labeled with: Upper branch, Upper branch, Lower branch, NCP, N, Horizon, S, SCP

TIME AND THE CALENDAR

Time of Day

Let us imagine a great circle that passes through the celestial poles and the zenith; this circle is called the **celestial meridian**, or meridian for short (Figure 2.12). The meridian crosses the horizon at points that are due south and due north of the observer. The meridian is cut into two half-circles by the celestial poles. The half that contains the zenith is called the upper branch of the meridian; the other half is called the lower branch. Note that part of the lower branch of the meridian may be above the horizon: the part extending from the north celestial pole to the northern horizon is above the horizon for a northern-hemisphere observer, for instance.

The concept that the Earth is round is ancient; it was known to the Greek philosophers of Aristotle's time. The reason why the concept is relevant here is as follows: at any given instant, each person on Earth has his or her own zenith (see Figure 2.6). The stars at the zenith are different for a person in the U.S.A., in China, and in Australia. Since the zenith is different for each person and the meridian passes through the zenith, the meridian is different for each person—with one stipulation. Since the meridian is a complete circle, everyone at a given longitude on the Earth has the same meridian; that is, one circle passes through the celestial poles and all their zeniths simultaneously. With this bit of preliminary ground-work, we are now prepared to talk about time of day.

The apparent solar day begins when the Sun crosses the *lower* branch of the meridian; and it is 12:00 apparent solar time when the Sun crosses the *upper* branch of the meridian. The time interval from one crossing of the lower branch to the next crossing is divided into 24 equal hours. Time measured by the Sun in this way is called **local apparent solar time**. Local apparent solar time is not convenient for timekeeping.

Earlier, when we talked about the annual motion of the Sun, we pointed out that the Sun moves roughly 1° per day eastward along the ecliptic, which makes the day as determined by the Sun 4 minutes longer than the day determined by, say, a star crossing the meridian. One problem arises because the motion of the Sun along the ecliptic is not uniform. In January the Sun moves faster among the stars than it does in July. If you work out the exact figures, you find the solar day is 16 seconds longer in January than in July. Thus, to keep apparent solar time, clocks would have to run at different rates in different seasons.

Like the Earth, the celestial sphere has an equator. It is a great circle midway between the celestial poles. Astronomers have invented a **mean sun**, an imaginary sun which moves eastward around the celestial equator at a rate of roughly 1° per day, but at the *same* rate each day. Since the rate of motion is constant, it is easy to calculate where the mean sun is each day. Mean solar time is kept by using the mean sun and its meridian crossings, and it runs at a constant rate all year.

Now we come to a second problem; this one arises from the fact that people at different longitudes have different meridians. Let's think about the case of two cities that are not too far apart in longitude, for example, New York and Philadelphia, which differ in longitude by about $1\frac{1}{4}°$. When the mean sun is on the lower branch of the meridian of New York City and a new day begins, the same is not true for Philadelphia. In fact, the celestial sphere must rotate another $1\frac{1}{4}°$ before the day starts in Philadelphia. We have already pointed out that the celestial sphere rotates 1° in 4 minutes; so it rotates $1\frac{1}{4}°$ in 5 minutes. Thus the day begins in Philadelphia 5 minutes after it began in New York City; that is, the local mean solar time in Philadelphia is 5 minutes earlier than the local mean solar time in New York City.

Now, this isn't a very convenient state of affairs. One can easily drive from Manhattan to Philadelphia in two hours, except at rush hour. To have to reset your watch by 5 minutes after such a drive is not acceptable. If everyone kept local mean solar time, you would

FIGURE 2.13 Time-zone map of the United States.

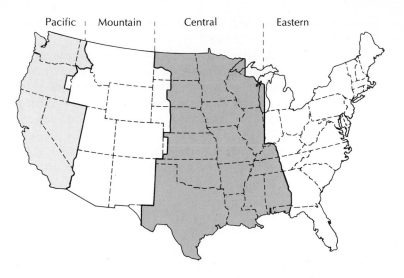

Pacific | Mountain | Central | Eastern

have to constantly reset your watch as you traveled around even for very short trips. To avoid this nuisance, people have adopted **zone** time. The local mean solar times within a band of longitude 15° wide are all within one hour of one another. So the convention could be this: all locations between longitude 67°.5W and 82°.5W adopt the local mean solar time for the central meridian 75°W, and call their time Eastern Standard Time. All locations between 82°.5W and 97°.5W adopt the local mean solar time for the central meridian of 90°W, and call their time Central Standard Time, which is one hour earlier than Eastern Standard Time; and so on. This is not exactly the convention, however.

The difficulty with the simple picture presented above is that the longitude line of 82°.5W, where the time changes by one hour, passes through the middle of states and cities. It is not very convenient to have a time change in the middle of a state and especially in the middle of a city. So instead of using exactly the 82°.5W longitude line for the time change, a zigzag line is drawn north and south across the country, keeping largely to state borders, to serve as the boundary for the time zones (Figure 2.13). Only in the oceans do meridians like 82°.5W, 97°.5W, etc., serve as boundaries of time zones.

Astronomers keep two kinds of clocks in their observatories. One clock keeps mean solar time, typically not for the local zone, but rather for the zone that is centered on the **prime meridian** (0° longitude), which passes through Greenwich, England. This mean solar time is called **universal time** (UT), or **Greenwich mean time** (GMT), and is useful when observatories in two zones want to compare observations. The universal time is the same everywhere on the Earth at a given instant.

The second clock keeps **sidereal time**. One sidereal day is complete when a point fixed among the stars goes from the upper branch of the meridian around the sky and back to the upper branch. This fixed point is the vernal equinox in the constellation Pisces. The sidereal day is 24 sidereal hours long, but 24 sidereal hours take only 23 hours and 56 minutes to complete by the solar clock. Thus the sidereal clock gains 4 minutes on the solar clock each day. Sidereal time is useful for pointing telescopes to fixed points among the stars.

The Calendar

There has been a fundamental need for a calendar since people began to settle down and take up agriculture; and no doubt even preagricultural nomads needed some knowledge of the seasons to time their wanderings. The Babylonians, the Hebrews, the Egyptians, and the Greeks each had their own calendars. In general, the year was divided into units like our months today. Each month began on the day of the new moon. Since the period between successive new moons is about $29\frac{1}{2}$ days, ancient months were either 29 or 30 days long, probably in some alternating pattern.

The year of the seasons—the **tropical** year—is 365.2422 days; whereas 12 lunar months are $12 \times 29\frac{1}{2}$ or 354 days. To keep the months in step with the seasons, the ancients occasionally threw in an "intercalated" or extra month. The basic problem is that there is no way to divide the tropical year into an even number of lunar months; that is, the two time periods are incommensurable. Julius Caesar made a first step at solving this problem—with the help of leading astronomers of his day.* He decided to drop the dependence of the months on the lunar phases, and to divide the year into 12 months whose lengths are 30 or 31 days, with the exception of short February, to give a 365-day year. Thus he arrived at our modern months. He then made the length of the year 365.25 days by adding one day to February every fourth year—that is, every **leap year**. But the **Julian calendar** of 365.2500 days was still 0.0078 days longer than the tropical year of 365.2422 days. This may seem like a small difference. However, between Caesar's time

---

*He also borrowed some basic ideas from the Egyptian calendar, in which the year consisted of twelve 30-day months plus an extra five days. Oddly, the Egyptians made no provision for leap years, and let their calendar drift against the seasons, coming back to its starting point only after every 1,460 years.

and the sixteenth century, the difference amounted to almost two weeks. This meant in effect that Easter—which traditionally falls on the first Sunday after the first full moon after the beginning of spring—fell earlier and earlier in the year, a situation which the church fathers found unacceptable. So, in 1582, Pope Gregory XIII announced a calendar reform. He decreed: first, that the Friday after Thursday, October 4, would be October 15; and, second, that only those century years divisible by 400 would be leap years. By this rule 1600 was a leap year, but 1700, 1800, and 1900 (which would be leap years in Caesar's calendar) would not be leap years. In this **Gregorian calendar**, the year is 365.2425 days—very close to the correct value.

The Gregorian calendar was accepted reluctantly in most Catholic countries. However, the Protestant and other nations did not accept the change immediately. The German states came around in 1700, England in 1752, and Russia in 1918. The historian is plagued by the fact that two calendars were in use in the world between 1582 and 1918.

In this section we have discussed two important topics: time and the calendar. We explained why it is necessary to introduce mean solar time and zone or standard time. We then discussed the calendar. The modern, Gregorian calendar is based on a leap-year rule where every year divisible by 4 is a leap year except for century years, which must be divisible by 400.

REVIEW QUESTIONS

1. Describe or define: (a) zenith; (b) circumpolar stars; (c) ecliptic; (d) retrograde motion.

2. Venus and Mars have distinctively different motions among the stars. Describe their motions.

3. Why do sidereal days and solar days differ in length? What is the magnitude of the difference?

4. Why is it not practical to base our months on lunar phases?

5. Describe the Gregorian calendar, especially the leap year rule. Why was that calendar introduced?

6. How can you determine your latitude at sea? Why is longitude more difficult to determine?

3

THE HISTORICAL DEVELOPMENT OF GRAVITATIONAL ASTRONOMY

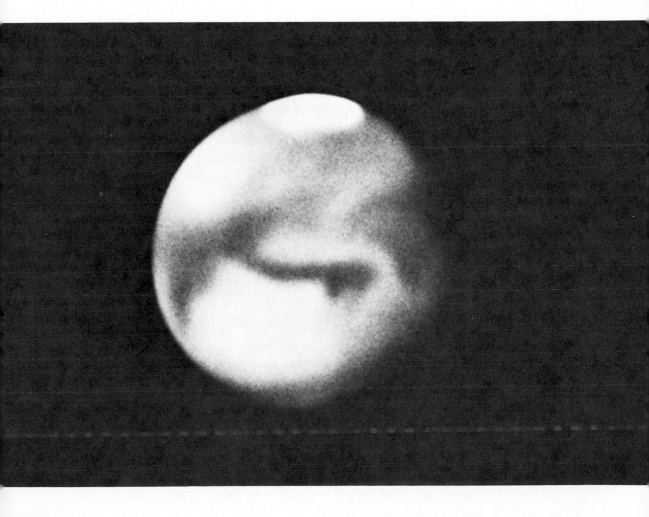

In the last chapter, we talked about the phenomena that we could discover by watching the sky night after night. We described the rising and setting of the stars; we talked about the annual motion of the Sun and the associated annual progression of the constellations across the night sky; and we discussed the motions of the planets. How do we explain these observed facts? The goal of this chapter is to answer this question, in an historical context. We will show how our concept of the sky developed from a simple knowledge of celestial motions to an understanding of the physical foundations of celestial mechanics.

Since the observed facts we wish to explain can be discovered without a telescope or sophisticated observing equipment, they have been known for a long time. Our story begins at the dawn of recorded history. We use the word **cosmology** to mean the branch of astronomy which deals with the structure of the universe; it can also mean the collection of ideas on this structure held by an individual or group.

THE MOTIONS OF THE PLANETS

The earliest astronomers were interested in studying the sky for very practical reasons: they needed to know when to plant and harvest their crops and when to hold religious festivals.

The cosmological theories of the Babylonians and Egyptians are historically interesting, and those of the former are surprisingly similar to Old Testament cosmology (Figures 3.1 and 3.2). The theories of the Greek philosophers, from the earliest times to Aristotle, are also important, since these concepts had a significant influence on later thought. However, for our story there are only

FIGURE 3.1 Egyptian universe. (Yerkes Observatory; from G. Maspero, *The Dawn of Civilization*. S.P.C.K., 1894.)

FIGURE 3.2 Babylonian universe. (Yerkes Observatory; from G. Maspero, *The Dawn of Civilization*. S.P.C.K., 1894.)

two points from the early theories that are important. First, most ancient cosmologies were geocentric; that is, the Earth was assumed to be at the center of the universe. Second—and this is Plato's philosophical contribution to cosmology—the motions of the planets were assumed to be combinations of uniform circular motions

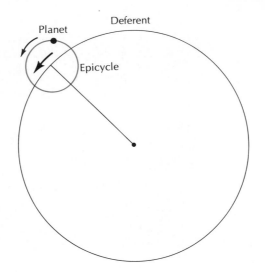

FIGURE 3.3 Basic units of the Ptolemaic system.

about the Earth as center. Most theories of planetary motions from Plato's day (fourth century B.C.) to the Renaissance were dominated by these two concepts. The culmination of these theories is the Ptolemaic system.

The Ptolemaic System

Ptolemy (Claudius Ptolemaeus) lived in the second century A.D. The details of his life are largely unknown. However, the details of his contributions to astronomy are carefully spelled out in his great book, the *Almagest*. Ptolemy's astronomical work is an outgrowth and extension of the work of Hipparchus of Nicaea, who lived in the second century B.C. and whose writings are all lost. We know of Hipparchus' work because Ptolemy frequently cites him and describes his earlier contributions. Ptolemy obviously admired Hipparchus.

Ptolemy assumes that the Earth is at the center of the universe, and that each planet circles about it with two simultaneous, uniform circular motions. The planet moves around a circular path called an **epicycle**, while the center of the epicycle moves around a larger circular path called the **deferent**. The Earth is at the center of the deferent (Figure 3.3). The relative sizes of the epicycle and the deferent, and the rates at which the planet moves on the epicycle and the epicycle moves on the deferent, can all be adjusted. With the proper combination of sizes and rates, the observed motions of the planets can be matched quite well (Figure 3.4).

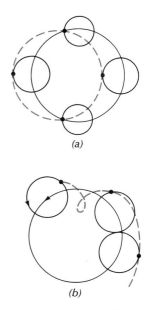

FIGURE 3.4 Different kinds of epicycles: (a) produces an offset circle; (b) produces a figure known as a cardioid.

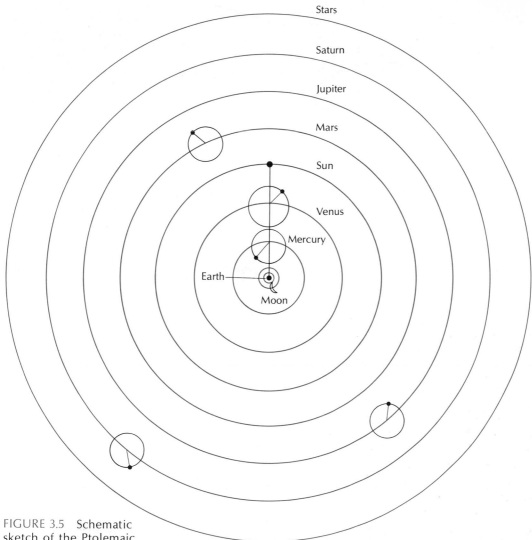

FIGURE 3.5 Schematic sketch of the Ptolemaic picture of the solar system.

The retrograde loops of Mars, Jupiter, and Saturn (described in Chapter 2) are easily produced by this scheme. The theory also accounts for the fact that Mercury and Venus never stray far from the Sun as seen from the Earth, by requiring that the centers of these two planets' epicycles always remain on a line joining the Earth and the Sun as they move around their deferents (Figure 3.5).

Ptolemy adjusted all the sizes and motions until his system agreed with observed planetary motions. As time went on, he made

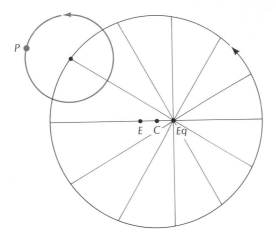

FIGURE 3.6 Fully detailed Ptolemaic concept of planetary motion. In this picture, Earth (*E*) does not lie at the center (*C*) of the deferent, but is offset from it. The center of the epicycle moves around the deferent in such a way that its angular motion is uniform as viewed from the equant (*Eq*). The planet (*P*) moves around the epicycle at a uniform rate. The center of the deferent is midway between Earth and the equant.

a few modifications in the details of the system to make it fit observations better (Figure 3.6). He was then able to predict future positions of the planets—with a considerable amount of computational effort, it should be mentioned. If, in a few years, the observed motions departed from the predictions, the scheme could be tuned slightly by changing a size or rate. This flexibility made the theory highly successful, and it was still being used 1,400 years after Ptolemy's death.

To obtain the diurnal motion of the stars, Ptolemy assumed that the stars hang on a sphere larger than the deferents of all the planets. The rotation of this sphere accounts for the rising and setting of the stars.

Even as early as the third century B.C., not everyone agreed with the geocentric concept. The idea that the Sun was the center of the universe and that the Earth moved around it was suggested by Aristarchus. It was, however, an idea whose time had not come. By the beginning of the sixteenth century A.D., with Europe in the midst of a great intellectual Renaissance, the time was right for the heliocentric theory to again appear on the scene.

The Copernican Universe

Nicholas Copernicus was born at Torun, Poland, in 1473. When he was 10, his father died, and he was adopted by his mother's brother, Lucas Waczenrode. Waczenrode was an intellectual priest who

FIGURE 3.7 Copernicus. (Yerkes Observatory.)

would rise to the position of Bishop of Ermland. These events set the tone for Copernicus' life: his uncle saw to it that he obtained the best education possible, and that he would enter the service of the church.* Copernicus officially studied medicine and church law, but his passion was for astronomy and mathematics—to which he devoted all his spare moments throughout his life. The result of his study is summarized in the treatise *On the Revolutions of the Heavenly Spheres,* which finally reached print the day Copernicus died, in 1543.

The fundamental difference between the cosmology of Copernicus and the cosmology of Ptolemy is the place of the Earth. Copernicus sets the Sun at the center of the universe, and makes the Earth move as another planet (Figure 3.8). Why did Copernicus doubt Ptolemy's cosmology? The answer to that question is complex, and scholars differ in detail about it, but basically it is this: observations of planetary positions were becoming more precise, and it was becoming increasingly more difficult to adjust the Ptolemaic system to account for discrepancies. In addition, as we mentioned in Chapter 2, it was becoming clear that the year of the Julian calendar was too long, since Easter came earlier than the church fathers liked, and the possibility of calendar reform had astronomers attuned to the problem.

In Book One of the *Revolutions*, Copernicus asserts that the universe and the Earth are both spherical. He accepts Plato's idea that the movements of celestial bodies are uniform and circular, or compounded of circular motions, which go on forever. He then assumes that the planet Earth also moves. He argues that risings and settings of the stars and other heavenly bodies are due to Earth's rotation, according to the Pythagoreans; so why not go one step further and make our planet also move around something else at the center of the universe?

Among the arguments leveled against the concept that Earth rotates was the idea that its rotation would tear it apart. Copernicus replied by pointing out that the immense sphere of the stars—the celestial sphere—would have to rotate much more rapidly than Earth to achieve the observed daily motion and would therefore be even more likely to be torn apart by its motion.

Copernicus next discusses the ordering of the planets in the universe. He states that the Sun occupies the center of the universe, then presents arguments to show that, from the Sun outward, the

*To put Copernicus in historical perspective, we can note that he was a student at the University of Cracow when Columbus landed in the New World.

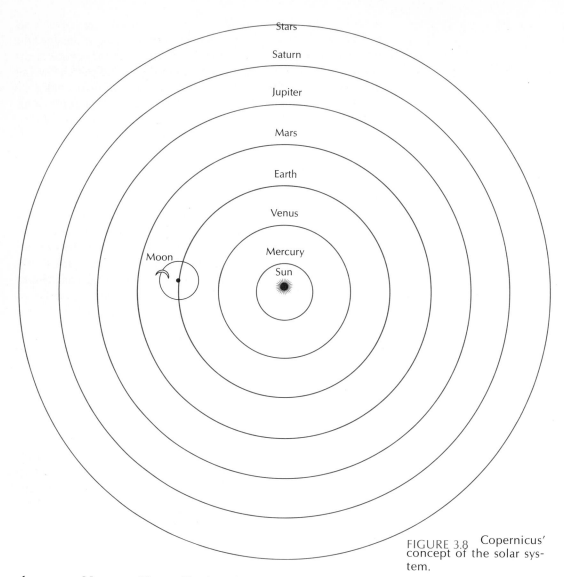

FIGURE 3.8 Copernicus' concept of the solar system.

planets are Mercury, Venus, Earth (with the Moon circling the Earth), Mars, Jupiter, and Saturn.

The fact that the orbits of Mercury and Venus are inside Earth's orbit is clear from the observation that each of these planets does not wander far from the Sun in the sky, as can be seen in Figure 3.9. Furthermore, the retrograde motion of Mars, Jupiter, and Saturn results from the relative motions of Earth and each of these planets, as can be seen in Figure 3.10. To describe how Copernicus came up with the exact ordering of the planets, we have to discuss

FIGURE 3.9 Why Mercury and Venus do not appear to stray far from the sun. The maximum distance these planets can be from the Sun is the angular sizes of their orbits as seen from the Earth.

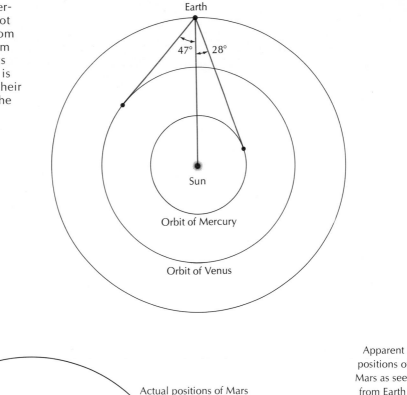

Earth

47° 28°

Sun

Orbit of Mercury

Orbit of Venus

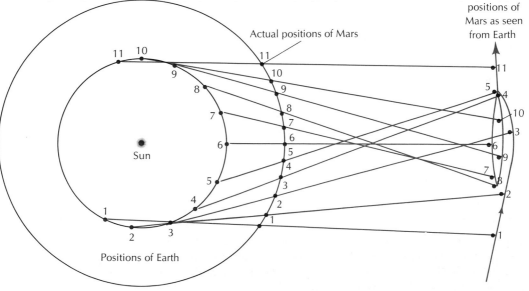

Actual positions of Mars

Apparent positions of Mars as seen from Earth

Sun

Positions of Earth

FIGURE 3.10 The retrograde motion of Mars. The modern explanation is that the "retrograde motion" results from the changes in the sighting line from Earth to Mars as Earth passes Mars in their orbits.

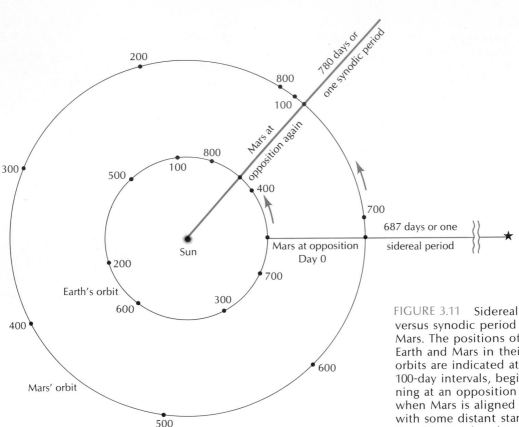

FIGURE 3.11 Sidereal versus synodic period of Mars. The positions of Earth and Mars in their orbits are indicated at 100-day intervals, beginning at an opposition when Mars is aligned with some distant star. Roughly 687 days later, Mars has returned to the same point in its orbit, and Earth has made just under two circuits of its orbit. Earth overtakes Mars, and another opposition occurs, about 780 days after the initial opposition.

two concepts that are used today: the **sidereal** revolution period and the **synodic** revolution period of a planet.

The sidereal revolution period of a planet is the time required for it to complete one full circuit around its orbit relative to the stars. (The term sidereal comes from the Latin *sidus,* meaning star.) The synodic period is the time required for the planet to move from some Sun-Earth-planet arrangement back to the same arrangement. For instance, the time it takes Mars to go from one opposition— that is, the Sun, Earth, and Mars on a straight line, with Mars on the opposite side of the celestial sphere from the Sun—to the next opposition is its synodic period (Figure 3.11). The synodic and sidereal periods differ in length because the motion of the Earth and the planet together determine the synodic period, whereas the motion of the planet alone determines its sidereal period. We can observe the sidereal period of the Earth and the synodic period of the other planets. From these facts, we can calculate the sidereal periods of the planets. Copernicus said that the distances of the planets from the Sun are related to their sidereal periods. Mercury, with the shortest period, he assumed to be closest to the Sun; Saturn, with the longest period, he assumed to be farthest away.

Copernicus' theory seems at first glance to be somewhat simpler than Ptolemy's, since he explains retrograde motion, for instance, as being due to the relative motion of both the Earth and the other planet. However, with simple circular motions around the Sun, Copernicus could not explain the observed details of planetary motions and in the end he had to introduce epicycles again. This does not take away from his great leap forward in placing the Sun at the center of the universe and recognizing the Earth to be a planet.

Copernicus' thought gained several early disciples, including the great philosopher René Descartes and the monk Giordano Bruno. The latter was burned at the stake in 1600 for his heretical beliefs, which included the Copernican theory, prompting many thinkers like Descartes to decide discretion is the better part of valor and to hold their peace on Copernicanism. But the work that would provide the proof of Copernicus' hypotheses was in progress.

Tycho and Kepler

Tycho Brahe was born in Denmark in 1546. He developed an early interest in astronomy and spent many nights studying the sky rather than the books of law his family expected him to study. This interest in the sky was motivated in part by a belief in astrology.

Tycho's brilliance was brought to the attention of King Frederic of Denmark, who offered him a small island near Copenhagen—called Hven—on which to build an observatory (Figure 3.13). The income provided by the serfs on the island was available for Tycho's support. He designed and built several instruments for measuring the positions of stars and planets in the sky that were far superior to any previous instruments, and then spent the next two decades carefully making precise measurements.*

Tycho was not a good manager, despite his brilliance. For instance, he channeled far too much of the resources of his fief to the support of his studies. In the end, he was forced out of Denmark, and moved to Prague, where in 1599 he came under the patronage of Emperor Rudolph II.

FIGURE 3.12 Tycho. (Yerkes Observatory.)

---

*Tycho's work was done before the invention of the telescope. All his instruments were built without lenses or mirrors to aid his observations.

FIGURE 3.13   Tycho's observatory at Hven. (Yerkes Observatory.)

To compound his lack of management skills, Tycho was also an arrogant individual, often at odds with his peers. Legend has it that as a young man he was involved in a duel with swords, and had his nose sliced off. Supposedly, he had a gold replica sculpted, which he wore in place of the original.

Tycho's careful observations were mere numbers unless put to good use. Fortunately, a young man came along who would be able to take full advantage of them. Johannes Kepler was born in 1571 in Wurtemberg. He studied at the University of Tubingen, where he became a disciple of Copernican cosmology. In 1594 he moved to Styria (part of modern Austria), where he taught mathematics and wrote about Copernican science. Like Tycho, Kepler was partly motivated by a belief in astrology. Tycho became acquainted

FIGURE 3.14   Kepler. (Yerkes Observatory.)

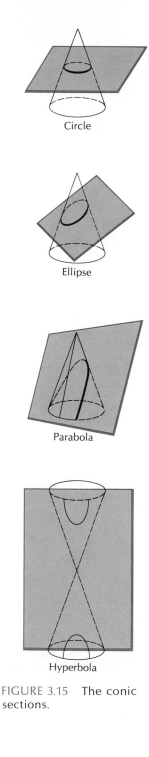

Circle

Ellipse

Parabola

Hyperbola

FIGURE 3.15 The conic sections.

with Kepler through the latter's writings. Furthermore, Kepler knew of Tycho's great observations, and wished to study them. Fate pushed the two men together. Styria was strongly Catholic and Kepler was a Protestant. Religious intolerance was finally too much for Kepler, and he left Styria for Prague in 1600. Here he and Tycho joined forces, and Kepler began an analysis of Tycho's measurements, which continued long after Tycho's death in 1601.

Kepler's most detailed studies were on the motions of the planet Mars. In 1609 he published *Commentaries on the Motion of Mars,* in which he makes two general statements that today are called Kepler's first two laws of planetary motion. They are:

> The planets move around the Sun in orbits that are elliptical in shape, with the Sun at one focus;

and (the **law of areas**)

> A straight line between the moving planet and the Sun sweeps out equal areas in equal intervals of time.

These laws, which will be explained below, were demonstrated by Kepler only for Mars. He then argued that they should apply to the other planets. This process of arguing from specifics to generalities happens so frequently in science that it is given a name. It is called **induction** or **inductive reasoning**.

What do these laws mean? The ellipse is a closed figure obtained by cutting a circular cone with a plane. It is therefore called a **conic section**, one of a family of curves which includes the circle, **ellipse**, **parabola**, and **hyperbola** (Figure 3.15). When the plane cutting the cone is perpendicular to the axis of the cone one obtains, of course, the limiting case of the circle. When the plane is tipped at an angle to the axis, one obtains an ellipse. The greater the tipping of the intersecting plane, the more the ellipse is elongated, until a stage is reached when the curve resulting from the intersection is no longer a closed curve. At this stage the plane is parallel to a line on the surface of the cone which passes through the apex, like a pole in a teepee. The conic section resulting from this intersection is a parabola. Finally, if the plane is made parallel to the axis of the cone, the curve is called a hyperbola.

What are the foci of the ellipse? We can answer this question by looking at a method for drawing an ellipse (Figure 3.16). One needs two pins, a loop of string, pencil, paper, and a drawing board. Drive the pins through the paper and the board, loop the string around the pins, and draw a line, keeping the string taut. The figure you have drawn is an ellipse. Since the loop of string does not stretch, the sum of the distances from any point on the perimeter of the ellipse

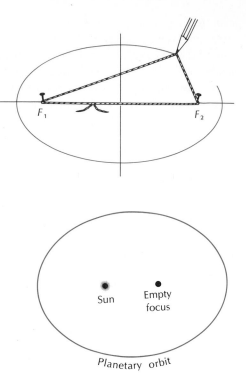

FIGURE 3.16 Drawing an ellipse. (From *New College Physics* by A. V. Baez. W. H. Freeman and Company. Copyright © 1967.)

FIGURE 3.17 Planetary orbit. The orbit is an ellipse with the sun at one focus.

to the two pins is the same at every point on the ellipse. The two pins are the foci of the ellipse.

According to Kepler's first law, each planet moves in an elliptical orbit (Figure 3.17). The Sun is at one focus, and the other focus is empty.

Kepler's second law, the law of areas, is illustrated in Figure 3.18, where the elliptical shape of a planet's orbit is exaggerated to show the idea. The time it takes the planet to move from a to b, from c to d, and from e to f are all equal. Thus the planet moves fastest when it is closest to the Sun, and slowest when it is farthest away.

A decade after he published his first two laws of planetary motion, Kepler found a third relationship, which is now called Kepler's third law of planetary motions: the ratio of the squares of the periods of revolution of any two planets is equal to the ratio of the cubes of their mean distances from the Sun. If we take the Earth as one of the two planets, this law allows us to find a simple relationship between a planet's period of revolution in years and its distance from the Sun in **astronomical units** (see Glossary). The relationship is shown in Figure 3.19.

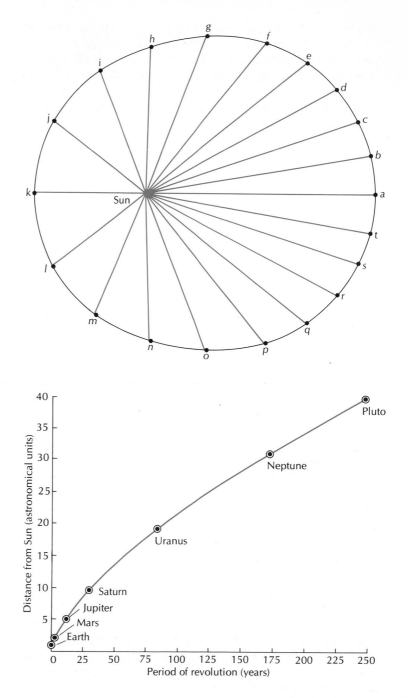

FIGURE 3.18 Kepler's law of areas. The planet moves from a to b, from b to c, etc. in equal time-intervals. The areas of the roughly triangular-shaped segments are all equal.

FIGURE 3.19 Kepler's third law leads to the plotted relationship between the planets' periods of revolution and distances from the Sun. Mercury and Venus also fall on the curve.

Kepler's three laws are clearly Copernican in nature. They all imply that the Sun is the center of the universe. Kepler believed in Copernicus' ideas and spelled out his belief in the book *The Epitome of Copernican Astronomy,* which was fully published in 1621.

In this section, we have discussed the work of Tycho and Kepler. We have described how Tycho's measurements led Kepler to his three laws of planetary motion. The final nails were driven into the coffin of the Ptolemaic system by a slightly older contemporary of Kepler, Galileo.

THE BEGINNING OF MODERN SCIENCE

Galileo

While Kepler was analyzing Tycho's observations at Prague, another chapter of our story was unfolding in Italy. Galileo Galilei was born at Pisa, in 1564. In 1581 he entered the University of Pisa, supposedly to study medicine. However, he soon showed a special talent for physics and mathematics. For example, in his first year at the university, he demonstrated that the period of time required for a pendulum to make one swing depends only on its length. During his early years, Galileo spent his time in a study of the motions of bodies and the strength of material bodies.

One of Galileo's famous experiments during this time was a study of falling bodies. Aristotle had stated that a brick will fall faster than a pebble because the brick was heavier. Galileo dropped many bodies from the Leaning Tower of Pisa and found that, no matter what the weight, all objects fall the same distance in the same time. He recognized, too, that a dropped feather would fall more slowly because of air resistance. Thus he made one of the first important discoveries about gravity.

In 1609, he heard of a remarkable discovery by a Flemish spectacle maker, Jan Lippershey. By placing a concave lens at one end of a tube and a convex lens at the other end, Lippershey had made a "telescope." Galileo immediately made one himself and found that if he looked through the tube, with the concave lens nearest his eye, distant objects appeared to be nearer. After several trials, he made a telescope that made distant objects appear about 30 times nearer (Figure 3.21).

When Galileo turned this telescope toward the heavens, he made one astonishing discovery after another. The sky appeared crowded

FIGURE 3.20   Galileo. (Yerkes Observatory.)

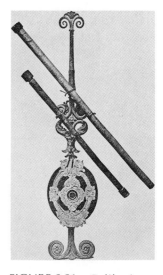

FIGURE 3.21   Galileo's telescope. (Yerkes Observatory.)

with fixed stars. The Milky Way, which to the unaided eye appeared to be a nebulous band stretching across the sky, was seen to be a myriad of stars through the telescope.

When he observed Jupiter, Galileo found it accompanied by four stars which changed their positions relative to the planet from night to night without straying far from it (Figure 3.22). His observations showed that these stars were satellites which revolve about Jupiter just as Jupiter revolves about the Sun. Later Kepler would show that these satellites also obey his third law. Galileo also showed that Venus exhibits phases, just like the Moon.

Galileo decided early in his life that Copernicus' ideas of planetary motions were correct. He also knew of Kepler's work, and his own observations convinced Galileo further of the truth of Copernicus. In 1632, Galileo published his *Dialogue on the Two Chief World Systems,* spelling out his pro-Copernican arguments. Unlike some of the earlier books mentioned, this work is very readable.

On the rotation of the Earth, Galileo argues that it is easier to make the small terrestrial globe rotate than the immense sphere of the fixed stars. In addition, he points out that the sidereal periods of the planets increase with increasing distance from the Sun. Mars completes one orbit in two years, Jupiter in twelve, Saturn in thirty. The same is true of Jupiter's satellites: the larger the orbit of the satellite, the longer its orbital revolution period. This relationship would be destroyed if the starry sphere were made to rotate in one day, for then this largest sphere would have the shortest period. If, on the other hand, the sky were fixed, then its rotation period would be infinite; that is, the largest possible. In this case the smallest sphere, the Earth itself, would have the short, one-day rotation period, in harmony with the other motions.

On the Copernican vs. Ptolemaic picture of the position of the Earth, Galileo offers many arguments. One of the most compelling is based on his discovery of the phases of Venus. In Ptolemy's planetary system, Venus always lies almost exactly on the line joining the Sun and Earth, and must appear as a thin crescent or a dark disc, if we admit that its light is reflected sunlight. However, in Copernicus' theory, both Earth and Venus orbit the Sun (Figure 3.23). At one time Venus is between the Earth and the Sun, and would appear as a dark disk. At another time it is on the far side of the Sun, and would appear fully illuminated. Between these extremes, Venus is off the Sun-Earth line, and could exhibit any phase between full and dark.

Six months after the first copy of Galileo's *Dialogue* was sold, the Inquisition banned its further sale and confiscated all remaining copies. Galileo was summoned to Rome, where he was eventually

FIGURE 3.22 Galileo's drawing of Jupiter and its moons. (Yerkes Observatory.)

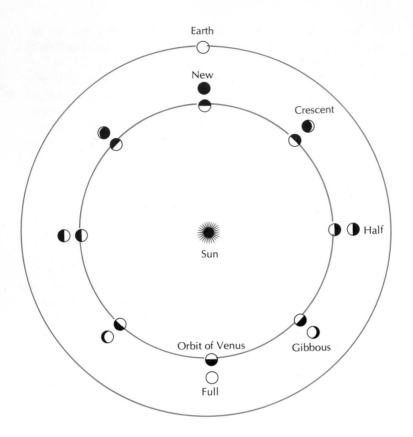

Earth

New

Crescent

Sun

Half

Orbit of Venus

Gibbous

Full

FIGURE 3.23 Phases of Venus according to Copernican theory.

found guilty of heresy for his strong defense of Copernicanism. However, he was offered absolution if he disavowed further belief in Copernican cosmology. Galileo recanted his belief before the Inquisition. But legend has it that, as he left the room, he muttered under his breath, "But the Earth does move." The remainder of his life he remained under house arrest at his villa in Arcetri. During this time he devoted his efforts to the study of mechanics that had occupied his younger years. His final work, a masterful and readable description of his lifelong studies, was the *Dialogue Concerning Two New Sciences*.

In this section we have summarized briefly some of Galileo's contributions to science. Although he personally was silenced by the church, his observations shouted the truth from the hilltops.*

---

*The events of Galileo's confrontation with the church were complex. Even today, certain points of controversy remain. A true perspective can be obtained only from the reading of a detailed history of the incident, such as de Santillana's *The Crime of Galileo*.

The brilliant work of Kepler, added to Galileo's, ended all doubt. The heliocentric nature of the solar system, was fully accepted at centers of learning.

Newton

Isaac Newton was born on Christmas day in 1642, in England. He entered Trinity College at Cambridge in 1661, where he studied natural philosophy. In 1665 and 1666 the University was closed because the plague was ravaging London, and Newton spent two years at relative leisure at his family home. In fact, his leisure was more apparent than real. He used the time for quiet contemplation, which led to the discovery of the differential calculus, the theory of colors, and the quantitative science of mechanics. Newton's intellectual achievement in those two years was phenomenal.

Most of Newton's discoveries were not published until 1687, when a friend, Edmund Halley, persuaded him to write them down. The result is the book *Philosophiae Naturalis Principia Mathematica,* or the *Principia* for short. In the *Principia,* Newton first presents a theory of the motions of bodies, then states the principle of universal gravitation. As we shall see, mechanics and gravitation tie together the earlier work of Kepler, Galileo, and others into a quantitative and logically consistent whole. In the following paragraphs, we will describe a few of Newton's contributions.

Let us begin by stating Newton's first two laws of motion:

(1) Every body continues in its state of rest, or of uniform motion in a straight line, unless it is compelled to change that state by forces impressed upon it.

(2) The acceleration of a body is proportional to the force acting on it and is in the direction in which the force acts.

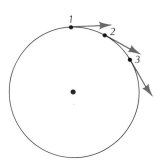

FIGURE 3.25 Circular motion. The speed of the body moving in the circular path remains constant. However, the direction in which it moves is constantly changing, and therefore its velocity changes. The body can be said to be accelerated because of the change in velocity. Thus acceleration does not necessarily mean a change in speed alone.

We have two concepts to define: **acceleration** and **force**. Let's consider them in order. To understand acceleration, we must understand **velocity**. To specify the velocity of an object, we must know both its *speed* and its *direction* of motion. An acceleration is a *change* in velocity. Thus a car accelerates from a stop sign as the driver increases its speed from zero to 35 mph on a straight road. A race car going 150 mph around a circular track is accelerating, even though its speed remains constant, because its direction of motion and therefore its velocity are changing (Figure 3.25). A planet orbiting the sun in an elliptical orbit is constantly accelerating, for two reasons: Kepler's law of areas shows that the speed changes; and, since the planet moves on a curved path, its direction of motion changes. Therefore the planet must experience a force;

otherwise it would fly off into space on a straight line, at constant speed. We can define force as a push or pull that causes a body to be accelerated.

The three quantities, acceleration, force, and velocity, are examples of **vector quantities** or, more simply, **vectors**. A vector has both magnitude and direction. The push or pull of a force has a certain magnitude or strength, and the force is exerted in a certain direction. To calculate the effect a force will have on the motion of an object, we must know both the strength and direction of the force. A quantity that has only magnitude is called a **scalar**. The speed of an object is the magnitude of its velocity, and is a scalar quantity.

When Galileo dropped bodies from the Leaning Tower of Pisa, he found that they accelerated. The force pulling the bodies down to the Earth is the force of gravity. Gravity acts on all bodies on the Earth. How, then, did Newton manage to formulate his first law? Look especially at the part which says that a body continues to move uniformly in a straight line *unless* a force is acting on it. He could never truly make a direct test of this law. The answer is, the tests are indirect. All Newton's laws together permitted him to make a series of predictions. These were found to agree with observations; therefore the laws are correct.

In his *Dialogue Concerning Two New Sciences,* Galileo presents an argument that makes Newton's first law plausible. He considers a ball rolling down an inclined board (Figure 3.26). He argues that if we start with the board at a very steep angle and roll the ball down the board, it will accelerate rapidly. If we make the angle less steep, the rate of acceleration will decrease until it becomes zero for a horizontal board. On the other hand, if we try pushing the ball up the inclined plane, it will experience a negative acceleration (or **deceleration**), which decreases as the angle of the board decreases until it becomes zero for horizontal boards. For a horizontal board, the ball is neither speeded up (accelerated) nor slowed down (decelerated); therefore its velocity must remain constant. If it were not for friction, the ball would roll forever on the horizontal board. Friction, however, is also a force which acts to slow down the ball.

Newton's second law is a quantitative statement of the effort of a force. If we apply a force to a body, it accelerates. If we double the force, the acceleration doubles; if we halve the force, the acceleration halves; and so on. To make the law even more quantitative, we must define **mass**. Mass is a measure of the amount of matter contained in a body. Now, experiment shows that for a constant force, acceleration depends inversely on mass; that is, if you double the mass, you halve the acceleration, and so on. Also, we know from

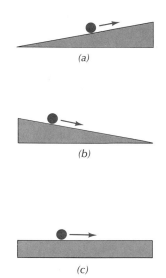

FIGURE 3.26 Ball on an inclined plane. (a) The ball slows down as it moves up the incline. (b) The ball speeds up as it moves down the incline. (c) The ball neither slows down nor speeds up as it rolls along a horizontal plane (ignoring friction), and its speed remains constant.

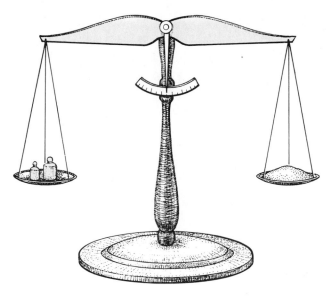

FIGURE 3.27 Use of balance to measure mass. When the pans are balanced, the known mass on the lefthand pan equals the unknown mass on the righthand pan.

Galileo's experiments that a falling body experiences the same acceleration no matter what its mass is. If you break a brick in half, each half falls with the same acceleration as the whole brick did. This observation tells us that the force of gravity acting on a body is proportional to its mass. If the whole brick and the half-brick experience the same acceleration, the force on the whole brick must be double the force on the half-brick.

This observation gives us a way to measure mass. The standard of mass is a cylinder of platinum kept at the *Bureau International des Poids et Measures* near Paris. Replicas of this standard mass are used as standards throughout the world. This standard mass is by definition **one kilogram**. How do we divide the kilogram into smaller units? Taking an equal-arm balance, we put the standard kilogram on one pan and, say, a pile of sand on the other pan until a balance is reached—then the force of gravity on both sides is the same, and the mass of sand is one kilogram (Figure 3.27). If we want to have half a kilogram of sand, we put the standard kilogram away, and divide the kilogram of sand between the two pans until a balance is found. Then each pan holds a half kilogram. We can go on to get a quarter kilogram and other fractions in this way. One-thousandth of a kilogram is called a gram. This way of measuring mass compares the force of gravity on a known mass with the force of gravity on the unknown mass. When the forces are equal, the pans balance and the masses are equal. This technique works any-

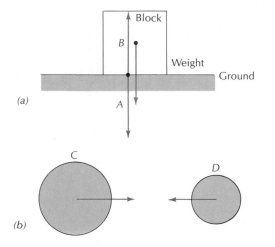

FIGURE 3.28 Action-reaction concept. (a) The block exerts a downward force A on the ground due to its weight. The ground exerts an equal upward force on the block. Forces A and B are an action-reaction pair. (b) Planet C exerts a gravitational force on planet D, and planet D exerts a gravitational force on planet C. The two forces are equal and opposite.

where: on Earth, on the Moon, where gravity is less, and on another planet where gravity is greater, the results are always the same.

The **weight** of an object is the force that gravity exerts on the object. Weight depends on the object's mass and on the strength of gravity. Astronauts who explored the Moon found their weight there to be one-sixth of what it is on Earth. Weight changes as gravity changes, but mass stays the same. We can measure weight using a spring scale. We hang a mass whose weight we wish to discover on one end of a spring and attach the other end to a fixed object. The force of gravity, acting on the mass to be weighed, stretches the spring by an amount that is proportional to its weight. In the metric system the unit of weight is also the kilogram. A kilogram mass weighs one kilogram on the Earth. We can calibrate the spring scale by weighing known masses and measuring the stretch of the spring for each. It is important to remember the distinction between weight and mass.

Newton also stated a third law (illustrated in Figure 3.28):

> (3) To every action there is always opposed an equal reaction; the mutual actions of two bodies on each other are always equal and oppositely directed.

Think about a brick falling toward the Earth. The brick is accelerating because the Earth exerts a gravitational force on it. The brick also exerts a gravitational force on the Earth. The forces the brick and the Earth exert on one another are equal in strength but are in opposite directions.

Without doubt, Isaac Newton's greatest contribution is his study of gravitation. We can infer some of the characteristics of gravity from what has been discussed before. We have already seen that the pull of the Earth's gravity on any object is proportional to the mass of the object. Furthermore, Newton's third law says there must be a reaction to this tug: the object pulls back on the Earth. Newton generalized and said: every body in the universe exerts a gravitational force on every other body that is proportional to the product of the masses of the two bodies. If the bodies are perfect spheres, the force of gravity acts along a line joining their centers.

The next step in arriving at the universal law of gravitation is not so easy. Newton showed that for the orbits of the planets to be ellipses, with the Sun at one focus, the force of gravity must depend on the square of the distance between the planet and the Sun. The simplest demonstration of this relationship requires Newton's mathematical invention, calculus—so we will not dwell on it here.

Put the parts together, and you get the **universal law of gravitation**:

> Every body in the universe attracts every other body with a force that is proportional to the product of their masses and inversely proportional to the square of the distance between their centers.

One final point should be mentioned here: Kepler's law of areas is the statement of an important physical principle known as the law of conservation of **angular momentum**. The angular momentum of a small moving mass is the product of its mass, its distance from the point about which it moves, and its speed about that point (Figure 3.29). The angular momentum of any body remains constant if the only forces acting on it are along the line between it and the point about which it rotates. Since gravity acts along the line joining two bodies, angular momentum is conserved in the orbital motion of the bodies around one another.

A familiar situation in which the conservation of angular momentum plays a role is the spinning ice skater. At the finale of her act, the lovely performer presses her arms tightly against her body and goes into a rapid spin. Her axis of rotation is a vertical line through the center of her body. When the skater extends her arms, her rate of spin slows down. This is due to the fact that, when she extends her arms, the mass of her arms is moved farther from the axis about which she is rotating. Since the distance increases, the speed must decrease to conserve angular momentum.

The same phenomenon occurs for planets orbiting the Sun (or for any two bodies orbiting one another). The force acting on the planet is gravity, which acts along the line joining the planet and the Sun. As the planet moves in its elliptical orbit, its angular momen-

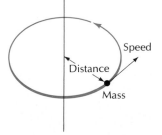

FIGURE 3.29 Angular momentum equals mass times distance times speed.

tum remains constant. Thus its speed must increase as it moves closer to the Sun and decrease as it moves farther away. This effect is just what is described by the law of areas.

One can take Newton's three laws of motion and his law of gravitation and predict the motions of the planets, quantitatively. The very massive Sun dominates the much less massive planets, and they orbit the Sun on elliptical paths. In addition, the law of gravitation says that the Earth tugs on Venus, Jupiter tugs on Mars, and so on; that is, the planets **perturb** each other's motions slightly. As a result, the paths depart slightly from perfect ellipses. Newtonian mechanics accounted for the observed motions of the known planets remarkably well.

In this section, we have described Newton's contributions to science, particularly his laws of motion and his universal law of gravitation. In the next section we shall describe the later events that demonstrated the validity of Newtonian cosmology.

FIGURE 3.30 Edmund Halley. (Yerkes Observatory.)

The Triumph of Newtonian Cosmology

Periodically we see in the sky a wispy object known as a comet. We will discuss these bodies further in Chapter 8. Tycho Brahe showed that comets are celestial bodies, and Newton assumed that they moved around the sun in very elongated orbits. Newton then found a method for calculating the orientations of these orbits in space from a series of observations. His protégé, Edmund Halley, applied the method to several historical comets whose positions had been carefully recorded, and discovered that three of these comets—the comets of 1531, 1607, and 1682—followed nearly the same orbit. He concluded that these were in fact reappearances of one and the same comet, that it had an orbital period of 75 years, and that it would return in 1758. The comet returned in 1759, soon after Halley's death (Figure 3.31). In the meantime, a French astronomer, Alexis Clairaut, showed that the comet would pass close to Jupiter and Saturn as it approached the Sun and would therefore have its motion perturbed by their gravitational influence. When the comet did return, it made its closest approach to the Sun within a month of Clairaut's prediction. The comet is known as Halley's comet, in honor of Edmund Halley's pioneering work.

In 1781, William Herschel was scanning the sky with one of his excellent telescopes and chanced upon a new planet beyond Saturn. It was ultimately named Uranus (Figure 3.32). As it turns out, Uranus is just at the limit of naked-eye visibility at times, and it had

FIGURE 3.31 Halley's Comet. (Hale Observatories.)

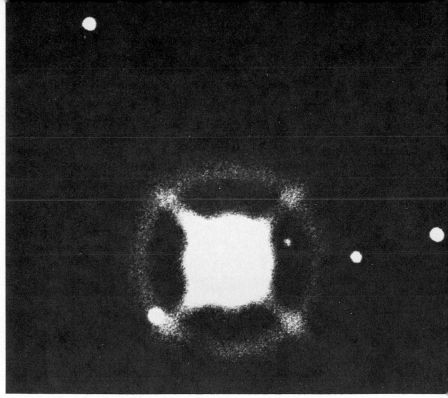

FIGURE 3.32 Uranus and its five known moons. From left to right, they are Oberon, Ariel, V, Umbriel, and Titania. The small fifth satellite was discovered in 1948. (Yerkes Observatory.)

FIGURE 3.33 William Herschel. (Yerkes Observatory.)

FIGURE 3.34 John Couch Adams. (Yerkes Observatory.)

actually been plotted, unrecognized, on charts of the sky several times between 1690 and its discovery. Once Herschel discovered the planet, these earlier observations were easy to locate, and they could be used for calculating the new planet's orbit. It was quickly noted that Uranus' motion did not follow the path that Newton's mechanics predicted, even allowing for the influence of Jupiter and Saturn. Some scientists suggested that Newton's theory might begin to fail over such great distances, and that perhaps some other law might better explain Uranus' motion.

A young English mathematician, J. C. Adams, who had just finished his degree at Cambridge University, took a different point of view. He assumed that Newton's theory was correct, but that there existed another planet beyond Uranus. Uranus' erratic behavior, according to Adams, was due to perturbations by this hypothetical planet. Adams calculated the position of the suspected planet and sent his predictions to the Royal Greenwich Observatory. Unfortunately, the Astronomer Royal was busy and chose to ignore the results of so young a mathematician.

Adams' idea was a pregnant one, to be sure, for soon a Frenchman, Urbain Leverrier, published similar results. Using Leverrier's results, astronomers at the Berlin Observatory found the planet

FIGURE 3.35 Neptune and its two satellites. (Yerkes Observatory.)

FIGURE 3.36 Urbain Leverrier. (Yerkes Observatory.)

immediately. The planet was named Neptune (Figure 3.35). Adams' and Leverrier's use of Newton's mechanics proved its great value once and for all. The only cloud in the entire episode was the squabble between the English and the French about who deserved credit for the discovery.

After the triumph of Newtonian science, all seemed well for a long time. However, by the beginning of the twentieth century, observations were being made that did not seem consistent with Newtonian mechanics. Another giant appeared on the Earth to help unravel the problems.

Einstein

Albert Einstein was born in 1879 at Ulm, Germany. Since his adulthood was spent in the twentieth century, much is known about Einstein. Recently, two excellent biographies of Einstein have been published.*

There is an interesting parallel between the life of Einstein and the life of Newton. As a young man, Newton had two years without any responsibilities while Cambridge University was closed because of the plague. Einstein could not get an academic post when he finished his formal education, and had to take a job as a technical officer in the Swiss Patent Office, where he worked for seven years.

FIGURE 3.37 Albert Einstein. (Yerkes Observatory.)

---

*Ronald W. Clark, *Einstein: The Life and Times* (New York: World, 1971), and Banesh Hoffman, *Albert Einstein, Creator and Rebel* (New York: Viking, 1972).

The responsibilities of the job were undemanding, and there are indications that Einstein could finish his day's work in less than the alloted time. This left time on the job, plus all his free time off the job, for intellectual pursuits. He used the time well, for the papers he published in 1905 alone constitute an awe-inspiring flood of creativity, equalled by that of only a handful of thinkers in the history of mankind.

Einstein's contributions to gravitational astronomy began with a small paper in 1913 and culminated in 1915 with the paper "The Foundation of the General Theory of Relativity."

Einstein kept Newton's first law of motion, but rephrased it to say that *bodies will move in straight lines if not disturbed*. He rejected the idea that gravity is a force which acts at a distance to disturb the straight-line motion of a body. Instead, he assumed that the presence of a massive body changes the geometry of space itself: it warps space. Bodies will move along straight lines in this warped space, as Newton said they must, but the straight line warps with the space.

The mathematics of general relativity can be complex. But in the end they produce equations for the motion of a planet around the Sun that are almost precisely like those of Newtonian mechanics, except for a tiny correction. It has been observed that in each century the long axis of Mercury's orbit rotates 43 seconds of arc *more* than Newtonian theory predicts (Figure 3.38). That is a small effect, but an important one: Einstein's theory accounts for it, whereas Newton's theory cannot.

There is one very important difference between Newtonian gravitation and Einsteinian gravitation, in how the propagation of light is affected. A light beam can be considered a stream of massless particles, called photons. According to Newton, a gravitational field cannot effect a photon, since its mass is zero. According to Einstein, however, the light beam will follow a warped "straight line" and will appear to be deviated. As Figure 3.39 shows, starlight passing near the Sun should be deflected. Observations made at total solar eclipses have shown that this deviation does exist.

Einstein's theory of general relativity makes predictions that Newtonian mechanics cannot make. Does this fact mean that Newton was wrong? The philosopher of science says, *no*. For most problems that we deal with on Earth and in the universe, Newtonian mechanics is the correct system to use. It is simpler than relativity and gives precise predictions. In a few situations—for example, for bodies moving very near a massive body or for a light beam passing close to a massive body—Newton's theory is inadequate, and we must use general relativity. Both theories have their

Perihelion advance

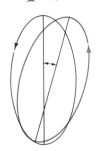

FIGURE 3.38 Precession of Mercury's orbit. Its perihelion advances 1° per century, of which 43″ are not accounted for by Newtonian theory.

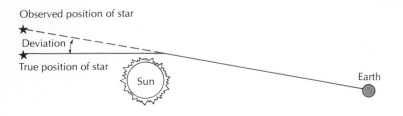

Observed position of star

Deviation

True position of star

Sun

Earth

FIGURE 3.39 The direction of motion of photons is apparently deviated by the intense gravitational field of the Sun, if the photons pass very near the Sun, according to Einstein's general theory of relativity. A star will appear to be located a greater angular distance away from the Sun because of the deviation.

strengths and weaknesses, and both theories are the invention of the human mind. They are a way to view the real world. Someday a discovery may be made that is not predictable by either Newton or Einstein—and another great thinker will have to come up with a new idea. To put it in the simplest terms: we can never know the real world fully; we can only approximate it in our theories.

We have now surveyed the development of thought that led to our limited understanding of gravity. We now know why we observe the things described in Chapter 2. With one or two very rare exceptions, everything we know about the universe outside the Earth comes to us on light beams. Our next order of business is to find out enough about light to help us understand what we see.

REVIEW QUESTIONS

1. Compare and contrast the Ptolemaic and the Copernican pictures of the solar system.

2. Describe the contributions of Tycho and Kepler. How are they intertwined?

3. What is an ellipse? Describe the motion of a planet in its elliptical orbit.

4. Describe three discoveries Galileo made with the telescope.

5. What were Newton's contributions to the study of the solar system?

6. Define: (a) mass; (b) velocity; (c) acceleration; (d) force.

7. How did discoveries by Halley and Herschel confirm Newton's theories?

8. How does general relativity describe gravity?

9. Why do the motions of Venus and Mars among the stars differ?

4

LIGHT AND THE TELESCOPE

How do we find out about the bodies that make up the universe? Except for the few situations so far where we have direct contact with the bodies, such as by men visiting the Moon, we learn about the heavenly bodies by studying the light we receive from them. In this chapter we will describe some basic information about light and about telescopes that will help us understand the material in the rest of the book. In particular, we wish to address the question: how are observations of celestial objects made? The question of how to analyze light will be addressed in Chapter 7.

THE NATURE OF LIGHT

In the previous chapter, we pointed out that a light beam can be viewed as a stream of massless particles known as photons. That this view is correct is proved by the photoelectric effect, which we will discuss later in this chapter. One of the great mysteries in physics is the fact that light *also* behaves like a wave. Light is a **transverse wave**.

Two examples of transverse waves that are familiar to all of us are waves moving (or propagating) along a rope and waves in the open ocean. If you shake one end of a long rope, a wave will seem to run along the rope (Figure 4.1). The material of the rope moves perpendicular to the rope's length as the wave moves along the rope. This is the fundamental characteristic of the transverse wave. As the wave propagates in a particular direction, the disturbance— in this case, the motion of the material making up the rope—is in a direction perpendicular to the propagation direction. The same thing is true of water waves. If you have ever sat in a boat in the ocean with long swells moving by, you will know that the boat moves up and down with the water as the wave crest passes by

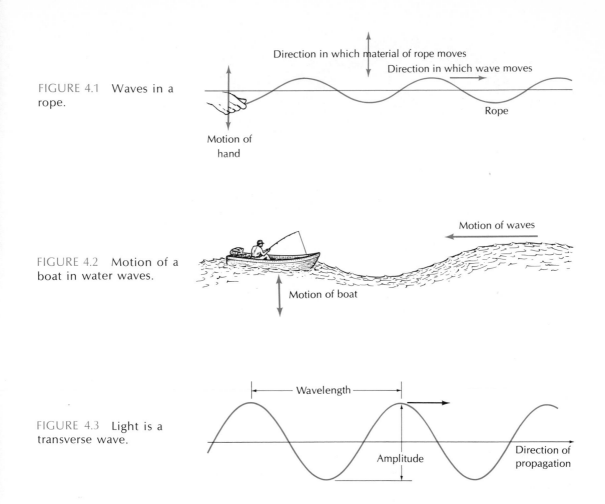

FIGURE 4.1 Waves in a rope.

Direction in which material of rope moves

Direction in which wave moves

Rope

Motion of
hand

FIGURE 4.2 Motion of a boat in water waves.

Motion of waves

Motion of boat

FIGURE 4.3 Light is a transverse wave.

Wavelength

Amplitude

Direction of
propagation

(Figure 4.2). The boat does not move along the surface of the ocean with the wave.

Light is a transverse wave, just like the water wave or the wave in the rope (Figure 4.3). However, for light the disturbance is electromagnetic. The electromagnetic disturbance takes place at right angles to the direction of propagation of the wave. We define **amplitude**, which is the total magnitude of the electromagnetic variation in the light wave. The **wavelength** of the light is the distance from one crest to the next in the wave. Sound is an example of a longitudinal wave (Figure 4.4).

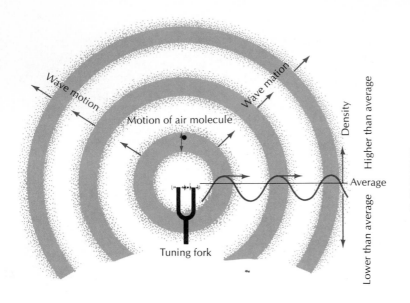

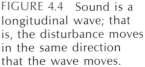

FIGURE 4.4 Sound is a longitudinal wave; that is, the disturbance moves in the same direction that the wave moves.

Interference of Light

The wave nature of light is best illustrated by the phenomenon known as interference, which can only be produced by waves. When two light waves cross one another, they add together or **interfere**. If the wave trains are exactly **in phase**—that is, if the crests in both waves occur at the same location at the same time—the two waves add together to give a final wave whose amplitude is equal to the sum of the amplitudes of the two waves. This is called **constructive interference** (Figure 4.5). On the other hand, if the two wave trains are exactly one half-cycle out of phase—that is, if the crests of one wave fall in the troughs of the other—the two waves add together to give a wave with an amplitude equal to the *difference* between the amplitudes of the two waves. This is called **destructive interference**. If the two initial amplitudes are equal, the difference is zero—there is no light.

In the early nineteenth century, an English physicist named Thomas Young carried out an experiment to show that light waves can be made to interfere. His experimental layout is shown in Figure 4.6. Light from a source such as a bulb passes through a narrow slit in a piece of metal, then through a pair of slits (S_1 and S_2) in a second metal sheet. Light from the two slits combines at a screen where it can be observed. At a point on the screen where the light from both slits arrives in phase, constructive interference occurs and the screen is bright. At a point where the light from the

Relative amplitudes

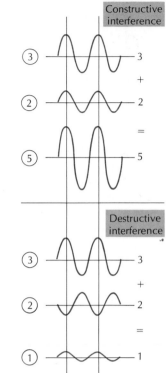

Constructive
interference

③ ——————— 3

+

② ——————— 2

=

⑤ ——————— 5

Destructive
interference

③ ——————— 3

+

② ——————— 2

=

① ——————— 1

FIGURE 4.5 Interference
proves the wave nature
of light.

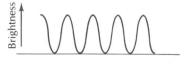

Brightness

Screen

Dark

FIGURE 4.6 Layout of
Young's double-slit ex-
periment (see text for
explanation).

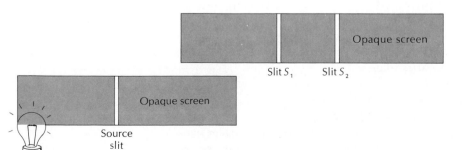

Opaque screen

Slit S_1 Slit S_2

Opaque screen

Source
slit

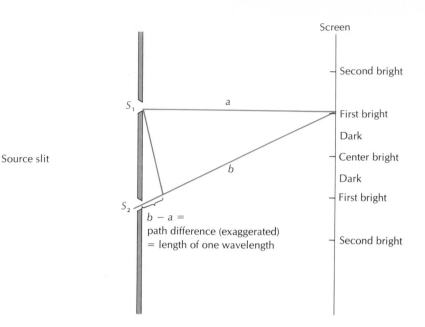

FIGURE 4.7 Wavelength calculated using Young's double-slit experiment. The distances from the slits S_1 and S_2 to the first noncentral bright fringes differ by one wavelength of the light; the distances to the second bright fringes differ by two wavelengths; and so on.

two slits arrives out of phase, destructive interference occurs and the screen is dark. The bright and dark stripes are called **fringes**.

The central bright fringe on the screen is equally distant from the two slits. The next bright fringe, say, on the right, is closer to one slit than to the other, but the waves from the two slits still arrive at the location of the bright fringe in phase. Therefore, the fringe must be closer to the nearest slit by a distance that is equal to one wavelength of the light (Figure 4.7). If we measure all the appropriate distances, we can calculate the difference between the distances to the two slits and thereby find the wavelength of the light.

If the experiment is carried out with white light, the calculation yields a wavelength of 5×10^{-5} centimeters, a very tiny distance. When astronomers talk about the wavelengths of visible light, they use a unit of length known as the **Angstrom** (Å), where 1 Ångstrom is 10^{-8} centimeters. The wavelengths of white light lie in a range around 5000 Ångstroms.

Velocity of Light

In the last sections we talked about the fact that light behaves like a wave which propagates through space just as a water wave propagates over the surface of the ocean. How fast does the wave move?

FIGURE 4.8 Römer measures the speed of light. In the six months in which Jupiter moves from A to B, Earth has moved from one side of its orbit to the other. The eclipses of Jupiter's moons appear to be late by the amount of time it takes light to cross Earth's orbit.

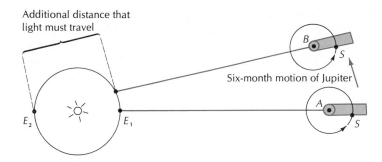

Additional distance that light must travel

Six-month motion of Jupiter

S is position of satellite at beginning of eclipse

Many people, including Galileo, tried to measure the speed of light with little success. They could conclude only that the speed was great. Finally, in 1675, the Danish astronomer Ole Römer found a method for making the measurement, using Jupiter's satellites. Beginning with Galileo's discovery of the moons of Jupiter, astronomers had carefully studied their motions, and the period of time that it takes each moon to orbit around the planet was measured. Based on this well-known orbital period, astronomers made predictions of the times that the moons would appear to pass behind or in front of the planet; that is, the time of eclipses of these moons by Jupiter. The observations on which the predictions were based were the times of eclipses measured when the moons were easiest to see, which was, of course, when the Earth and Jupiter were closest to one another. On the other hand, when Jupiter appeared to be near the Sun in the sky, that is, when it and the Earth were on opposite sides of their orbits, the eclipses were more than ten minutes late.

Römer explained the fact that the eclipses were late by saying that it took ten minutes for the light from the eclipse to travel across the Earth's orbit (Figure 4.8). Using the size of the Earth's orbit as known at that time, Römer arrived at a velocity of roughly 200,000 kilometers per second.

Modern measurements of the velocity of light are made in a different way that yields more precise results. The best modern value is 299,792.4562 kilometers per second. Römer was as close as could be expected, given the data he had to work with. In most of our work, it is sufficient to remember that the speed of light is 300,000 kilometers per second or 3×10^{10} centimeters per second.

We now know the wavelength of white light and the speed of light. We are thus ready to calculate another parameter, the **fre-**

quency of light. If you watch a light beam go by and count the number of crests that pass you in a second, you will have measured the frequency. In one second a section of light beam 3×10^{10} cm long will pass you, and there will be one wave crest every 5×10^{-5} cm in the beam. Therefore, the number of crests (or cycles) that go by is

$$\frac{3 \times 10^{10}}{5 \times 10^{-5}} = \frac{30 \times 10^9}{5 \times 10^{-5}} = 6 \times 10^{14} \text{ cycles/second.}$$

Of course, you really couldn't count the crests. If you counted one, two, three at the rate of one number per second, counting to 6×10^{14} would take 20 million years.

We now know that light behaves like a wave, and we know its speed, wavelength, and frequency. We know how to measure each of these properties. We will next look at the polarization of light.

Polarization

A natural light beam contains light waves with electric and magnetic disturbances in planes that are in all possible orientations (Figure 4.9). Each wave in the beam is a transverse wave with its electric field varying in a specific plane. The natural light is unpolarized. If we remove certain of the planes, so that there are more light waves in a certain orientation than in any other orientation, the light is **partially polarized** in that direction. If all the electric field planes are in the same orientation, the light is **fully polarized**. Certain crystals pass only one plane of polarization; they thus polarize natural light.

The material known as polaroid is made of a layer of needle-like crystals, which are all lined up and imbedded in plastic (Figure 4.10). When a natural light beam passes through a polaroid filter, it is fully polarized in a given direction. The light will then pass through a second polaroid that is aligned in the same way as the first. However, if the second polaroid is rotated a quarter turn, it will only pass light waves that are oriented perpendicular to those passed by the first polaroid. These latter waves were not passed by the first polaroid, of course; so nothing gets through the second polaroid.

These details are important because the light from some heavenly bodies is polarized, and measuring the polarization tells us something about the body.

FIGURE 4.9 Polarization of light: (a) The electric portion of light's electro-magnetic disturbance takes place in one plane, the plane of polarization. (b) Natural light is a bundle of light waves with all possible planes of polarization present.

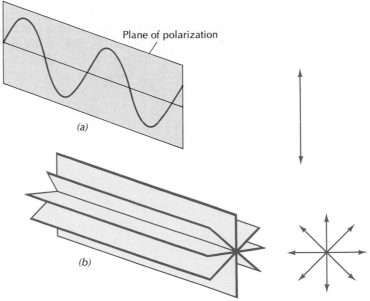

Plane of polarization

(a)

(b)

Natural radiation contains many planes of polarization

FIGURE 4.10 Polaroid filter is like a picket fence; it lets only one plane of polarization pass.

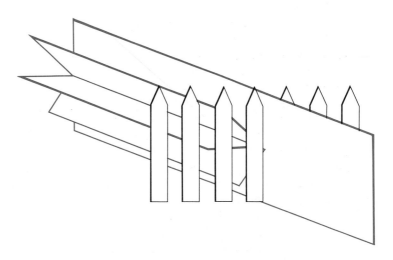

The Electromagnetic Spectrum

Isaac Newton made many fundamental experiments in light and optics. One of the most important was his demonstration that white light is, in fact, a combination of all colors. Newton darkened his study and let a beam of sunlight enter through a small hole in a shade. He passed the light through a prism and found that the prism spread the light out into a band of colors or **spectrum**, which he projected onto the wall (Figure 4.11). He found the order of colors in the spectrum to be *r*ed, *o*range, *y*ellow, *g*reen, *b*lue, *i*ndigo, and *v*iolet. (To remember this order, remember the name of the colorful

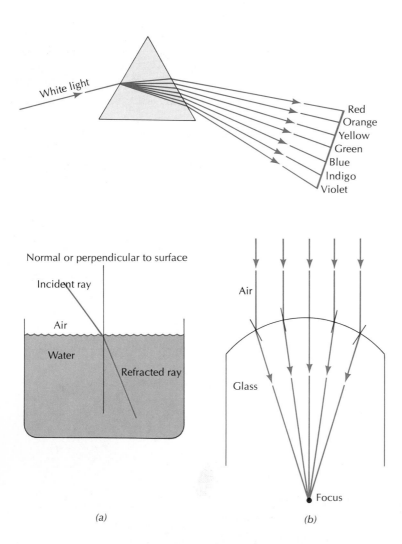

FIGURE 4.11 White light is dispersed into its constituent colors by a prism.

(a)

(b)

FIGURE 4.12 When light passes from a rare medium (for instance, air) to a dense medium (for instance, water or glass), the light ray is refracted toward the **normal** to the surface. When light passes from a dense to a rare medium, the refraction is exactly the reverse. (a) Light passes from air to a tank of water. (b) Light passes from air into glass through a convex spherical surface. The initially parallel rays are refracted and meet at a single point, the **focus** of the surface. Note that there is a different normal at each point on the entrance surface.

Table 4.1 Relationship Between Color, Wavelength, and Frequency

Color	Wavelength Range (in Ångstroms)	Frequency Range (in cycles/second)
Red	7500 to 6400	4.0×10^{14} to 4.7×10^{14}
Orange	6400 to 5900	4.7×10^{14} to 5.1×10^{14}
Yellow	5900 to 5600	5.1×10^{14} to 5.4×10^{14}
Green	5600 to 5000	5.4×10^{14} to 6.0×10^{14}
Blue	5000 to 4400	6.0×10^{14} to 6.8×10^{14}
Violet	4400 to 3800	6.8×10^{14} to 7.9×10^{14}

elf, Roy G. Biv—each letter in the name is the first letter of a color, in proper order.)

The prism spreads white light out into separate colors because of two facts. First, a light beam is bent or **refracted** whenever it passes from one substance (such as air, glass, or water) into another (Figure 4.12). Second, the amount of refraction depends on the color of the light. Red light is refracted the least, blue light the most. This effect is called **dispersion**.

The color of the light depends on its wavelength, as we could discover by carrying out interference experiments (like that of Young) using light of different colors. Table 4.1 gives the wavelengths of the various colors. White light behaves as if it had roughly the middle wavelength in the table, that is, 5000 Ångstroms.

The colors of the rainbow are just a small part, the visible part, of the **electromagnetic spectrum**. Figure 4.13 shows the rest of the electromagnetic spectrum. All the radiations, from the short-wavelength gamma rays, X rays, and ultraviolet, to the long-wavelength

FIGURE 4.13 The electromagnetic spectrum. (From *New Horizons in Astronomy* by J. C. Brandt and S. P. Maran. W. H. Freeman and Company. Copyright © 1972.)

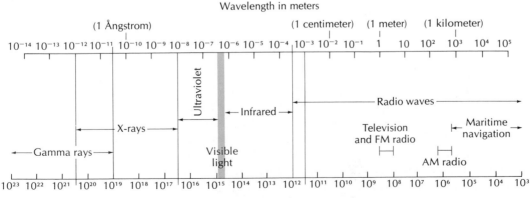

infrared and radio radiation, are electromagnetic waves just like visible light. The only difference is the wavelength, which spans the fantastic range of 10^{22}.

We have briefly discussed the basic nature of light. Let us now proceed to describe how we collect and study electromagnetic radiation.

TELESCOPES

To collect light or other electromagnetic waves, we need a telescope. In this section we will concentrate on a discussion of visible-light telescopes; then in the next section we will talk about what kind of telescopes are used for other parts of the spectrum. To begin, we need some background material on how lenses focus images.

Let's imagine a lens mounted in a device which holds it vertically. The optical axis of the lens, which is a line through the center of the lens normal to its plane, is then a horizontal line. If we shine a narrow beam of light parallel to the optical axis and through the lens, it will be bent or refracted and will cross the optical axis. All beams that enter the lens parallel to the optical axis pass through the same point after refraction (Figure 4.14a). This point is called the lens' **focal point**. The distance from the center of the lens to the focal point is the lens' **focal length**.

If we shine a beam of light on a lens in such a way that it appears to come from the focal point (Figure 4.14b), it will be refracted by the lens and leave the lens parallel to the optical axis—just the reverse of the first case. A beam directed through the center of the lens (Figure 4.14c) will pass through the lens without being bent.

Each object is seen by reflected or emitted light. Every point on the object is a light source, beaming light waves in all directions.

Let's place a simple object like an arrow upright on its feathered end some distance from a lens (Figure 4.15). The pointed tip of the arrow reflects light in all directions. The rays it gives off will include one that is parallel to the optical axis of the lens, another that is heading toward the center of the lens, and a third that goes through the lens' focal point that is nearest the arrow. These three rays are refracted by the lens in the manner described in the previous paragraphs, then all intersect at one point, which is the **image** of the top of the arrow. We can next follow the same process with rays coming from every other point on the arrow and find the image

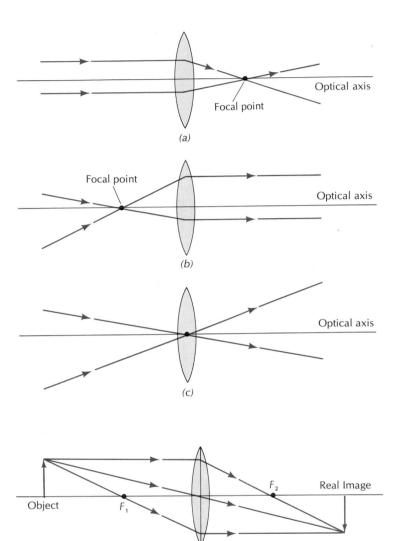

FIGURE 4.14 Effect of a lens on light beams. (a) A beam of light that is parallel to the optical axis of a lens is refracted so that it passes through the lens' focal point. (b) A beam of light that appears to come from the focal point of a lens is refracted by the lens so that it leaves parallel to the optical axis. (c) A beam of light that passes through the center of a thin lens is not refracted.

(a)

Optical axis

Focal point

(b)

Focal point

Optical axis

(c)

Optical axis

FIGURE 4.15 Formation of a real image by a lens. Illustrated are the three types of beams of Figure 4.14, coming from the tip of an arrow. These three beams are refracted by the lens and meet at the image of the arrow tip. F_1 and F_2 are the focal points of the lens.

Object

F_1

F_2

Real Image

of each point. In this way we build up an image of the entire arrow. This image is a real image. If we place a sheet of paper at the position of the image, we will see the image of the arrow projected on the paper.

A concave mirror forms an image in the same way a lens does (Figure 4.16). The three types of rays behave the same for a mirror as for a lens. The chief difference is the fact that the object and image are both on the same side of the mirror.

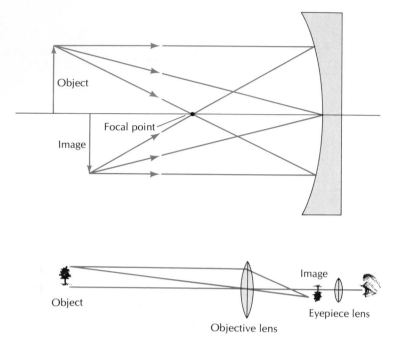

FIGURE 4.16 Formation of a real image by a mirror.

Object

Focal point

Image

Object

Image

Objective lens

Eyepiece lens

FIGURE 4.17 A simple telescope. The objective lens forms a real image, which can be examined in detail with a magnifying glass. The eyepiece lens of the telescope is the magnifying glass.

What if we form a real image with a lens or mirror, then carefully examine it with a magnifying glass? We will see a close-up image of the object with our eye. This combination of two lenses is a **telescope** (Figure 4.17). The main lens or mirror of the telescope, often called the **objective**, forms an image that we examine with the **eyepiece**, which is in effect a magnifying glass.

We have now described how a telescope works by discussing how lenses and mirrors form real images. In the process, we have left out several details that one would have to know to make quantitative calculations of the properties of any telescope. The next step is to look at specific types of telescopes used in astronomical research. Before we do so, it is useful to talk about the characteristics of telescopes that make them useful to us.

Parameters of the Telescope

The telescope allows its user to study distant objects in more detail than is possible with the unaided eye. The usefulness of a telescope depends on two characteristics: its ability to resolve fine detail, and its ability to collect large quantities of light. Let's look at these two abilities in turn.

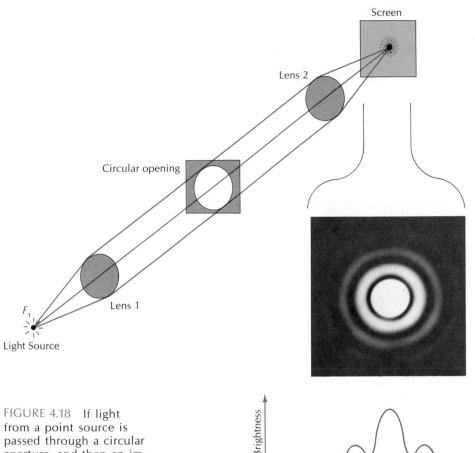

FIGURE 4.18 If light from a point source is passed through a circular aperture, and then an image if the source is formed on a screen, the image is no longer a point. The image is a pattern of light and dark rings known as the diffraction pattern of a circular aperture.

Resolving Power. Scientists describe the resolving power of an optical instrument in terms of the angular distance between two objects, such as stars, that are just discerned as being separate by the instrument.

If the light from a point source is passed through a circular aperture as shown in Figure 4.18, the image of the source is not a point, but is spread out into a tiny disk as shown. Experiments show that the size of the image increases as the size of the circular aperture decreases, and vice versa. This physical phenomenon is called **diffraction**. The lens of a telescope is a circular aperture and produces diffraction images of stars. Furthermore, each point of an extended object such as a planet acts like a point source, and the image of the planet is made up of the diffraction images of all of these point sources superimposed on one another.

The fact that each point on an object is smeared out on the image tells us that there is a limit to the amount of detail that can be seen. Figure 4.19 sketches the images of two stars that are not resolved, that are just resolved, and that are obviously resolved. We find experimentally that a telescope one inch (2.5 centimeters) in diameter can resolve two details that are about five seconds of arc apart, and that the bigger the objective lens (or mirror) is, the smaller can be the distance between the closest two points that the telescope can resolve. The actual theoretical resolving power of telescopes of various sizes is listed in Table 4.2. Incidentally, an object 1 cm in diameter seen at 2 km (about 1.2 miles) appears to be one arc second in diameter.

The 5-meter (200-inch) telescope at Palomar Observatory theoretically should be able to resolve details 0.02 arc seconds apart on a celestial object such as a planet, but it cannot. Why not? Because it has to look at the object through the Earth's seething, turbulent atmosphere. Just like water or glass, the gas in the Earth's atmosphere refracts light. But because the gas is moving, the refraction changes rapidly; as a result, the image "dances" around rapidly. Our eye cannot follow the rapid dance, and instead sees the image of the star blurred out over an area that may be much larger than the diffraction image produced by the telescope. This effect of the Earth's atmosphere is called **seeing**.

On the very best nights at mountain-top observatories, the size of an image of a point source is spread out by the seeing over $\frac{1}{4}$ arc seconds. Poor nights have a seeing of 1 arc second. Under the very best conditions, then, a 0.5-meter (20-inch) telescope will be able to resolve details at its theoretical diffraction limit. Any larger telescope will be limited by atmospheric seeing to the same $\frac{1}{4}$ arc second resolution. On relatively poor nights, any telescope with more than about a 12-centimeter (5-inch) aperture will be limited by the atmosphere.

We have now described the resolving power of a telescope, and have talked about the limitations placed on the resolution by the Earth's atmosphere. If, even under the best conditions, a 0.5-meter telescope is the largest that can resolve to its theoretical limit, why do we build instruments as large as the 5-meter (200-inch) telescope? The answer lies in the second characteristic that makes telescopes useful, their light-gathering ability.

Light Gathering Power. A telescope can be considered to be a funnel for light. All the light which is collected by the large objective is focused through the eyepiece to the viewer's eye. The amount of light which enters the objective is proportional to its

FIGURE 4.19 The images of stars produced by telescopes are tiny diffraction patterns, because the objective of the telescope is a circular aperture. Depending on how the patterns of two stars overlap, it may or may not be possible to tell that two stars are present.

Table 4.2   Resolving Power of Telescopes

| Diameter of Objective | | Finest Detail That Can |
Inches	Meters	Be Resolved
1.0	0.025	4″.5
6.0	0.15	0.75
12.0	0.3	0.4
24.0	0.6	0.2
40.0	1.0	0.1
80.0	2.0	0.05
200.0	5.1	0.02

area, that is, to the square of its diameter. Thus a 5-meter telescope gathers $10^2 = 100$ times more light than a 0.5-meter telescope, and can therefore help the observer see objects 100 times fainter.

The largest telescopes are not typically used for direct viewing of celestial objects by an observer. Instead, astronomers will use a photographic plate or photoelectric system to make observations. Long-exposure photographs permit the study of even fainter objects than is possible with the eye. We will return to the use of photography later.

Types of Telescopes

Telescopes can be divided into three general types: **refracting** telescopes, which have a lens as objective; **reflecting** telescopes, which have a mirror as objective; and telescopes which have a combination of lens and mirror as the main optical element. We will describe each type of telescope in turn.

Refracting Telescopes. The objective of a refracting telescope is a lens, typically one whose focal length is long compared to its diameter. The objective lens is usually a compound lens, made of two pieces of glass in a way that corrects for **chromatic aberration**. Such a lens is called an **achromatic lens** or **achromat** for short.

Chromatic aberration can be seen in inexpensive telescopes as a red or blue halo around bright objects (Figure 4.20). Since glass refracts blue light more strongly than red light, just as we saw happen with the prism, the focal length of a single lens will be shorter for blue light than for red light. We can illustrate the problem by forming the image of a star with a simple lens. When the red light is in focus, the blue light is not; so the star is surrounded by a blue halo. However, a carefully designed combi-

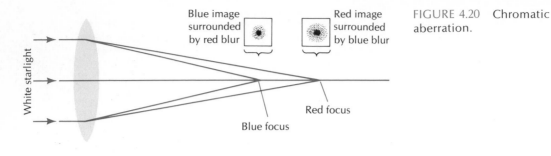

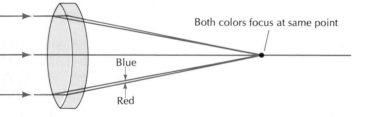

FIGURE 4.20 Chromatic aberration.

FIGURE 4.21 The achromatic doublet lens consists of two elements constructed of different kinds of glass. Typically the positive element is crown glass, and the negative element (with one concave surface) is flint glass.

nation (Figure 4.21) of a convex lens of crown glass and a concave lens of flint glass will focus light of all colors at almost the same point. Such a compound lens minimizes chromatic aberration.

Refracting telescopes have traditionally been used in applications where great magnification is needed, for example, in studying fine details on the Moon and the planets, and in looking for tiny relative motions between stars. The long focal lengths of refractors make it easy to achieve high magnification.

The largest refracting telescope in the world is the 1-meter (40-inch) instrument at Yerkes Observatory, Williams Bay, Wisconsin (Figure 4.22). The large objective lens has a focal length of 19 meters or over 60 feet. The tube of the telescope is very long and requires a tall, massive mounting.

As the telescope is swung from looking at the horizon to looking overhead, the massive lens sags under its own weight, distorting its shape and changing its optical properties. Because of this problem, it has not proved feasible to build an acromat over one meter in diameter. The lens would have to be too thick to be practical. It is much easier to build large reflecting telescopes.

FIGURE 4.22 Yerkes 1-meter refractor. (Yerkes Observatory.)

Reflecting Telescopes. A mirror used in a reflecting telescope is ground and polished on its front surface, and then has a thin layer of aluminum vaporized on it (Figure 4.23). Therefore, light does not penetrate the glass of the mirror: it reflects from the front surface. This fact gives large mirrors one advantage over large lenses: the mirror can be made very thick to give it rigidity. In addition, a ribbed pattern can be molded into the back of the mirror for added strength, with minimum weight. In Figure 4.24, the ribbing of the 3-meter (120-inch) mirror at Lick Observatory can be seen through the front of the unaluminized mirror. Furthermore,

FIGURE 4.23 The 3-meter mirror of the Lick Observatory after aluminizing. (Lick Observatory photograph.)

FIGURE 4.24 The 3-meter mirror of the Lick Observatory on the grinding machine. The ribbing on the back of the mirror provides structural rigidity with low weight. (Lick Observatory photograph.)

FIGURE 4.25 The 3-meter reflecting telescope at Lick Observatory. (Lick Observatory photograph.)

the mirror can be supported in a cell that is designed to prevent sagging.

There are several telescopes operating in the United States with mirrors in the 3- to 5-meter range, including the 3-meter (120-inch) telescope at Lick Observatory (Figure 4.25), a new 4-meter (158-inch) telescope at the Kitt Peak National Observatory (Figure 4.26), and the 5-meter (200-inch) telescope at Palomar Observatory (Figure 4.28). The Soviet Union has built a 6-meter (236-inch) telescope.

FIGURE 4.26 The 4-meter Mayall Reflector at Kitt Peak National Observatory. (Kitt Peak National Observatory. Copyright © 1975 by the Association of Universities for Research in Astronomy, Inc.)

FIGURE 4.27 The building that houses the Mayall Reflector. For an idea of the scale, note the door and catwalk at the base of the dome. (Kitt Peak National Observatory. Copyright © 1975 by the Association of Universities for Research in Astronomy, Inc.)

FIGURE 4.28 The 5-meter Hale Reflector on Palomar Mountain. (Hale Observatories.)

Several optical arrangements are possible in reflecting telescopes, with a second or even a third mirror used to fold a long focal length into a relatively short tube. The most familiar optical arrangement for small telescopes, such as an amateur astronomer might use, is the **Newtonian** focus (Figure 4.29a). A flat mirror is placed at 45° to the optical axis of the main mirror, and reflects the light beam, which has been focused by the main mirror alone, to the side of the telescope tube, where it can be observed. The surface of the main mirror of a Newtonian telescope is paraboloidal in shape. That is, its cross section is a parabola.

In the **Cassegrain** system (Figure 4.29b), a convex secondary mirror reflects the converging beam from the primary mirror back through a central hole in the main mirror. The secondary mirror also decreases the convergence of the beam, making it appear to come from a mirror with a much longer focal length. The main mirror of a Cassegrain is also a paraboloid, but the secondary mirror has a hyperboloidal shape. A modern variant of the Cassegrain is the **Ritchey-Chrétien** system, in which both the primary and the secondary mirror are hyperboloidal; this system gives a larger field of view than the classical Cassegrain.

The largest telescopes use the **prime focus**, where an image is formed inside the tube of the telescope (Figure 4.29c). The astronomer sits inside the prime-focus cage and makes his observations (Figures 4.30 and 4.31). The prime-focus cage blocks a portion of the main mirror, but the center of the mirror is already cut out for

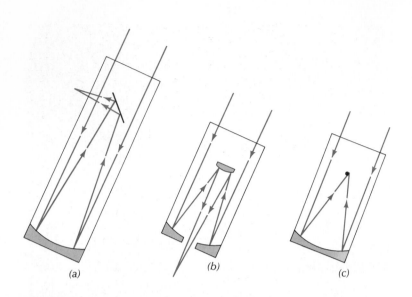

(a) (b) (c)

FIGURE 4.29 Focus arrangements in reflecting telescopes; (a) Newtonian focus. (b) Cassegrain focus. (c) Prime focus.

FIGURE 4.30 The 5-meter Hale Reflector, showing an observer using an elevator to approach the prime-focus cage. This picture gives one a good feeling for the scale of the immense optical device. (Hale Observatories.)

FIGURE 4.31 Observer at work at the prime focus of the Hale Reflector. (Hale Observatories.)

the Cassegrain focus so it does not matter. To go from Cassegrain-focus to prime-focus operation, the Cassegrain secondary mirror must be removed.

The **Coudé** focus is a system designed to allow very large instruments to be used with the telescope, for making measurements. The Coudé focus remains fixed in the observatory as the telescope is moved about.

Catadioptric Telescopes. The mirrors of reflecting telescopes must be ground to a precise paraboloidal shape in order for beams that hit the center of the mirror and those that hit the edge of the

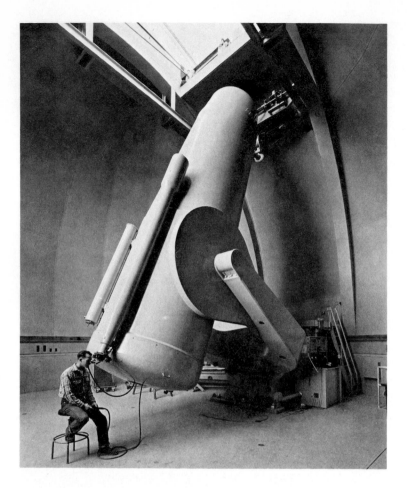

FIGURE 4.32 Schmidt telescope at Palomar Mountain. (Hale Observatories.)

mirror to focus to the same point. A spherical mirror does not have this property: it suffers from spherical aberration. On the other hand, spherical mirrors are easier to build than paraboloidal ones. Catadioptric systems combine a spherical mirror and a thin lens that corrects for spherical aberration. One type of catadioptric system with a wide field of view is the Schmidt optical system. The Schmidt telescope at Palomar Observatory can photograph a $6° \times 6°$ area of the sky (Figure 4.32).

We have now discussed the three types of telescopes: reflectors, refractors, and catadioptric systems. We next want to take a brief look at mountings for telescopes.

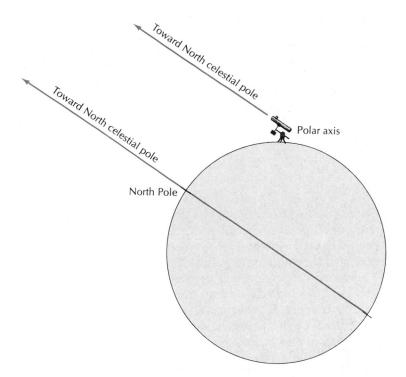

FIGURE 4.33 The polar axis of a telescope and its relationship to Earth's rotation axis.

Mountings

The mounting of a telescope has two functions: to allow the telescope to be pointed easily at any spot in the sky; and to allow the telescope to be driven easily to compensate for the rotation of the Earth. Most mountings are of a type known as **equatorial mounts**.

The **polar axis** of an equatorial mount is parallel to the Earth's axis of rotation (Figure 4.33). A telescope at the equator has a horizontal polar axis; at 40°N latitude the axis is tipped up 40° above the horizontal. As the Earth rotates, causing the stars to rise, move across the sky, and set, a small clock motor drives the telescope in the opposite direction, keeping it pointed at the same spot in the sky. Different types of mountings can be seen in the photographs illustrating this chapter. The mounting of the Schmidt telescope at Palomar or the 3-meter Lick reflector is called a fork mount. The "handle" of the fork is the polar axis. The giant "horseshoe" of the 5-meter Hale reflector at Palomar actually floats on a thin film of oil. All these telescope mounting systems are very delicately balanced. For example, the 530-ton mass of the Hale reflector can be driven by a $\frac{1}{12}$-horsepower motor.

In the discussion so far, the human eye has been the chief light detector mentioned. However, in modern astronomical research the human eye is seldom relied on. Astronomers use either photography or electronic devices to detect and record images. When a device such as a photographic plate is used, the eyepiece is removed from the telescope, and the real image is focused on the plate, where it is recorded. What are the advantages of photography?

Photography

A photographic plate consists of a glass or celluloid base covered with a thin emulsion layer of gelatin that contains compounds of silver. When light strikes these silver compounds, it causes them to break up and deposit small grains of silver in the emulsion layer. The process of development of the photographic plate simply removes all the unaffected compounds and leaves behind the exposed silver grains. The amount of darkening on the photographic plate depends on the number of silver grains on a given area of the plate, which in turn depends on the amount of light that fell on the area when it was exposed.

The photographic plate has two obvious advantages over the eye: it is a permanent record of an observation which can be studied again and again; and it can integrate. The ability of a photographic plate to integrate can be understood as follows. Roughly speaking, the blackening of an area of a plate depends only on the total amount of light that strikes it. It does not matter whether the plate is exposed to bright light for a fraction of a second or to the light of a faint celestial object for minutes or even hours. To record fainter and fainter objects, one need only make the exposure time longer and longer to have the same amount of light reach the plate.

There is a limit to exposure time, however. The Earth's atmosphere itself glows faintly—a phenomenon known as **airglow**. If an exposure is too long, airglow will blacken a plate so much that other objects cannot be discerned easily. As long as the astronomer does not exceed the limits imposed by airglow, he can photograph objects much fainter than he can see through the eyepiece.

The galaxy in Andromeda (see Figure 13.25) is a stunning object in a photograph. Most people that see it for the first time through an eyepiece are disappointed. It looks like a featureless smudge of light among the stars. The great detail in the photograph is brought out by long exposure.

Photoelectric Effect

When a metal is illuminated by light with the right range of wavelengths, it will emit electrons. The speeds of the emitted electrons depend on the wavelength of the light: the shorter the wavelength, the greater the speed; the longer the wavelength, the less the speed of the electrons, until a wavelength is reached when the speed goes to zero. If light of even longer wavelength illuminates the metal, no electrons are emitted. The number of electrons depends on the brightness of the light: the brighter the light, the more electrons are emitted. These two observations comprise the photoelectric effect.

In 1905 Albert Einstein explained the photoelectric effect by saying that light consists of a beam of particles he called **photons**. The energy of a photon depends on its wavelength (or frequency), in the sense that shorter wavelengths correspond to higher energies. When a photon strikes the metallic surface, it gives its energy to an **electron** that is bound to the metal. If the electron is given enough energy to overcome the binding, it may leave the metal. More-energetic, short-wavelength photons may be able to remove an electron from the metal, but less-energetic, long-wavelength photons may not be able to. Any energy that is in excess of the amount needed to tear the electron from the metal appears as energy of motion (kinetic energy) of the electron. This explains the relationship between speed of the electron and wavelength of the light.

The brightness of light is a measure of the number of photons in the beam: the brighter the light is, the more photons there are in the beam. Increasing the brightness of light hitting a metal increases the number of photons that can remove electrons, and, if the wavelength is right, increases the number of electrons removed.

When Einstein explained the photoelectric effect, he talked about the frequency of photons rather than about the wavelength. Either way, a characteristic usually associated with a wave is ascribed to a particle. In Young's double-slit experiment, light behaves like a wave; in the photoelectric effect, light behaves like a particle stream. In time it became clear that this wave-particle duality of light applied to other particles, such as electrons, as well—and the science of quantum mechanics was born.

Photoelectric Detectors

The photoelectric effect can be used to measure the brightnesses of celestial objects. A strip of metal is placed inside a glass tube from which the air is partially evacuated, and the image from the telescope is formed on the strip—now called the **photocathode** of the tube—which emits electrons at a rate that depends on the brightness of the image. To measure the brightness, we must be able to count the number of emitted electrons. The emitted electrons are multiplied into a measurable current inside the tube.

A fast-moving electron that hits a metallic surface can knock off several additional electrons. In a photomultiplier tube there are several metallic surfaces called **dynodes** connected to successively higher voltages (Figure 4.34). When an electron is emitted from the photocathode, it strikes dynode 1, where several electrons are emitted and accelerated toward dynode 2 by the difference in electrical voltage between the dynodes. At dynode 2, each electron from dynode 1 causes several electrons to be emitted, and these are accelerated toward dynode 3. After several such stages, a million electrons are collected for each electron emitted by the photocathode. The resultant current is easily detected and recorded.

The modern trend is to use a device similar to a television camera which can work with very low levels of light to detect images of astronomical objects. Brightness at each point in a two-dimensional image can be read off the tube numerically, and can be written on computer tape for later processing. The great advantage of such a technique over photographic methods is the numbers can be turned into images by automatic computer processing.

We have talked about the detection of light by using optical telescopes and either the eye, a photographic plate, or a photoelectric device. All the discussion has been about visible wavelengths. Let's now move into the longer wavelengths, and talk about radio telescopes.

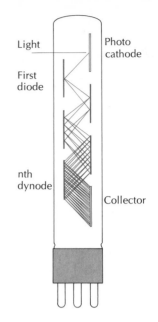

FIGURE 4.34 Photomultiplier tube. (From *New Horizons in Astronomy* by J. C. Brandt and S. P. Maran. W. H. Freeman and Company. Copyright © 1972.)

INVISIBLE RADIATION

Not all electromagnetic radiation of interest to astronomers can be seen by the human eye. The full range of the electromagnetic spectrum has already been described earlier in this chapter, and astronomers study the sky at almost every wavelength represented. Here we want to talk very briefly about telescopes used in three wavelength bands: the radio; the infrared; and the very short ultraviolet wavelengths, often called the extreme ultraviolet or EUV. But first, let us say a few words about why we study these other wavelength regions at all. In general terms, the answer is quite simple. The radiation that reaches Earth in different wavelength regions from a specific celestial object is produced by different

FIGURE 4.35 The 140-foot (43-meter) radio telescope at NRAO. (National Radio Astronomy Observatory, Greenbank, West Virginia.)

physical processes. Thus, to learn about all the processes that occur in the object, we must observe it in all possible wavelength regions. An example of what we mean is provided by the Sun (Chapter 8). The visible radiation from the Sun is radiated in the relatively hot gas of its photosphere. However, the extreme ultraviolet radiation that we observe from space satellites is emitted by the ultra-hot gas in the corona. To learn about the different gaseous layers of the Sun, we must observe it in different wavelength regions. On the other hand, some objects emit almost all their radiation in some wavelength band other than the visible. For instance, very cool, young stars (Chapter 15) are brightest at infrared wavelengths. To learn about the early life histories of stars, we will have to carry out our studies in infrared wavelengths.

Radio Telescopes

The branch of astronomy called **radio astronomy** began in 1931, when Karl Jansky, a radio engineer at Bell Research Laboratories, proved that some of the static picked up by his radio receivers came from a cosmic source. Impetus was given to the field by the tremendous amount of research on radio and radar during World War II. Armed with the knowledge that cosmic sources emit radio waves, and with constantly improving radio-receiver technology, astronomers began to build radio telescopes.

The easiest way to collect and focus radio waves is with a mirror—called a dish for short—that reflects the radio waves to a radio detector (Figure 4.35). The resolving power of a telescope is pro-

FIGURE 4.36 The Culgoora radioheliograph (solar radio telescope) in New South Wales, Australia. The 96 antennas lie along a circle 3 kilometers across. (Courtesy J. P. Wild, Division of Radiophysics, C.S.I.R.O.)

portional to the wavelength being observed. A visible-light telescope receives light with a wavelength of about 5×10^{-5} cm, whereas a radio telescope may receive radiation with a 10-cm wavelength. To have the same resolving power as an optical telescope of a given size, the radio telescope must have a mirror 200,000 times larger. Therefore, to observe the sky with a resolution of 1 arc second at the 10-cm wavelength would require a dish almost 23 kilometers (14.7 miles) in diameter.

The fully steerable radio telescopes—that is, radio telescopes on mountings that can point them to any point in the sky—are smaller than 75 meters (250 feet) in diameter. Since these mirrors are much less than 20 kilometers in diameter, they certainly cannot resolve objects 1 second of arc apart when they are used to observe at the 10-centimeter wavelength. Nevertheless, radio astronomers have found a way to observe finer details on the sky by a technique called **aperture synthesis**. Here, the signals from several radio telescopes are combined in a computer to produce a resultant signal that appears to have been made with a telescope whose diameter equals the largest distance between the telescopes being used (Figures 4.36 and 4.37).

FIGURE 4.37 Three-element radio interferometer at NRAO. (Courtesy National Radio Astronomy Observatory, Greenbank, West Virginia.)

We chose 10 centimeters as a representative wavelength for radio astronomy. In fact, observations are being made today at wavelengths ranging from less than a millimeter to several meters. As we shall see, each wavelength range (millimeter, centimeter, decimeter, and meter) contributes something to our understanding of the cosmos.

Infrared Astronomy

How would you like to try to take photographs with a camera that was coated inside and out with phosphorescent paint, so that it glowed? This is the problem that astronomers studying the infrared must face. Warm objects emit infrared: human bodies do, telescopes do. Although ordinary telescopes can be used to collect infrared radiation, special techniques must be used to get rid of the glow of the telescope and its surroundings.

Infrared wavelengths cover the range from 1 micron (10,000 Ångstroms or 10^{-4} centimeters) to about 1,000 microns (1 millimeter), where they merge into the domain of radio astronomy. At the shortest infrared wavelengths, observations can be made from the ground with ordinary telescopes. The telescope may first be pointed at the object being studied, then at the nearby sky. The difference between the measured signals is due to the radiation from the object

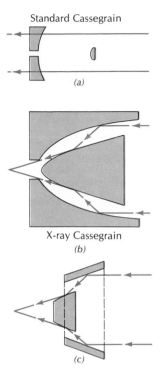

Standard Cassegrain

(a)

X-ray Cassegrain

(b)

(c)

FIGURE 4.38 Wolter-type telescope for X-rays. (a) X-rays that strike the mirror perpendicularly just pass through the mirror. (b) However, X-rays that strike the mirror at a glancing angle are reflected. (c) Removing extraneous mass leaves us with an X-ray telescope.

under study. At longer wavelengths, the observing equipment must be cooled with liquid nitrogen to about $-320°$ F, or with liquid helium to even lower temperatures. The telescope also must be flown above the atmosphere on high-flying aircraft or space satellites, since at long wavelengths the atmosphere is no longer transparent.

Short-Wavelength Astronomy

The Earth's atmosphere is opaque to all radiation with wavelengths shorter than roughly 3000 Å. To observe celestial objects at shorter wavelength ranges, for example, at the extreme ultraviolet, X-ray, or gamma-ray wavelengths, the telescope must be flown above the atmosphere on a space satellite. Telescopes similar to our ground-based reflectors can be used at wavelengths as short as 1000 Å. However, at 300 Å the reflectivity of the mirrors is very small, and a different design must be used (Figure 4.38).

We have described very briefly the problems that astronomers face when they observe objects outside the visible-light region of the spectrum. The solutions that have been found to these problems are ingenious and complex.

THE INVERSE-SQUARE LAW

Before we end this chapter, there is one more important fact that we must introduce. How does the brightness of a light source vary as a function of distance between the source and the observer?

The answer to this question is relatively easy to understand if we know that the apparent brightness of a source is measured by the amount of light energy that strikes each unit of area of our detector. As light travels away from a source, it spreads out as illustrated in Figure 4.39. All the radiation from the light bulb that passes through the surface labeled 2 first passes through surface 1. However, surface 2 has four times the area of surface 1; so the amount of light that passes through each unit area of surface 2 is one-fourth the amount that passes through the unit area of surface 1. If we place a light meter at surface 2, it measures one-fourth the brightness it would measure at surface 1. The observed brightness of a source decreases with the square of the distance from the source. If we measure the brightness of a source at a distance of 1 meter, then at 2 meters, 3 meters, 4 meters, and so on, we will find its brightness

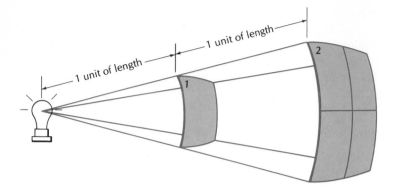

FIGURE 4.39 Inverse-square law (see text).

at 2 meters to be $\frac{1}{4}$ its brightness at 1 meter, its brightness at 3 meters to be $\frac{1}{9}$ its brightness at 1 meter, its brightness at 4 meters to be $\frac{1}{16}$ its brightness at 1 meter, and so on. This is known as the inverse-square law of brightness as a function of distance.

Armed with the knowledge we have gained in this chapter, let's begin our survey of the observable universe, beginning with the Earth and working our way outward. We will have to digress on occasion and discuss some needed basic physics. We will, from time to time, also talk about other types of astronomical instruments.

REVIEW QUESTIONS

1. Describe or define: (a) transverse wave; (b) wavelength; (c) amplitude; (d) interference; (e) polarization; (f) refraction; (g) focal length; (h) image; (i) seeing.

2. What experiments show that light behaves as a wave? As a particle?

3. Describe reflecting and refracting telescopes. How does a telescope work? Why were reflecting telescopes invented?

4. What are the advantages of a telescope over the unaided eye?

5. What is the purpose of an equatorial mounting for a telescope?

6. How do astronomers detect light?

7. How does the brightness of an object change as its distance from the observer changes?

5

THE EARTH AS A PLANET

The Earth is a planet, orbiting the Sun on a path between the orbits of Venus and Mars. It took a revolution in philosophy for us to be able to make this simple statement. The revolution started by Copernicus removed the Earth from the center of the universe, and, at first, gave the Sun that honored position. Today we know that the Sun is not at the center of the universe. The Sun is merely an ordinary star, occupying no special place in the universe.

The Earth and the Sun are the center of our existence, however. Virtually all the power we consume to make our lives comfortable comes from the Sun, indirectly, and we will continue to lead lives of high quality only so long as we keep the Earth a good place to live. We will begin our survey of the universe here at home, on Earth, and work our way outward.

In this chapter we will consider three basic topics: the Earth's motions and their consequences; the dimensions of the Earth; and the internal structure of the Earth. We will talk about the Earth's atmosphere and near environment in space later, after we have talked about the Sun. This plan of attack will allow us to discuss the strong influences of the Sun on the Earth's environment.

We have already mentioned the motions of the Earth in Chapters 2 and 3. The opening sections of this chapter will focus more closely on these motions, and describe some of their consequences. It is these consequences that showed that the Earth does move.

THE REVOLUTION OF THE EARTH

The Earth's Orbit

The Earth orbits the Sun on a path that is an ellipse with the Sun at one focus. How big is the Earth's orbit, and just how elliptical is it? The second question is the easiest to answer; so we will answer it

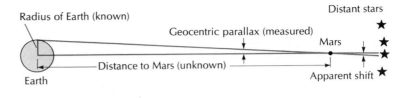

first. We could measure the shape of the Earth's orbit by measuring the size of the Sun at various times of the year. When the Earth is nearest the Sun, it would appear largest and brightest; when the Earth is farthest from the Sun, it would appear smallest and faintest. In this way, we would find that when the Earth is nearest the Sun, it is 1.17% nearer than average, and when it is farthest from the Sun, it is 1.17% more distant than average. In short, the Earth's orbit does not depart much from a circle. How do we find the size of the Earth's orbit?

From studies such as Kepler carried out to arrive at his famous laws of planetary motion, we can calculate *relative* distances in the solar system. The distance unit is the average distance between the Sun and the Earth, a unit we call the **astronomical unit**. We know, for instance, that Mars' average distance from the Sun is 1.5 astronomical units, and Venus' average distance is 0.72 astronomical units. If we could measure in kilometers any distance that we already know in astronomical units, then we will know how many kilometers there are in an astronomical unit, which is equivalent to knowing the Earth's average distance from the Sun in kilometers.

Of course, we cannot use a yardstick to measure distances in the solar system; so we must use an indirect means. Triangulation is one of the best indirect means available. For instance, we might measure the distance to Mars when it is closest to the Earth, as follows. We measure the position of Mars relative to the background stars when it is just rising and then again when it is directly overhead (Figure 5.1). Between the two measurements, the observer has moved a distance equal to the radius of the Earth because of the Earth's rotation. This distance is the base of the triangle. The amount by which Mars shifts relative to the stars between the measurements is called its **geocentric parallax**, because it is measured as the observer moves around the center of the Earth. Mars' geocentric parallax is small—around 50 seconds of arc—and the measurement is subject to large errors. The skinny triangle that must be solved to obtain a numerical value for Mars' distance is discussed further in Appendix A.

One modern method of measuring a distance in the solar system involves bouncing radar signals off the planets—especially Venus.

Since we know the speed of light with great precision, and we can measure the time-interval between transmission of the radar signal and receipt of the signal reflected by Venus also with great precision, we can calculate the round-trip distance with high precision. The result gives us our best value for the astronomical unit; it is close to 149,598,000 kilometers (92,960,000 miles). For our purposes, it is usually sufficient to remember that the astronomical unit is 150,000,000 kilometers or 93,000,000 miles.

Earth-Moon System

The Moon orbits around the Earth much as the Earth orbits the Sun. The Moon's orbit, too, is an ellipse with the Earth at one focus. The motion of the Moon that would be seen by a space traveler hanging motionless above the solar system is complicated, since the Earth continues to move around the Sun as the Moon moves around the Earth.

There is a fact about orbital motion that we have not mentioned before. If two bodies are attracted by gravity, one appears to orbit the other. However, the pair of bodies obeys Newton's first law and travels in a straight line unless acted upon by an external force (external to the mutual gravitational force between the two bodies). The thing that moves in a straight line is the **center of mass** of the two bodies. Imagine a giant rod connecting the two bodies, and imagine a giant planet with a fulcrum attached to it, so that we can try to find the balance point on the rod (Figure 5.2). The balance point is the center of mass. In fact, as seen from the center of mass, both bodies perform elliptical orbits, with the center of mass at one focus. In astronomical situations the center of mass is sometimes called the **barycenter**.

The Sun is the most massive object in the solar system, by far. The center of mass between the Sun and any other planet lies close to the center of the Sun. (That is why this complication did not come up when we were discussing planetary orbits.) In the Earth-Moon system, the center of mass is roughly 1,600 kilometers below the surface of the Earth. That point moves around the Sun in an elliptical orbit. In the roughly one-month orbital period of the Earth-Moon system, the center of the Earth describes a small ellipse around the barycenter. So the actual motion of the Earth, too, is complicated.

We must continue to think about the Earth-Moon system as a unit, for we shall discover later that the Moon has other important effects on the Earth. Additional details about the system will be discussed in Chapter 6.

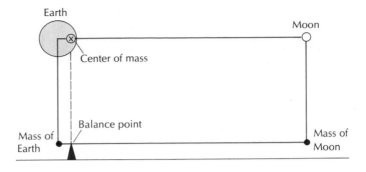

FIGURE 5.2 The center of mass of the Earth-Moon system is the point where it would balance if the bodies were connected by a rod and set on a giant fulcrum.

Seasons

Because of the motion of the Earth around the Sun, the Sun appears to move among the stars from day to day. It is like sitting in a smooth-riding car watching the scenery pass. One could almost imagine that the car was stopped and that the scenery was rushing by the car window. The path the sun follows among the stars (see Chapter 2)—the ecliptic—is actually the intersection of the plane of the Earth's orbit and the celestial sphere. We often call the plane of the Earth's orbit the plane of the ecliptic.

The Earth's rotation axis is tipped $23\frac{1}{2}°$ away from the normal to the plane of the ecliptic. As the Earth revolves around the Sun, the axis of rotation always points to the same spot in the sky (Figure 5.3). This fact can be stated in another way: as the Earth revolves around the Sun, its axis of rotation remains parallel to itself.

The unchanging orientation of the Earth's rotation axis is the basic cause of the seasons. During summer in the northern hemisphere, the northern end of the Earth's axis is tipped toward the Sun; so the northern hemisphere receives more direct sunlight than the southern hemisphere (Figure 5.4). When it is winter in the northern hemisphere, the northern end of the rotation axis is tipped away from the Sun. On the first day of spring and of fall, the axis is perpendicular to the direction from the Earth to the Sun. (The orientation of the Earth's axis relative to the Sun described above is what would be seen by a space traveler looking at the Earth from a great distance. What do we see here on Earth?)

In Chapter 2 we talked about the celestial poles. These are the points on the sky where the Earth's rotation axis intersects the celestial sphere. The **celestial equator** is a great circle on the celestial sphere midway between the poles. If one extended the plane of the Earth's equator until it intersected the celestial sphere,

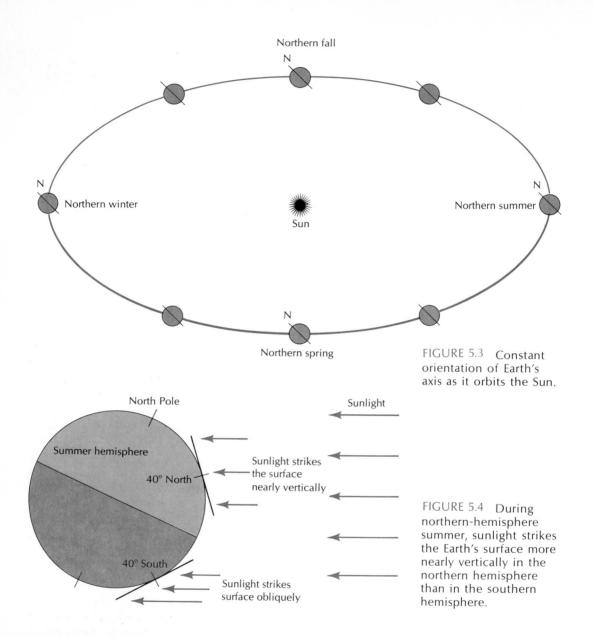

Northern fall

N

Northern winter

N

Sun

Northern summer

N

Northern spring

N

FIGURE 5.3 Constant orientation of Earth's axis as it orbits the Sun.

North Pole

Sunlight

Summer hemisphere

40° North

Sunlight strikes the surface nearly vertically

40° South

Sunlight strikes surface obliquely

FIGURE 5.4 During northern-hemisphere summer, sunlight strikes the Earth's surface more nearly vertically in the northern hemisphere than in the southern hemisphere.

it would intersect the celestial sphere along the celestial equator. Because the Earth's axis is tipped $23\frac{1}{2}°$ to the ecliptic, the celestial equator is tipped $23\frac{1}{2}°$ to the ecliptic. That is to say, there is an angle of $23\frac{1}{2}°$ between the plane of the ecliptic and the plane of the celestial equator (Figure 5.5). These two circles on the sky intersect at two points.

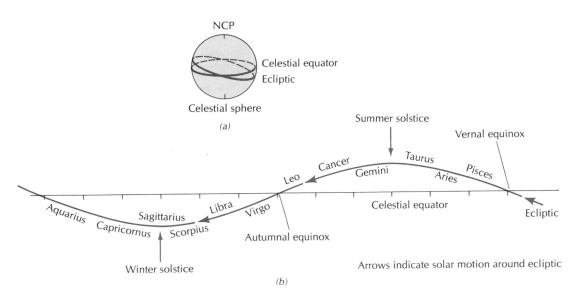

NCP

Celestial equator
Ecliptic

Celestial sphere

(a)

Summer solstice

Vernal equinox

Cancer
Leo
Taurus
Gemini
Aries
Pisces

Aquarius
Capricornus
Sagittarius
Scorpius
Libra
Virgo

Celestial equator

Ecliptic

Autumnal equinox

Winter solstice

Arrows indicate solar motion around ecliptic

(b)

FIGURE 5.5 Relationship between celestial equator and ecliptic: (a) on the celestial sphere; (b) on the sky.

As the Sun moves 1° per day eastward along the ecliptic, it also moves relative to the celestial equator. On about March 21, the Sun crosses the celestial equator, going from south to north. The point where this occurs is one of the two points at which the ecliptic and the celestial equator intersect; it is called the **vernal equinox**. The Sun continues along the ecliptic, moving further northward from the equator, until it is as far north as it will go, on about June 22. This day is the first day of summer, and this point on the ecliptic is called the **summer solstice**. From June 22 until about September 23, the Sun moves back toward the equator. On September 23 the Sun crosses the **autumnal equinox**, the other intersection point. It then moves toward the **winter solstice**, about December 22, where it is as far south of the equator as it will reach. The Sun then moves back toward the vernal equinox, where the process begins again.

If June 22 is the day we, in the northern hemisphere, receive the most direct radiation from the Sun, why is it not the hottest day of the year? In simplest terms the answer is that many factors other than the solar-energy input control our climate. Another important factor is the temperature of the ground and the oceans. The surface of the Earth continues to absorb the solar heat and continues to warm up until it reaches a temperature where it radiates as much energy back to the atmosphere in the form of infrared radiation as it receives in all radiations from the Sun. At that point it can get no warmer, of course. The ground reaches this point three or four weeks after summer begins, and the ocean reaches it fully two months after summer begins. Thus the hottest weather of the year arrives earlier in the midwest than in the coastal states, but in both places it lags after the opening of summer.

Of course, location effects climate, too. In high mountain meadows in the Rockies, the last snows of the previous winter do not fully melt until August; so spring flowers bloom very late in the year.

Incidentally, Earth is at perihelion (the nearest point to the Sun in its orbit) in January and at aphelion (the farthest point from the Sun in its orbit) six months later, in July. The changing distance from the Sun affects our seasonal temperature variations slightly: our northern-hemisphere winters are a bit warmer, our summers a bit cooler, than they would be if the Earth moved in a circular orbit. Of course, the tilt of the Earth's axis has a far larger effect on the seasonal temperatures than the planet's changing distance from the Sun.

In this section, we have described the orbital motion of the Earth and its main effect, the seasons. We know that the seasons come about because Earth's rotation axis is tilted to the plane of its orbit and keeps a fixed orientation in space as the planet circles its orbit. We will now look in detail at Earth's rotation.

THE ROTATION OF THE EARTH

In Chapter 3, the arguments in favor of the fact that the Earth (rather than the celestial sphere) rotates were rather unscientific. The arguments made the Earth's rotation seem reasonable, but were not hard and fast proof. Proof of the Earth's rotation was provided by a French physicist, Jean Foucault, using a simple, ingenious method. In 1851, Foucault suspended a free-swinging pendulum from the interior of the dome of the Pantheon in Paris. The dome was very high, allowing the pendulum to be over 60 meters (200 feet) long.

Foucault drew the heavy weight of the pendulum to one side, tied it securely with a thin cord, then burned the cord to set the pendulum swinging smoothly in one plane.

Gravity pulled the pendulum toward the center of the Earth; it could not act to change the plane in which the pendulum swung. Thus, as the Earth rotated, the pendulum swung in a plane fixed in space, and the Earth rotated under it. To the spectators watching Foucault's demonstration, the plane in which the pendulum swung

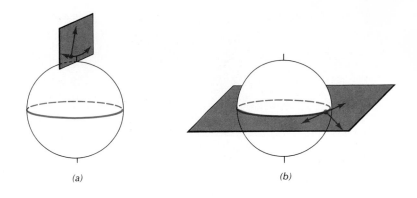

FIGURE 5.6 Foucault pendulum at the North Pole and at the equator. As Earth rotates under a pendulum at pole (a), its plane of swing appears to rotate. In contrast, as Earth rotates, plane of swing of pendulum at equator (b) remains fixed.

(a)   (b)

appeared to rotate slowly, opposite in direction to the Earth's rotation. The rotation of the Earth was proven.

The Foucault pendulum is easiest to understand if it is set swinging at the north pole (Figure 5.6). Then, as the Earth rotates under the pendulum, its plane appears to make one rotation in 24 hours. At the equator, the plane of the pendulum does not appear to rotate relative to the Earth. Between the pole and the equator, the apparent period of rotation of the plane of the pendulum increases from 24 hours, through 37 hours at 40° latitude, to infinity (non-rotation) at the equator.

The Shape of the Earth

The knowledge that the Earth is a sphere goes far back in time, probably to the sixth century B.C. One argument used by the ancients to prove the Earth's spherical shape has been discussed in Chapter 2. As one travels north or south on the Earth, different stars are seen overhead (Figure 5.7). As Aristotle put it in his treatise "On the Heavens," there are some stars seen from Egypt which are not seen from more northerly regions, and some stars which are circumpolar in the north rise and set as seen from Egypt. If the Earth were flat, the same stars would be overhead no matter where one stood on the surface and looked at the sky (Figure 5.8).

Today we need no such indirect arguments. Men have stood on the surface of the Moon and looked back at the finite Earth. They have seen with their own eyes that it is a sphere.

However, very careful measurements show that the Earth has an equatorial bulge: its equatorial radius is 21 kilometers more than its polar radius. The equatorial bulge is a consequence of the rotation of the Earth. We call the shape of the Earth an oblate spheroid.

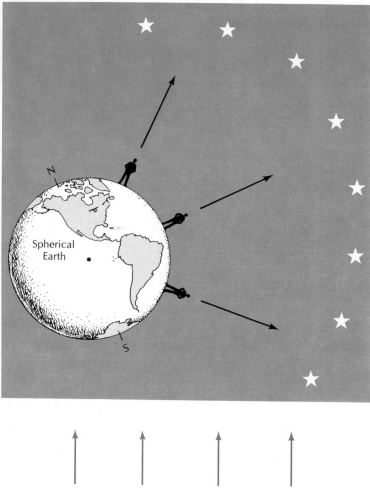

FIGURE 5.7 On the Earth, as a person travels along a north-south line, the stars that are near the zenith change. This change is due to the Earth's spherical shape.

Spherical Earth

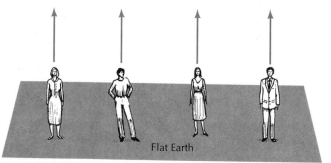

Flat Earth

FIGURE 5.8 If the Earth were flat, the same stars would always be near the zenith. Lines to different person's zeniths are all in the same direction, and all point to the same, distant stars.

Tides

If you have ever visited an ocean beach, you are undoubtedly familiar with the ocean tides. The level of water at the seashore (and throughout the ocean) goes through two high tides and two low tides each day, the average time between high tides being roughly 25 minutes longer than twelve hours. The range of the tide is the

FIGURE 5.9 Resultant net tidal force on the Earth's oceans (and rocks). (a) The total forces on different particles of water are indicated by arrows. The direction of the arrow is the direction of the force, and the length of the arrow indicates its size. The vector (arrow) **A** represents the average force on the Earth. (b) If we subtract the average force on the Earth from all the total forces, we find so-called net forces as illustrated. Note that the net effect of the forces is to push downward on the water at the poles and along the meridian *B* whose plane is perpendicular to the Earth-Moon line, and to pull outward on the water nearest to and farthest from the Moon.

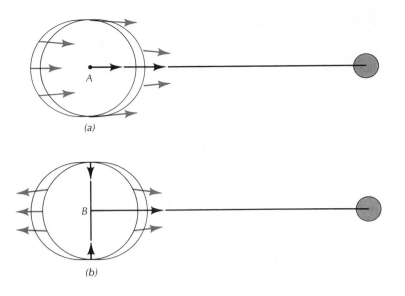

(a)

(b)

difference between the levels of successive high waters and low waters.

The greatest range of the tide on Earth is measured at the head of the Bay of Fundy, between New Brunswick and Nova Scotia, Canada, where it reaches 15 meters (50 feet). At islands in the mid-Pacific Ocean, the range is usually less than a meter.

The time of high tide is not usually uniform along a stretch of coastline, but will vary from place to place. In tidal rivers, such as the Potomac near the District of Columbia, the time of high tide becomes increasingly later with distance from the ocean.

Tides are caused by the gravitational pull of the Moon on the Earth and are modified by local coastal conditions. Let's think about a very idealized situation, a perfectly smooth, rigid Earth completely covered by a deep ocean. The gravitational pull of the Moon acts on each particle of water in the ocean with a force that is proportional to the mass of the Moon times the mass of the particle of water, and is *inversely proportional to the square of the distance from the particle to the center of the Moon.* The latter phrase is all important. The force on the particles of water on the side of the Earth nearest the Moon is greater than the force on the water on the side of the Earth farthest from the Moon. If we calculate the average force of the Moon on the whole Earth, then subtract this average from the force of the Moon on each particle of water, we get resultant **net forces** as shown in Figure 5.9. Force is a vector quantity. The arrows in the figure show the net force of the Moon on the water in the ideal ocean. The net force on the water nearest the Moon is directed toward the Moon, whereas the net force on the water on the opposite side of the Earth is directed away from the Moon. On a great circle around the Earth midway between the extreme points, the net force is toward the center of the Earth. These net forces tend to pull water toward and away from the Moon at the extreme near and far points and to push the water toward the center of the Earth along the great circle between the extremes. The result is a bulge in the ocean, increasing its depth toward the Moon and away from the Moon.

In the real situation, the ideal case remains true on the average. But the fact that there are continents with jagged coastlines modifies the heights and times of high and low tides. The Earth rotates under the Moon, causing the bulge to appear to move around the Earth. As the Earth makes one rotation on its axis, the Moon moves one-thirtieth of its orbit or 13°, and the Earth must rotate for an additional 50 minutes of time to catch up. That's why high tides are 12 hours and 25 minutes apart—as seen from the Moon, the Earth rotates in 24 hours and 50 minutes.

The Earth itself is not perfectly rigid. Its solid body can be deformed slightly, just like its oceans. Thus there are tides in the solid body of the Earth, too. These tides amount to about 20 centimeters (9 inches). They can be detected, for instance, by a change in the direction of a very accurate plumb line. Incidentally, the Earth's atmosphere has tides, too. These can be detected as a tiny oscillation in barometric pressure superimposed on the changes due to weather patterns.

Having talked about tides, we will now look at another motion of the Earth that occurs because of the Earth's rotation and the equatorial bulge.

Precession

Figure 5.10 illustrates a phenomenon that can be demonstrated with a toy gyroscope. The wheel of the gyroscope is made to spin rapidly, and the device is suspended horizontally from a pivot. If the wheel were not spinning, gravity would cause the toy to fall to the floor. However, with the wheel spinning, the gyroscope will swing around in a direction perpendicular to the direction of the force of gravity. The axis of rotation of the wheel rotates around the pivot, a motion called **precession**.

Like the toy gyroscope, the whole Earth precesses. The axis of rotation of the Earth swings around, keeping an angle of $23\frac{1}{2}°$ to the normal to the plane of the ecliptic. The axis describes a cone in

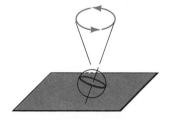

FIGURE 5.10 The precession of a top. As the top spins, the upper tip of its axle describes a circle as the lower tip remains fixed.

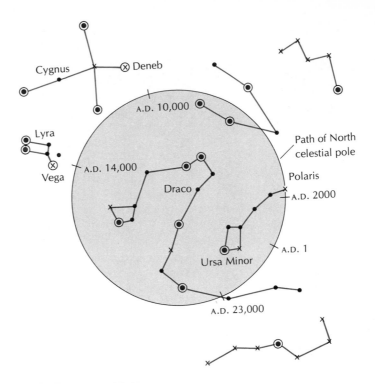

space. In fact, then, the axis of the Earth does not always point strictly in the same direction in space as it orbits the Sun. But since it does take 26,000 years for the axis to precess once around the normal, the change in a year (or even a century) is miniscule.

Figure 5.11 shows the path that the north celestial pole traces through the stars because of precession. Note that in about 13,000 years, the bright star Vega will be near the pole. The vernal equinox (and the autumnal equinox, too) move westward among the constellations. At the time of Christ, the vernal equinox was in the constellation Aries: today it is in the constellation Pisces.

Since the vernal equinox moves westward, it meets the Sun before the Sun appears to complete its circuit around the sky relative to the stars. Thus the tropical year, the time it takes the Sun to circuit the sky from vernal equinox to vernal equinox, is about 20 minutes shorter than the sidereal year, the time it takes the Sun to circuit the sky relative to the stars. The tropical year is the year on which our calendar is based.

Why does the Earth precess? The Sun always lies in the ecliptic plane, and the Moon is never far from the ecliptic. The gravitational force of these two bodies on the Earth's equatorial bulge tries to

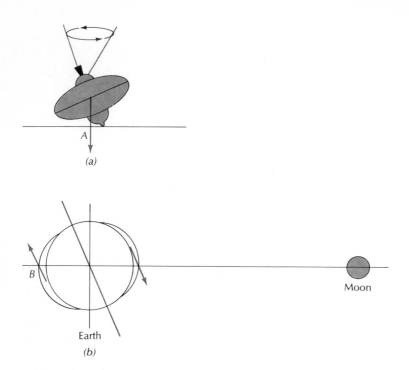

(a)

(b)

B

Earth

Moon

FIGURE 5.12 The cause of precession. (a) The force of gravity A on the spinning top tries to tip it over. The result is a precession. If the top were not spinning, it would simply topple over. (b) The net force of the Moon on the Earth's tidal bulge tries to pull the planet's equatorial plane into the ecliptic plane. The result is a precession. Like the top, if the Earth were not rotating, its equator would correspond to the ecliptic.

pull the bulge nearer to the plane of the ecliptic from its $23\frac{1}{2}°$ tilt. That is, as the Sun and the Moon try to pull the part of the bulge nearest to it toward its center, the net force on the Earth is a "tipping force" (Figure 5.12). Just as happens with the toy gyroscope, the axis of the spinning Earth reacts to the tipping force by moving in a direction perpendicular to the direction in which the force acts—that is, it precesses.

So far in this chapter we have discussed the motions of the Earth, revolution and rotation, and some of the implications of these motions. It is now time to take a closer look at the planet itself, and discuss its size, mass, density, age, and structure.

DIMENSIONS OF THE EARTH

The Size of the Earth

To the ancients, one proof that the Earth is round is the fact that, as one travels north or south on the Earth, different stars are seen overhead. This fact also provided a way to measure the size of the

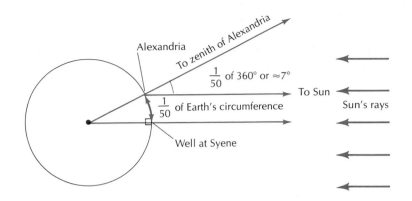

FIGURE 5.13 Eratos-
thenes measures the
Earth.

Alexandria

To zenith of Alexandria

$\frac{1}{50}$ of 360° or ≈7°

To Sun

$\frac{1}{50}$ of Earth's circumference

Sun's rays

Well at Syene

Earth. Eratosthenes, who lived in Alexandria in the third century
B.C., was the first known philosopher to make use of the idea. He
noticed that at the city Syene in Egypt, the Sun was directly
overhead at noon on the first day of summer. (The story says that
sunlight illuminated the bottoms of vertical wells at noon on that
day, showing the Sun to be exactly overhead.) At the same time,
however, the Sun was not overhead when viewed from Alexandria;
it was south of the zenith by a distance equal to one-fiftieth of a
circle around the whole sky (Figure 5.13). Eratosthenes concluded
from these observations that the distance from Alexandria to Syene
was one-fiftieth of the circumference of the Earth. Now, the distance from Syene to Alexandria had been measured to be about
5,000 stadia; so the circumference of the Earth worked out to be
250,000 stadia, and its diameter about 80,000 stadia.

There are several sources of error in Eratosthenes' calculation of
the diameter of the Earth. The distance from Syene to Alexandria
was known only roughly: it probably had been paced off by survey-
ors. Furthermore, Syene was not due south of Alexandria as the
method requires. Finally, we do not know exactly how long a
stadium (plural, stadia) was.* It seems that several values were in
use in Greece at various times in history. One stadium, known as
the Olympic stadium, would make Eratosthenes' result about 20%
too large. However, the exact value he obtained is not as important
as the fact that the method *is* basically correct.

In round numbers, the modern value for the diameter of the
Earth is 12,800 kilometers (7,900 miles).

Now that we know the size of the Earth, the next step is to find
its mass and density. The **density** is the mass per unit volume; that

---

*How do you suppose we got our current meaning of "stadium" if the word
originally meant a unit of length?

is, the number of grams for each cubic centimeter that there is in the Earth, on the average. We have to say on the average because we know that different rocks on the surface have different densities, and we will see that the density increases with depth in the Earth. Let's look at how to calculate the Earth's mass.

The Mass of the Earth

A precise method for calculating the mass of the Earth was used in 1881 by Von Jolly. He constructed a balance like that described in Chapter 3 for measuring masses by balancing the force of gravity on an unknown mass with the force on known masses (Figure 5.14). He then placed identical, known masses on each pan of the balance and made sure it balanced. Next, Von Jolly rolled a large lead ball under one pan. The ball's gravitational force acted on the mass on the pan, and pulled it down toward the lead ball's center, destroying the balance. To achieve a new balance, an extra small mass was added to the other pan. Now there were two forces acting on each pan of the balance. On one was the force of the Earth's gravity on the known mass and the force of the lead ball's gravity on the known mass. On the other was the force of the Earth's gravity on the known mass and the force of the earth's gravity on the extra (also known) mass. Since the force of the earth's gravity on the equal, known masses was the same on both pans, which were balanced, the other two forces must have been equal too.

Von Jolly knew, then, that the force of gravity of the lead ball on the known mass was equal to the force of gravity of the Earth on the extra mass needed to balance the two pans. On the lead-ball side of the balance, he knew both masses and the distance between them. On the other side of the balance he knew one mass (the extra mass) and the distance from the mass to the center of the Earth (essentially the radius of the Earth). The only unknown quantity was the mass of the Earth, which he calculated from the known quantities. The result is 6×10^{24} kilograms, or 6×10^{27} grams.

Why was this elaborate scheme needed to find the Earth's mass? Go back to Chapter 3 and reread Newton's law of gravitation. It says that the force of gravity between two bodies is *proportional* to (not *equal* to) the product of the masses divided by the square of the distance between centers. There is also a constant, called the gravitational constant, that has to be applied to the result of multiplying the masses and dividing by the square of the separation. This constant has to be measured by experiments. We can now measure this constant with the information we have.

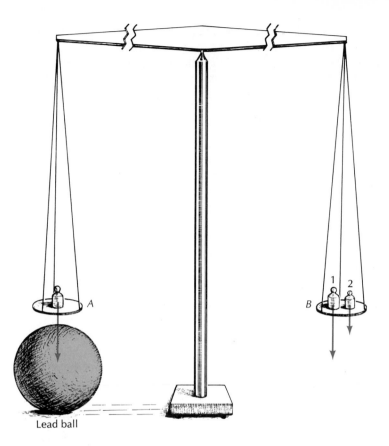

FIGURE 5.14 Apparatus used by von Jolly to find the Earth's mass. The balance arm is long enough that the lead ball effects only pan *A* and not pan *B*. The force on pan *A* is the sum of the Earth's gravity and the lead ball's gravity on the known mass. The forces on pan *B* are the force of the Earth's gravity on mass 1, which is identical to the mass on pan *A*, and the force of Earth's gravity on mass 2. When the pans balance, the force of the lead ball's gravity on the mass on pan *A* equals the force of Earth's gravity on mass 2.

Lead ball

Let's drop a body such as a brick from a high place. If we know the mass of the brick, we can calculate the force acting on it in two ways. First, we measure the acceleration that the brick experiences as it falls. According to Newton's second law of motion, the force on the brick is the product of its mass and its acceleration. Second, we know the brick's mass, the Earth's mass, and the Earth's radius; so we can calculate the force of gravity on the brick to be some number times G, the gravitational constant. The two ways of calculating the force must give the same answer; so we have an equation of the form (one number) $\times$ $G =$ (another number), and therefore we can find G. (The actual number, which we will never use, is $G = 6.67 \times 10^{-8}$ dyne cm^2/gm^2.)

Given the diameter of the spherical Earth to be 12,800 kilometers or 1.28×10^9 cm, we can find the Earth's volume to be 1.1×10^{27} cm^3 (this means "cubic centimeters"). The average density is then $(6 \times 10^{27}$ gm) $\div$ $(1.1 \times 10^{27}$ cm$^3) = 5.5$ gm/cm^3. The density of water is 1 gm/cm^3, to put that number in perspective. It is interesting to note that the average density of surface rocks is only about 2.7 gm/cm^3. Thus the density of the interior of the Earth must be much greater than 5.5 gm/cm^3. More will be said about this point when we talk about the Earth's interior, later in the chapter.

We have now arrived at the size, mass, and density of the Earth. There is one more fact that will prove useful in our studies: the age of the Earth. One reason the age of the Earth is interesting is the fact that it places a lower limit on the age of the universe: the age of the Earth is the youngest possible age of the universe.

The Age of the Earth

The earliest attempts to find the age of the Earth came from careful studies of Old Testament genealogies. In 1650, for instance, Archbishop Ussher placed creation in the year 4004 B.C. The general attitude in the seventeenth century was that the creation occurred sometime between 3900 B.C. and 4004 B.C. (In fact, this seems to be one of the few points on which the scholars of the Hebrew scriptures agreed in those days.) In the nineteenth century, the subject of prehistoric archaeology came into its own, and evidence began to pile up that the human race, and therefore the Earth, was far more ancient than the seventeenth-century theologians believed.

Then in 1896, natural radioactivity was discovered, and was soon applied to finding the ages of rocks. The nucleus of an atom is composed of two subatomic particles: positively charged **protons** and neutral **neutrons**. The charge of the nucleus—or, equivalently, the number of protons—determines which chemical element it is (Figure 5.15). The combined number of protons and neutrons determines the mass of the nucleus. Roughly, a nucleus contains equal numbers of protons or neutrons. However, different **isotopes** of atoms do exist. Isotopes are nuclei with the same charge but different mass; that is, the same number of protons but a different number of neutrons.

Uranium, for instance, has a nucleus with 92 protons and a few neutrons more or less than 146. The uranium isotope symbolized by $_{92}U^{238}$ (92 protons, 238 protons plus neutrons) is the most common isotope, but $_{92}U^{234}$ and $_{92}U^{235}$ also occur in nature.

Some isotopes are unstable: the nucleus breaks up or "decays," and emits a form of radioactivity. The rate of decay is measured by the **half-life** of the isotope. If one starts with a pure mass of the unstable isotope, half the nuclei will decay during a certain time interval. After the same period of time again elapses, half the

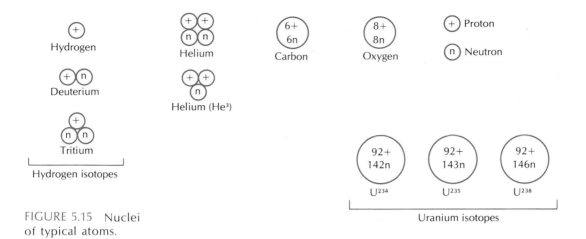

FIGURE 5.15 Nuclei of typical atoms.

remaining nuclei will have decayed, leaving only one-fourth the original number. After the next period, half those nuclei will decay, leaving one-eighth the original number, and so on. The time interval during which half the nuclei decay is the half-life, and it is a characteristic of the isotope. The half-life of $_{92}U^{238}$ is 4.5 billion years, and of C^{14} (carbon-14) is 5,580 years. There is a large range of half-lives of naturally radioactive nuclei.

Nuclei of $_{92}U^{238}$ decay through several intermediate steps and wind up as a stable isotope of lead, $_{82}Pb^{206}$. Basically, scientists measure the relative amounts of $_{92}U^{238}$ and $_{82}Pb^{206}$ in rocks of the Earth's crust. They assume that all the lead isotope there came from the decay of uranium. The relative amounts of the two isotopes tell how many half-lives have elapsed. If the number of uranium nuclei and lead nuclei are equal, then one half-life has elapsed in the rock's lifetime. The oldest rocks on the Earth that have had ages measured in this way are $3\frac{1}{2}$ billion years old.

This would be the end of the story if we had not landed men on the Moon. Some of the rock samples brought back to Earth were dated by radioactive methods and found to be as old as $4\frac{1}{2}$ billion years old. The question is, is the Moon older than the Earth or were the bodies formed at the same time? If the Moon and the Earth were formed at the same time, the Earth, too, is $4\frac{1}{2}$ billion years old. As we shall see later, this is near the probable age of the whole solar system.

Now that we have found the dimensions of the Earth—size, mass, age—we take a brief look at its interior structure.

If we had a giant saw and could cut the Earth in half, we would find a multilayered structure of rock and metal, with temperatures and densities increasing toward the core. Of course, we do not have such a saw. The deepest into the Earth that we can probe directly is into deep mines and oil wells, which are typically a few kilometers deep. Information on the deeper interior must be obtained by indirect means. Earthquake studies provide one such means.

When an earthquake occurs at some spot on the Earth, it sends waves out in all directions. There are basically two types of earthquake waves, the fast-moving *P waves* (*P* for primary) and the slower *S waves* (*S* for secondary). *P* waves can travel both through solid rock and through liquids, but the *S* waves travel only through solid rock. Seismographs all over the Earth pick up the waves from an earthquake after they have traveled through the body of the Earth. The arrival times and characteristics of the received waves tell the geophysicist the properties of the material through which the waves traveled (Figure 5.16). For instance, if the *S* waves are absent, the scientists knows the path from the quake to the recorder passed through liquid material.

Armed with the earthquake data from years of study and with theoretical calculations about the interior, scientists can come up with a reasonable picture of the interior. Figure 5.17 gives the current idea of the interior structure. The shell of the Earth is the outer **crust**, which is about 8 kilometers thick under the continents and somewhat thinner under the oceans. Beneath the outer crust is the lithosphere or rock sphere, which is about 100 kilometers thick.

Below these hard surface layers is the Earth's mantle, which is divided into two layers. The upper mantle or **asthenosphere** is about 650 kilometers thick. Its temperature is roughly 1,500° C (2,700° F); this heat is generated by the natural radioactivity in the rock and kept in by the insulating lithosphere. The material of the asthenosphere is above its normal melting point, but the pressure of the overlying rock keeps it in a soft solid state. Near cracks in the lithosphere, this pressure is relieved; the material liquifies and may flow to the surface as lava. The lower mantle or **mesosphere** is hotter than the asthenosphere. This 2700-kilometer-thick layer increases in temperature from 1,500° C at its upper edge to 3,000° C (5,400° F) at the bottom. The lower mantle, despite its higher temperature, is more rigid than the upper mantle, because of the great pressures. Finally, we come to the nickel-iron core, which is 3,000 km in radius and also consists of two layers. The outer core, with a density of 10 to 11 gm/cm³ and a temperature of 3,900° C

FIGURE 5.16 Propaga-
tion of P-wave earth-
quake waves through the
Earth. (From *Principles
of Geology* by J. Gilluly,
A. C. Waters, and A. O.
Woodford. W. H. Free-
man and Company. Copy-
right © 1959.)

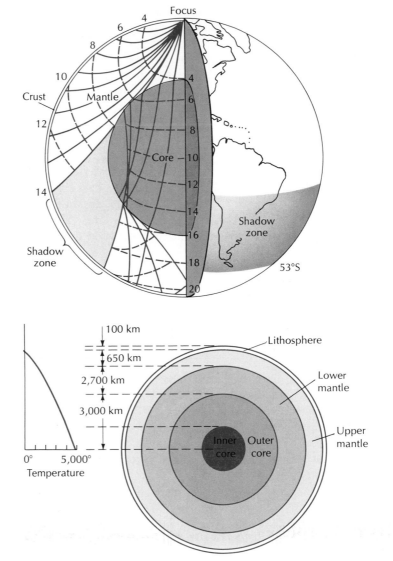

FIGURE 5.17 Cross sec-
tion of Earth's interior.
The temperature as a
function of depth into
the interior is indicated.

(7,000° F), is solid. The inner core, with a density of 12 gm/cm^3
and a temperature of 4,800° C (8,600° F), is liquid. The extreme
density of the core is not too surprising, since we found the mean
density of the Earth to be higher than the density of the surface
rock.

Most geophysicists accept the idea that the core of the Earth is a
mass of nickel and iron, though, of course, we have never been able
to sample it directly. The chemistry of the Earth's crust is more

complicated. More than 3,000 minerals, with a wide range of chemical compositions, have been identified among the crustal rocks. However, if we were to average the composition of all the rocks in the crust, we would find the main constituents, in order of abundance, to be oxygen, silicon, aluminum, iron, calcium, sodium, potassium, and magnesium. The oxygen of the crust is tied up in compounds, the most common being quartz, SiO_2.

As we will discover when we talk about the origin of the solar system, the composition of the Earth is a clue to its beginning.

We have described the interior structure of the Earth briefly and have seen that the temperature and density of the material increase toward the center. The central temperature may be as high as $4,800°$ C, and the central density as high as 12 gm/cm^3. In the last few years, much enthusiasm has been generated by a new theory of the lithosphere known as plate tectonics. Let's look at this theory next.

Plate Tectonics

The German meteorologist Alfred Wegener in 1912 put forward the theory that the continents drift about on the surface of the Earth. He presented several arguments to support his theory. For instance, he noted that the continents fit together like pieces of a jigsaw puzzle. He also pointed out that Greenland once had a tropical climate, and that Brazil was once heavily covered with glaciers. Wegener hypothesized that the continents were, at one time, all lumped together as a single supercontinent, which he called Pangaea (Figure 5.18). Many millions of years ago, Pangaea broke up and the continents moved to their present positions—very slowly, of course.

There was one serious problem: scientists could not imagine how the continents would plow through the solid lithosphere. Wegener's theory lay dormant for many years. In the meantime, more supporting evidence piled up. One very convincing bit of evidence was found in the fossil record. Paleontologists found fossils of land animals with similar ages and similar anatomies in African and South American rocks.

The clues that would provide the solution to the continental-drift mystery were found in the period from roughly 1955 to 1962. First, the ocean floor was carefully mapped, revealing several tantalizing features. On the floor of the ocean are a series of long mountain ridges, which, in all, extend almost 80,000 kilometers (Figure 5.19). The so-called **mid-Atlantic ridge**, for instance, runs virtually from pole to pole down the center of the Atlantic Ocean,

FIGURE 5.18 (*Opposite*) The fit of the continents. (From *The Confirmation of Continental Drift* by Patrick M. Hurley. Copyright © 1968 by Scientific American, Inc. All rights reserved.)

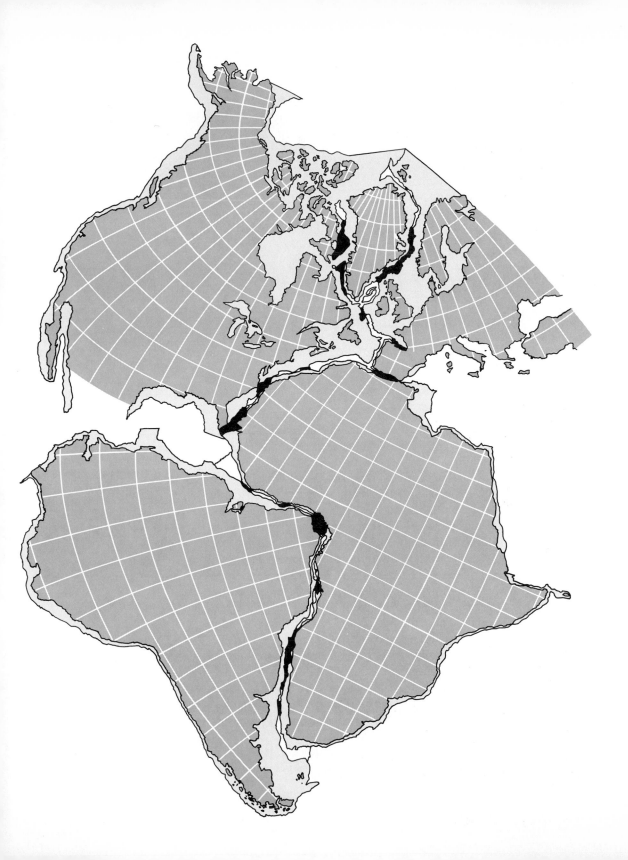

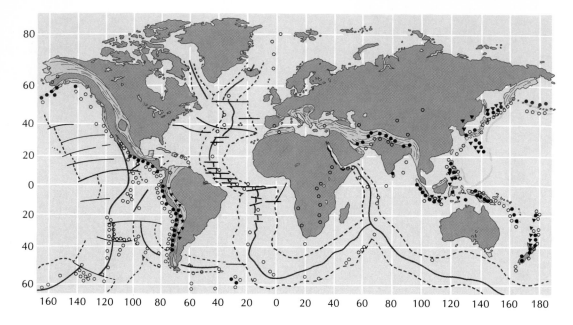

remaining equidistant from the continents on either side. Elsewhere
in the ocean there are deep trenches. One such trench—the Peru-
Chile trench—parallels the entire western coast of South America.

Another clue was provided by a worldwide network of sensitive seismographs designed to detect minute tremors in the Earth.
It was found that most earthquakes occur along the oceanic
trenches and the oceanic ridges. The earthquakes along the trenches
occur deep within the Earth's lithosphere. In addition, much of the
volcanic activity on the Earth falls along the oceanic ridges.

As the decade of the 60s opened, a new picture of continental
drift emerged, in the form of a theory called **sea-floor spreading**.
Molten rock from the asthenosphere wells upward through cracks
in the oceanic ridges where the lithosphere is very thin. The rock
solidifies, forming new sea-floor material, and pushes the continents
away from the oceanic ridges. As new sea floor is created, old sea
floor must be destroyed, since the Earth's surface area must remain
constant. Old sea floor is destroyed at the oceanic trenches, where
one moving mass of material plunges under another and melts in
the hot mantle. The sinking rock, rubbing against the material
above it, causes the deep-seated earthquakes.

The proof of the sea-floor spreading came quickly. It turns out
that, as the upwelling material at the midoceanic ridges solidifies, it

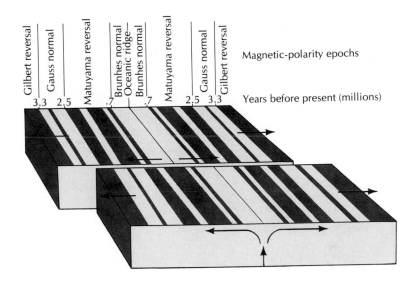

has frozen into it information about the Earth's magnetic field (Figure 5.20). Essentially, the magnetic particles in the rock are all aligned and point to the Earth's magnetic poles. However, for some unknown reason, the Earth's magnetic field reverses periodically. The north magnetic pole becomes a south magnetic pole, and vice versa. The magnetic structure of the sea floor was measured and found to consist of stripes of alternating magnetic polarity, all parallel to the midocean ridges. The reversals in the Earth's field were frozen into the sea floor as it spread. As Figure 5.20 shows, the magnetic patterns are symmetric on the two sides of the ridge.

The lithosphere is now pictured as a series of plates that slide about a few centimeters a year on the soft asthenosphere, growing at the oceanic ridges and being destroyed at the oceanic trenches or **subduction zones**. The general theory is called **plate tectonics**.

We have been able to hit only the high spots of the theory of plate tectonics. There are many other interesting aspects of the theory—for example, mountain building—that we cannot cover. There are many readable articles in the popular literature that contain more information on the subject.

We will delay further discussion of the Earth until Chapter 11. There, after we have covered some needed background material, we will discuss the Earth's near environment in space. In particular, we will cover the Earth's atmosphere and magnetosphere, and how the Sun affects them.

REVIEW QUESTIONS

1. What is the size and shape of the Earth's orbit?

2. Why are there seasons on the Earth?

3. Define: (a) vernal equinox; (b) celestial equator; (c) solstice.

4. How can one prove the Earth rotates on its axis?

5. Give one proof that the Earth is spherical.

6. What causes tides? What is the time interval between successive high tides?

7. How does one measure the Earth's (a) size; (b) mass; (c) age?

8. Describe the Earth's interior.

9. What is meant by plate tectonics?

6

THE MOON

Twenty years ago, everything we knew about the Moon was inferred from observations made by astronomers at their earth-bound, mountaintop observatories. Some things—such as the cause of eclipses and the phases of the Moon—had been understood since the time of Aristotle. On other topics, such as the origin of the lunar craters, disagreement was the order of the day.

In the meantime, a revolution has occurred in lunar science. In October 1959 the Soviet spacecraft Luna 3 made the first photo-graphs of the hidden lunar backside. In 1964 and 1965 the U.S. crash-landed three Ranger spacecraft at various sites on the lunar surface. As the craft sped toward the surface, they sent back photo-graphs of the terrain around the impact points. During 1966, 1967, and 1968, we soft-landed five Surveyor spacecraft, and the Russians soft-landed Luna 13. These automated devices took photographs and carried out analyses of the lunar surface soil at selected loca-tions.

Then at 3:17 P.M. EST on July 20, 1969, Apollo 11 landed at the edge of Mare Tranquillitatis, as the world held its breath, and United States astronauts began the brief series of manned explora-tions which ended in December 1972, with the departure for home of Apollo 17 (Figure 6.1).

Since then, we have sorted through much of the information that these missions collected. Our conception of the Moon has changed remarkably. Many old problems have been solved, though some still remain. Of course, new problems have replaced those that were solved. We can give only a brief survey of lunar science here. Let's begin by looking at some of the oldest ideas first.

FIGURE 6.1 Mankind achieves an age-old dream: the astronauts explore the surface of the Moon. (National Aeronautics and Space Administration.)

THE PHASES OF THE MOON

If we watch the Moon regularly, we can easily discover that it changes its apparent shape from night to night; that is, it goes through a series of phases.

Let's begin a series of nightly observations of the Moon on an early evening when we see it as a very thin crescent, not far from the spot where the Sun has just set. When we first notice the thin crescent, the red glow of sunset probably still colors the western sky. As it becomes darker, the entire circular disc of the Moon is illuminated with a faint, ruddy glow. This is known as the "old moon in the new moon's arms." The Moon then sets soon after the sun.

FIGURE 6.2 Crescent moon, four days after new moon. (Lick Observatory photograph.)

The next evening the Moon will appear higher in the sky and will be a thicker crescent (Figure 6.2). It sets 50 minutes later than it did the night before. The Moon moves its own diameter ($\frac{1}{2}°$) eastward among the stars each hour; so it covers 13° per day, and as a result rises and sets about 50 minutes later each day.

After we have been watching the Moon for a week, it will be at the phase called **first quarter** (Figures 6.3 and 6.4). Half the Moon's face toward the Earth is illuminated. At sunset, the Moon is roughly due south, near its highest point in its daily path across the sky. During the following week, the Moon passes through the **gibbous** phase as the face toward the Earth becomes more and more fully illuminated (Figure 6.5). At the end of the second week after we first saw the crescent, the Moon is **full** (Figure 6.6).

The full moon rises at sunset, and is in the sky for the entire night. On such nights the intense light of the full moon makes the sky so bright that all but the brightest stars are hidden in the glow.

For the next two weeks, the Moon goes through the phases in reverse: full, gibbous, third quarter, and crescent, until after four weeks (actually $29\frac{1}{2}$ days) the Moon is **new**. New moon is the opposite of full moon: the new moon is completely dark. During these two weeks the Moon still continues to rise later and later each night. The third-quarter moon rises around midnight, six hours

FIGURE 6.3 First-quarter moon. (Lick Observatory photograph.)

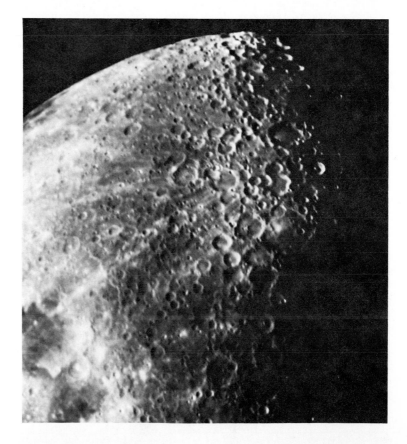

FIGURE 6.4 First-quarter moon, southern section, taken with a 16-inch reflecting telescope at Edinboro State College, Edinboro, Pa. (Courtesy James C. LoPresto, Edinboro State College, Edinboro, Pennsylvania.)

FIGURE 6.5 Moon at gibbous phase. (Lick Observatory photograph.)

FIGURE 6.6 Complete surface of the Moon. This is a composite of photographs taken at first and third quarters. At full moon, the lunar features cast imperceptible shadows; so there is little contrast in photographs. (Lick Observatory photograph.)

FIGURE 6.7 Crescent moon, as the lunar phase once again approaches new, almost 26 days after the last new moon. (Lick Observatory photograph.)

FIGURE 6.8 Lunar phases. The inner circle shows the Moon's phases as seen from Earth.

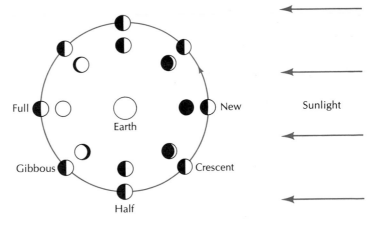

Full

Earth

New

Sunlight

Gibbous

Crescent

Half

before the Sun, and the thin crescent moon rises just before the Sun on the eastern horizon (Figure 6.7). You can check these facts for yourself: watch the Moon change shape and position for a month.

If we remember that the light we receive from the Moon is reflected sunlight, the progression of phases is easy to understand. In Figure 6.8, the outer circle of drawings shows the Moon's phases as they would look from a point far above the north pole of the

Earth, with sunlight coming from the right. At crescent phases the side of the Moon that is lit up is largely facing away from the Earth, whereas at gibbous phases it is the unlit side of the Moon that is largely facing away from the Earth. At first and third quarter, we see half the illuminated face. At new moon and full moon, we see either none of the illuminated face or all the illuminated face, respectively.

The "old moon in the new moon's arms" that is observed at crescent phases is also easy to understand. If you were an austronaut on the Moon looking back at Earth, you would discover that the Earth's phase is always opposite the Moon's phase. When the Moon is a crescent as seen from Earth, the Earth is gibbous as seen from the Moon. When the Moon is a thin crescent, the Earth, seen from the Moon, is nearly full and is extremely bright in the lunar sky. The Earth is so bright that it illuminates the moonscape. The old moon is simply the part of the Moon illuminated by earthshine.

In this section we have talked about the cause of the Moon's phases. We will now discuss its distance from the Earth, its size, and its motion.

DISTANCE, SIZE, AND MOTION OF THE MOON

The Distance to the Moon

The best measurement of the distance to the Moon has been made by bouncing radar signals off it (Figure 6.9). The time it took the signal to make the round trip was accurately measured; and the speed of the radar signal is the same as the speed of light, which is well known. The distance could then be calculated: it is 384,402 kilometers or 238,855 miles.

FIGURE 6.9 Radar distance to the Moon. A radar pulse reflected from the Moon traveled the round-trip distance (2 times D) in about $2\frac{1}{2}$ seconds. Since the pulse traveled 300,000 km in a second, it traveled about 750,000 km in the $2\frac{1}{2}$ seconds for the round trip. By taking half this distance, we find the distance D one-way between the surface of the Earth and the surface of the Moon. To find the distance between the center of the Earth and the center of the Moon, we need to add the radius of the Moon and the radius of the Earth to the one-way radar distance between their surfaces. (Notice that the radar installation is drawn out of scale. In comparison with the distance being measured, the radar antenna is right on the surface of the Earth.)

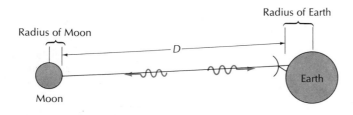

Radius of Moon

Radius of Earth

D

Moon

Earth

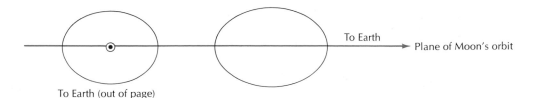

To Earth (out of page)

FIGURE 6.10 Three-axis ellipsoidal shape of the Moon. The shape of the Moon is exaggerated in the illustration. To the eye, the Moon appears spherical.

The Moon's Size

Given the distance to the Moon and its apparent angular size, we can calculate its true diameter. The diameter is 3,476 kilometers or 2,160 miles, along a diameter perpendicular to the line from the Earth to the Moon. Careful studies of the Moon, even before the Apollo missions, showed it to be an ellipsoid with three different diameters (Figure 6.10). The longest diameter is pointed toward the Earth, and its shortest diameter is along its polar axis.

The Moon's Mass

The modern way to find the Moon's mass is by studying the motion of an artificial satellite, like one of the lunar orbiters, that is circling it.

Let's think about a satellite orbiting the Moon in an elliptical orbit (Figure 6.11). If it were not for the Moon's gravitational force, the satellite would fly off into space on a straight line. However, the force of gravity accelerates the satellite and makes it travel on an elliptical path. We can calculate the acceleration from the size of the satellite's orbit and its revolution period; and then we can use the law of gravity to find the Moon's mass.

FIGURE 6.11 The Moon's mass. (a) The acceleration of an artificial satellite orbiting the Moon can be calculated from the size of its orbit and its revolution period. (b) The Moon's mass is calculated using the fact that it exerts a force on the artificial satellite that produces the observed acceleration.

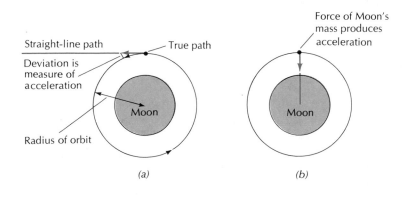

If one does that calculation using modern mechanics, the numbers that have to be measured are the satellite's period of revolution and its average distance from the Moon in its elliptical orbit. The result is the sum of the satellite's mass and the Moon's mass; and since the mass of an artificial satellite is tiny compared to the Moon's mass, we can take the whole sum to be the Moon's mass. Later, we will see how the same method is used to study the masses of binary star systems, where two stars orbit one another, and we will see what further measurements are needed to find the masses of the individual stars. The method also helps find the mass of a planet like Jupiter that has several natural moons.

Using the method, we find that the Moon's mass is $1/81$ of the Earth's mass. Using the diameter found earlier, we can next calculate the Moon's density, which turns out to be 3.3 gm/cm^3, somewhat less than the Earth's average density of 5.5 gm/cm^3.

Because the Moon's mass is so small, its gravity is weaker than the Earth's. The force of gravity at the lunar surface is only $1/6$ of that at the Earth's surface. Thus a man who weighs 180 pounds on Earth would weigh only 30 pounds on the Moon.

If we study the motion of a lunar orbiter very carefully, we will notice tiny irregularities in its path, which make it depart slightly from an ellipse. There seem to be concentrations of extra mass—now called **mascons**—beneath the Moon's circular maria. Some scientists feel that mascons may be the remains of huge meteorites that struck the Moon and created the circular maria.

The Moon's Orbit

The Moon's orbit is an ellipse with the Earth at one focus, as we would expect from Kepler's first law. The orbit is more elliptical than the Earth's orbit. Also, the plane of the Moon's orbit is tipped by about 5° to the plane of the Earth's orbit (Figure 6.12). Later, when we talk about eclipses, we will see that this 5° tip is important.

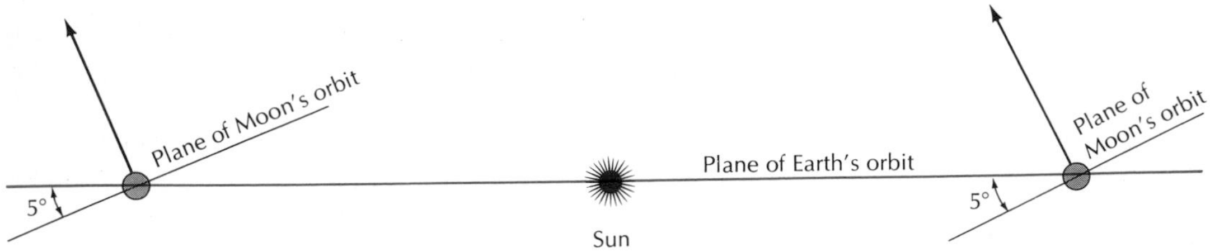

FIGURE 6.12 The Moon's orbit is tilted 5° to the plane of the ecliptic. As the earth orbits the sun, the perpendicular to the Moon's orbit maintains a fixed orientation in space.

Plane of Moon's orbit

Plane of Earth's orbit

Plane of Moon's orbit

5°

5°

Sun

FIGURE 6.13 Difference in length of sidereal and synodic months. At *A* the full moon is observed to be near some bright star in the sky. Roughly $27\frac{1}{3}$ days later, at *B*, the Moon is again near the same bright star. One *sidereal* month has elapsed from *A* to *B*. Meanwhile, the Earth has moved about $27\frac{1}{3}°$ in its orbit; so Sun, Earth, and Moon are not lined up. The Moon must move an additional $27\frac{1}{3}°$ in its orbit before it is again full; that is, before one synodic month has elapsed from time *A*. Since the Moon moves 13° per day, it must travel another $2\frac{1}{6}$ days before it is full. Thus the time between full moons (or first quarters, and so on) is $27\frac{1}{3} + 2\frac{1}{6} = 29\frac{1}{2}$ days.

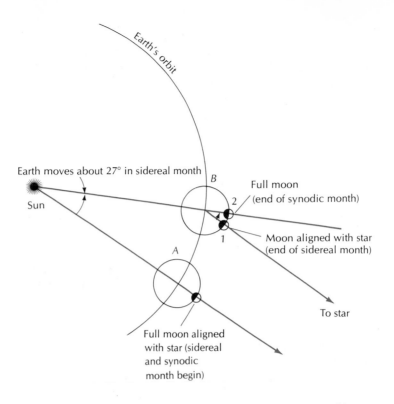

Earth moves about 27° in sidereal month

Sun

Earth's orbit

B

Full moon
(end of synodic month)

2

Moon aligned with star
(end of sidereal month)

1

A

To star

Full moon aligned
with star (sidereal
and synodic
month begin)

The Moon goes through a complete set of phases in $29\frac{1}{2}$ days, but it travels once around its orbit relative to the stars in $27\frac{1}{3}$ days. These two periods are called, respectively, the synodic period of revolution and the sidereal period of revolution. Why is the Moon's sidereal period shorter than its synodic period? Because (as can be seen in Figure 6.13) the Earth orbits the Sun as the Moon orbits the Earth.

The Moon's Rotation

The moon rotates once on its axis in $27\frac{1}{3}$ days, and therefore always keeps the same face toward the Earth (Figure 6.14). Until 1959 we had never seen the back side of the Moon. Now man-made satellites circling the Moon have revealed its entire surface to us in great detail. (By the way, the back side of the Moon is not the dark side of the Moon. At new moon, for instance, the entire back side is illuminated.)

Why does the Moon keep one face constantly toward the Earth? The Earth raises tides in the Moon's body just as the Moon raises tides in the Earth's oceans and rocks. As the Earth rotates, the tidal

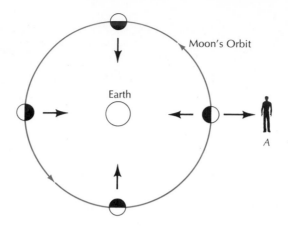

FIGURE 6.14 The Moon keeps the same face toward the Earth, but as seen by an outside observer *A*, the Moon does rotate on its axis.

bulge moves around the planet, causing tidal friction, which turns its rotation energy and the Earth-Moon orbital energy into heat. If the Moon had rotated more rapidly in the past, its excess rotational energy would have been lost to tidal friction in its interior.

The energy of motion in the Earth-Moon system is still being lost in the Earth's tidal friction. Even though the energy of the system is decreasing, its angular momentum remains constant. As a result, the Earth and the Moon will move apart as the Earth slows down. A few billion years from now, both the Earth and the Moon will rotate in about 45 of our present days, and the Moon will revolve about the earth in 45 days. The two bodies will therefore keep the same faces toward each other and will be about 540,000 kilometers apart. People in one hemisphere of the Earth will never see the Moon. A vacation treat might be a trip to see the Moon. (It is, of course, unlikely that people will be around in a few billion years to see the phenomenon.)

ECLIPSES OF THE MOON

At the time of full moon, the Sun, Earth, and Moon are nearly on a straight line. If the Earth passes directly between the Sun and the Moon, it prevents the sunlight from illuminating the Moon. The result is an **eclipse** of the moon (Figures 6.15 and 6.16). The eclipse can be either **partial** or **total**, depending on whether the Earth prevents a fraction of the sunlight or all of it from reaching the Moon.

At a total eclipse of the Moon, the disc of the Moon can be seen shining with a rusty red glow. The light reflected by the Moon is

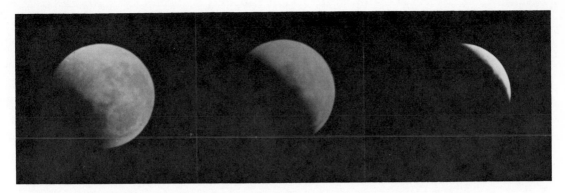

FIGURE 6.15 A lunar eclipse. (Yerkes Observatory.)

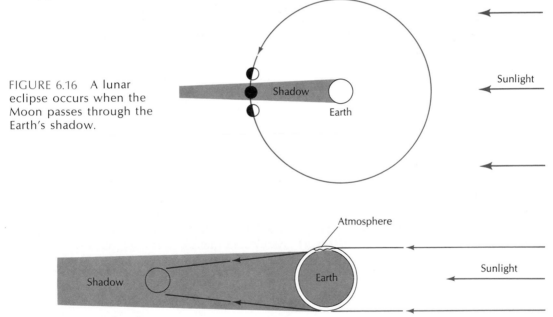

FIGURE 6.16 A lunar eclipse occurs when the Moon passes through the Earth's shadow.

FIGURE 6.17 At the time of total lunar eclipse, as seen from the Moon, the Earth would be dark except for the brilliantly illuminated atmosphere, which appears as a thin bright band encircling the Earth's disk. The light from the atmosphere casts a dim ruddy glow on the Moon.

sunlight that has been scattered by the Earth's atmosphere (Figure 6.17). A person on the Moon looking back at Earth would see a dark planet surrounded by a thin halo of brilliant atmosphere, the color of a sunset sky.

Eclipses of the Moon can only occur at full moon, for only then can the Earth pass directly between the Sun and the Moon. But why is there not an eclipse at every full moon? The answer lies in the 5°

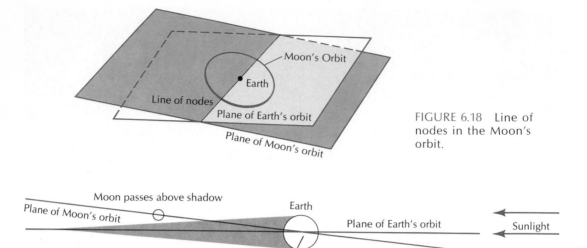

FIGURE 6.18   Line of nodes in the Moon's orbit.

Node: where Moon's orbit crosses the plane of the Earth's orbit

Moon at node encounters shadow; therefore eclipse occurs

Earth

Sunlight

FIGURE 6.19   A total eclipse does not occur at every full moon. Unless the Moon is near a node in its orbit, it will pass above or below the Earth's shadow.

tilt of the Moon's orbit to the ecliptic. At a typical full moon, the Moon passes "above" or "below" the Earth's shadow and there is no eclipse. However, if full moon occurs when the Moon is near one of the two points where its orbit crosses the plane of the ecliptic, an eclipse will occur (Figure 6.18). It is only then that the Moon will pass directly through the Earth's shadow.

The points where the Moon's orbit crosses the ecliptic are called **nodes**, and the line connecting the nodes of the Moon's orbit is called the **line of nodes**. If we draw a line through the Sun parallel to the line of nodes of the Moon's orbit, the line will cross the Earth's orbit at two points. Eclipses occur only when the Earth is near one of these two points (Figure 6.19). Thus, eclipses occur during **eclipse seasons** which are about six months apart.

So far in this chapter, we have talked about the Moon as an object in space. We have described its phases and explained their cause. We have found the Moon's distance, size, mass, and orbital motion. Finally, we discussed briefly lunar eclipses. We will return to eclipses later when we discuss solar eclipses.

Now we turn our attention to what has come to be called lunar geology, and discuss the Moon's surface features. (Lunar scientists use the term geology loosely, of course, for the word really means a study of the Earth.)

THE SURFACE OF THE MOON

Even to the unaided eye, it is clear that the Moon's surface is divided into two types of regions: the dark **maria**, so-called because the ancients thought they were seas (Latin, *mare*); and the lighter-colored surrounding areas. When Galileo turned his telescope toward the Moon, he found the light-colored areas to be extremely rough. Larger telescopes show the surface to be covered with innumerable **craters**. Large earth-based telescopes, as well as manned and unmanned satellite visits to the Moon, have revealed a wealth of information about the lunar surface features. In the following pages we will review these findings.

Maria

As the Apollo command modules passed over the lunar surface, the astronauts took measurements of the elevation of the landscape below using a laser altimeter. It turns out that the maria are basins, 3 to 4 kilometers below the surrounding areas. As a result, the heavily cratered lighter areas of the Moon are now called the **lunar highlands**. The entire surface of the Moon, therefore, is divided into the two regimes: maria and highlands.

Figure 6.20 is an outline map of the maria, with the poetic names given them by the ancients. With a little imagination, we can see the Woman-in-the-Moon in profile among the maria, with the crater Tycho forming her sparkling necklace, or the Man-in-the-Moon staring coldly at us (Figure 6.21).

The maria are not perfectly smooth. The basins are literally covered with craters, which are typically smaller and less densely packed than the great highland craters (Figure 6.22). Careful scrutiny of the maria also reveal **wrinkle ridges** (Figure 6.23). As the name implies, they look like wrinkles on the smooth maria material. The nearly circular maria, which are often called **ringed basins**, have systems of wrinkle ridges forming concentric rings on their floors.

FIGURE 6.20   Lunar map.

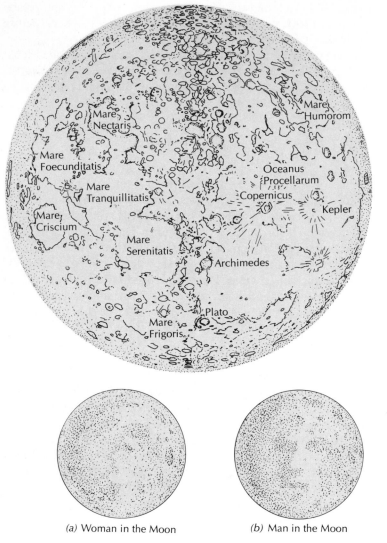

FIGURE 6.21   (a) Woman in the Moon. (b) Man in the Moon.

(a) Woman in the Moon   (b) Man in the Moon

The rock samples returned from the lunar maria are **basalts**. Studies of terrestrial basalts show that they are finely crystalized lava; so we assume the lunar basalts also originated from lava flows. More will be said about this later. The mineralogy of the lunar rocks, which has been studied in tremendous detail, will not be treated here. Ages of the basalt samples, as measured by radioactive dating methods, range between 3.2 and 4.0 billion years.

FIGURE 6.22 Mare Serenitatis. (Lick Observatory photograph.)

FIGURE 6.23 Wrinkle ridges in Oceanus Procellarum; photographs from Apollo 15 command module. (National Aeronautics and Space Administration.)

FIGURE 6.24 The giant crater Clavius. This crater is 235 kilometers in diameter. (Lick Observatory photograph.)

Craters

The lunar surface is pocked with an incredible number of craters. Under the best atmospheric seeing conditions, craters one kilometer across can be discerned easily on the lunar surface. Estimates indicate that the number of craters larger than 1 km in the Moon's earthward side exceeds thirty thousand. More than a thousand of those craters are larger than 16 kilometers in diameter. The craters are named for famous people.

The largest easily observed crater on the Moon's earthward side is Clavius, which is 230 km in diameter (Figure 6.24). Its area is roughly 40,000 square kilometers: Massachusetts, Connecticut, and Rhode Island would just fit inside the crater, to put the number in perspective.

Photographs of craters give one the impression that they are very deep; yet they are actually shallow features. Giant Clavius, for instance, is only 5 kilometers deep. If you stood at the center of the crater, its rim would be hidden beyond the lunar horizon. The floors of the craters are generally somewhat below the level of the surrounding surface (see Figure 6.25).

Many craters are like Tycho, with a central mountain peak (Figure 6.26). Others, like Plato, have seemingly smooth floors. The

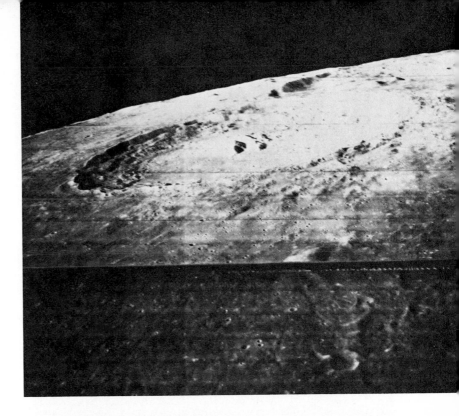

FIGURE 6.25 Oblique view of the crater Theophilus, made by Lunar Orbiter III. The central peaks in this 155-kilometer-diameter crater stand out clearly. (National Aeronautics and Space Administration.)

FIGURE 6.26 The crater Tycho. (National Aeronautics and Space Administration.)

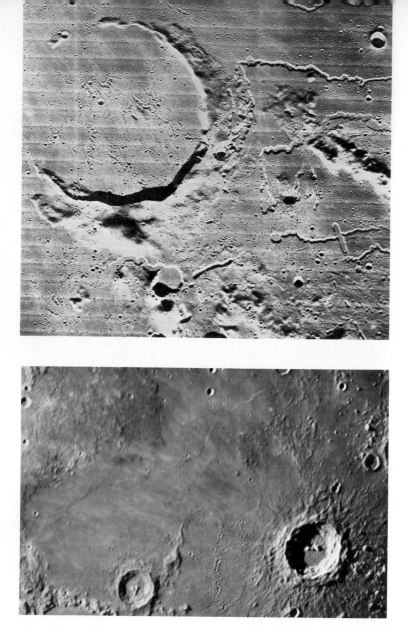

FIGURE 6.27 The partially rimmed crater Prinz and the channel-like rilles in its vicinity. Lunar Orbiter V took this photograph. (National Aeronautics and Space Administration.)

FIGURE 6.28 A photograph of the crater Copernicus made from Earth with the Hale 5-meter reflecting telescope on Palomar Mountain. (Hale Observatories.)

rims of some craters have obviously been destroyed somehow (see Figure 6.27). Several craters, Tycho and Copernicus being typical, are noted for their splash-like ray systems, most easily seen at full moon (see Figures 6.28 and 6.29). Each of these observations is a clue to the origin of lunar features.

The rock samples brought back from the lunar highlands are different from the maria rocks—but we cannot deal with their

FIGURE 6.29 Oblique view of Copernicus made by Lunar Orbiter II. (National Aeronautics and Space Administration.)

FIGURE 6.30 Lunar Orbiter III view of the Moon. Note the dome-shaped features in the foreground. (National Aeronautics and Space Administration.)

mineralogy in any detail. The highland rocks are mostly breccias: rocky material that was once broken into small pieces, then became cemented together again. The rocks contain much silicon and aluminum, like terrestial crustal rock. Their ages are in the range 3.9 to 4.0 billion years.

Figure 6.30 shows a lunar feature known as a **dome**. These domes may be of volcanic origin and are therefore of great interest.

FIGURE 6.31 The lunar Alps. The isolated peaks in Mare Imbrium are probably the highest parts of an old mountain chain that was buried by the lava flow that formed the mare. (Lick Observatory photograph.)

Lunar Mountains

There are many **mountain ranges** on the Moon. The astronomers who named these features gave them the names of terrestrial mountain ranges: the Alps (Figure 6.31), the Apennines (Figure 6.32), the Caucasus, etc. The highest mountains on the Moon tower 25,000 feet or more above the plains. Several isolated peaks can be seen on the lunar landscape (as in Figure 6.31).

An interesting feature of the Alps Mountains is the **Alpine Valley**, a cut through the mountain range (Figure 6.33). Along the midline of the Alpine Valley is a rille.

FIGURE 6.32 The lunar Apennines. (Lick Observatory photograph.)

FIGURE 6.33 Alpine valley photographed by Lunar Orbiter IV. (National Aeronautics and Space Administration.)

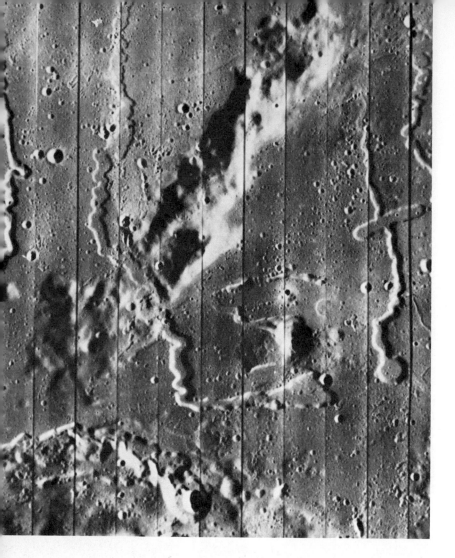

FIGURE 6.34 Rilles. (National Aeronautics and Space Administration.)

Rilles

Rilles are long, broad channels (up to 5 km wide) on the lunar surface that can take any of three forms: straight, curved, or sinuous (Figure 6.34). The rille on the floor of the Alpine Valley is straight. The sinuous rilles look very much like the dried bed of a meandering river, which prompted the speculation that the Moon once had liquid water on its surface in the distant past. As we will see, this is probably not correct.

The Apollo 15 landing site was near the 1-km-wide Hadley Rille which snakes across the mare at the foot of the Apennine Mountains. Several views of the rille are shown in Figures 6.35 and 6.36.

FIGURE 6.35 Photograph of the west wall of Hadley Rille made by the Apollo 15 crew. (National Aeronautics and Space Administration.)

FIGURE 6.36 Astronaut David Scott at Hadley Rille. (National Aeronautics and Space Administration.)

The Moon Through Binoculars

If you have available a pair of binoculars that provides a magnification of six or seven times (6X or 7X), you can observe some of the features we have discussed here. The best time to observe is near first (or last) quarter, because mountains and craters near the **terminator**—the line between lunar day and night—are in deep shadow at that time and are easiest to see. In addition, the thin crescent moon is lovely to observe, and the bright, full moon shows the ray systems of Tycho and Copernicus.

Let's look at the first-quarter moon through the binoculars, using the chart in Figure 6.20 as a guide. If you are using ordinary binoculars, then north on the Moon is up. The terminator is to the west, and the limb (edge) of the Moon to the east.

Almost in the center of the illuminated part of the Moon is Mare Tranquillitatis. The Apollo 11 landing site is in the southwestern part of the mare, near two moderately conspicuous craters. North of Mare Tranquillitatis is Mare Serenitatis (Figure 6.22). Two mountain ranges border this mare on the west: the Caucasus Mountains and the Apennines. Mount Hadley, near the base at which Apollo 15 landed, is in the Apennines near the terminator. North of Mare Serenitatis are two large craters, Aristotle and Eudoxus. Many large craters can be seen along the southern half of the terminator.

Wait two or three nights, then look at the Moon again. Mare Imbrium is clearly visible in the north, with the smooth-bottomed crater Plato at its northern edge. The crater Copernicus (Figure 6.28) is conspicuous south of Mare Imbrium.

A few nights later, when the Moon is full, it is almost uncomfortable to look at. A yellow filter actually makes for easier viewing. The main thing to note now are the rays, especially around Tycho in the south.

A week later, if you don't mind going out at 3 A.M., take a look at the last-quarter moon. Most of the surface area is covered in dark maria: Mare Imbrium, Oceanus Procellarum, and, to the south, Mare Humorum, Mare Cogitum, and Mare Nubium.

At the southernmost area of the Moon, near the terminator, are several giant craters, including Clavius.

With the detailed map of the Moon, you should be able to identify many lunar features as it changes phases. And you don't have to do your skywatching alone: consider being out on a dark night under the beautiful starry sky. . . .

FIGURE 6.37 Lunar Orbiter V view of the far side of the Moon. Note the absence of maria in this picture. (National Aeronautics and Space Administration.)

The Moon's Other Side

The side of the Moon that does not face the Earth is strikingly different from the earthward side. The biggest difference between the two faces is the almost complete lack of maria on the backside (Figure 6.37). One of the few maria, Mare Orientale, is a magnificent ringed basin with multiple rings of mountains surrounding it (Figure 6.39).

We have described many of the lunar features in the previous pages. Much of what we have stated was known prior to the Apollo

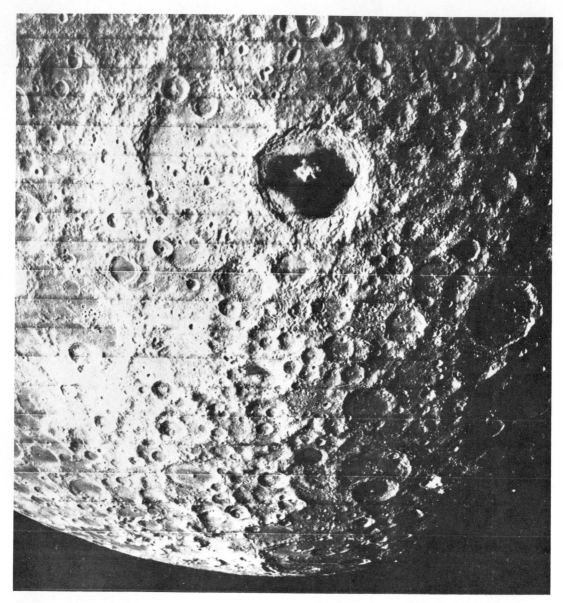

FIGURE 6.38 The unusual crater Tsiolkovsky on the far side of the Moon. (National Aeronautics and Space Administration.)

visits to the Moon, but some of the information comes from the manned exploration of our satellite. As we shall see, the major contribution of Apollo was to help answer the question, what do all these features tell us about lunar history? Let's look at the origin of lunar surface features according to our best estimates.

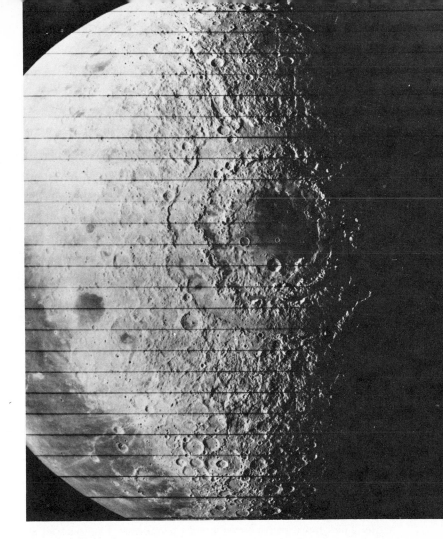

FIGURE 6.39 Orientale Basin on the far side of the Moon. Note the multiple rings of mountain ranges. These rings may be waves frozen in the lunar material after the splash of a giant meteor impact. (National Aeronautics and Space Administration.)

THE ORIGIN OF THE LUNAR SURFACE FEATURES

Two very important facts must be borne in mind as we discuss the history of the lunar surface. The Moon has no atmosphere, and the Moon has no water. These facts had been demonstrated before the visits by men, and were verified by experiments carried out by the Apollo crews. The fact that the Moon has no atmosphere shows it can have no liquid water. In the extreme vacuum any water would quickly boil away. (Later we will talk about how atmospheres escape from bodies like the Moon.) The reason these two facts are crucial, of course, is that without winds and rain there is virtually no erosion on the Moon. Once a lunar surface feature is formed, it will remain unchanged for many billion years.

The Origin of the Craters

What are the craters on the Moon? What is their origin? This is a question that was first asked when the craters were discovered in the seventeenth century, and the arguments continued for hundreds of years. The disagreement finally boiled down to two opposing hypotheses. Are the craters of volcanic origin or of impact origin?

The proponents of the volcanic theory argued that, since all the craters are circular, the impact hypothesis cannot be correct. If a large meteorite (see Chapter 10) were to strike a glancing blow to the lunar surface, they said, the resulting crater should be elongated in the direction of the motion. And besides, how could impacts produce central peaks in the craters? No, they argued, the craters must be the caldera produced by collapsed volcanoes, and the central peaks are the old volcanic vents. A giant mass of magma (liquid rock) would pour out on the surface, creating a typical volcano (Figure 6.40). However, a huge hollow chamber would be left inside the Moon as the eruption progressed. Finally the chamber would become so large that its roof could not support itself, and it would collapse, creating a crater with a central peak. Crater Lake in Oregon is an example of a collapsed volcano on Earth (Figure 6.41).

The arguments in favor of the impact hypothesis stressed the fact that meteorites strike the Moon with tremendous velocity. The impact would be explosive in nature, and the crater would be much larger than the body that created it. Experience with wartime bomb craters helped in the argument. Explosion craters are circular whether the bomb falls straight down or comes in at a glancing angle. The central peaks are rebound phenomena. The material at the point of impact is pushed downward so hard that it bounces upward after the impact, producing the central mountain. High-velocity projectiles fired from guns into sheets of steel produced craters with central "peaks," for instance.

The collapse of a volcanic caldera is not as violent as an impact that would produce the same size crater—not that volcanic collapse is a gentle event. For example, when Krakatoa, between Sumatra and Java, collapsed in 1883 the explosion was heard three thousand miles away in India; it produced a blast wave in the atmosphere which circled the Earth several times; and it produced a tidal wave which reached a maximum height of two hundred feet at the head of a narrow bay off the Sunda Strait. Still, the collapse could not throw great masses of material long distances to produce a ray pattern like that of Tycho or Copernicus. The explosion of Tycho sent material half the circumference of the Moon.

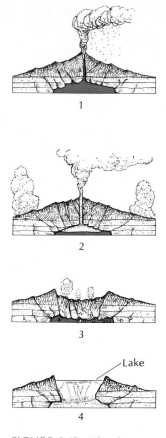

FIGURE 6.40 The formation of the caldera at Crater Lake. As lava flowed out from the volcano, it left an empty underground magma chamber. The entire volcano subsequently collapsed into the chamber. (From *Principles of Geology* by J. Gilluly, A. C. Waters, and A. O. Woodford. W. H. Freeman and Company. Copyright © 1968.)

FIGURE 6.41 View of Crater Lake, Oregon, from its east rim. The steep walls drop 2,000 feet to the lake surface. (Courtesy of National Park Service.)

Finally, studies of bomb craters revealed a relationship between depth and diameter that exactly fits the lunar craters (Figure 6.42). The fact that lunar highland rocks are breccias—that is, rocks that have been violently fractured, some of them several times—also supports an extremely violent origin for the craters.

The controversy has died down now. The impact origin of the lunar craters is almost universally accepted today. This is an important point because, as we shall see, Mercury, Mars and its moons, and probably Venus are also heavily covered with craters.

Several questions come to mind at this point. Why are there no craters on the Earth? Are there still giant meteors ready to strike the

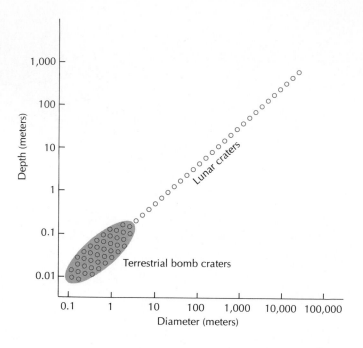

Moon—or worse, the Earth? Why are there areas on the Moon without many craters?

The major cratering on the Moon occurred long ago. The lunar highlands look today much as they did four billion years ago. At that time, the Earth was probably heavily cratered, too. But erosion has destroyed most evidence of the terrestrial craters. Look at Figure 6.43, showing the eastern shore of Hudson Bay just north of James Bay. Is the semicircular shoreline the remains of an ancient crater? Several possible old craters have been found on Earth, the most obvious one being Barringer Crater, Arizona (Figure 6.44).

In Chapter 10, when we talk about the origin of the solar system, we will have more to say about the bodies that struck the Moon to cause the craters. For now we can just say that until about four billion years ago, the solar system was heavily populated by asteroid-like bodies.

Today the asteroids lie between the orbits of Mars and Jupiter, for the most part. These bodies, sometimes called minor planets, are at most hundreds of kilometers in diameter, and the smallest are probably dust grains. Several asteroids travel in highly elliptical orbits which cross the Earth's orbit. In 1937 the asteroid Hermes came within 800,000 km (500,000 miles) of the Earth. Although it is extremely unlikely that any asteroid will strike the Earth, it is not impossible. A kilometer-sized body plunging into the Pacific Ocean would raise a wave several kilometers high.

FIGURE 6.43 The Eastern shore of Hudson Bay. Is the almost perfect segment of a circle in the Bay an old meteor crater?

FIGURE 6.44 Arizona meteor crater. This 1.2-kilometer-diameter crater is several kilometers south of Interstate 40, between Flagstaff and Winslow, Arizona. (Courtesy H. H. Nininger, American Meteorite Museum.)

The Origin of the Lunar Maria

The Orientale Basin on the far side of the Moon is a fascinating feature. It consists of several rings of mountains, the outermost of which is 900 kilometers (550 miles) in diameter. Some scientists believe that the basin was formed by an impact so huge that it actually raised waves in the solid rock of the moon, causing the concentric mountain rings. Four billion years ago, Mare Imbrium may have looked like Orientale Basin. The Caucasus Mountains and the Apennines are about all that is left of the outer mountain ring.

The Apollo 14 and 15 astronauts explored the lunar surface at points near Mare Imbrium, and collected material that is thought to be debris from the Imbrium collision. Roughly, the ages of the debris indicate an impact about 4 billion years ago. Then what happened? Why does Imbrium not look like Orientale Basin?

Remember that the maria rocks are basalts, indicating extensive lava flows, and also that the basalts have ages in the range 3.2 to 4.0 billion years. It seems that between 3 and 4 billion years ago, great masses of hot lava flowed from the interior of the Moon, filling large basins and obliterating the underlying terrain.

If you look at a photograph such as Figure 6.22, you can see many craters at the edges of the maria with rims that look relatively normal toward the highlands but are largely destroyed toward the maria. The crater Plato north of Mare Imbrium was filled by the lava from below, without ruining its rim. Just south of Plato are several isolated mountain peaks in the mare. These are probably the highest peaks of an old Imbrium inner mountain ring that were not buried by the lava flows. If only a few peaks were not buried, the lava must be very thick.

Some scientists believe that the wrinkle ridges on the maria mark the positions of faults where the lava flowed from the lunar interior. Rilles, which look like river beds, are probably just that, but not rivers of water: rather, rivers of very fluid lava flowed in them. Apollo studies of Hadley Rille support this idea.

Once the lava solidified and crystalized, some additional cratering took place, causing the craters now seen on the maria. Lunar scientists believe that Copernicus was formed only about a billion years ago, for instance.

Today the Moon seems to be a cold body. The outer thousand kilometers of its interior is solid, stable rock, like the Earth's lithosphere. The core of the Moon, roughly 700 kilometers in diameter, is material like the Earth's asthenosphere: a soft solid material. Of course, information about the lunar interior is derived from studies of moonquakes by seismographs left on the surface by Apollo crews. This discussion leads to one of the great remaining problems in lunar science. Where did the maria lavas come from? Could the heat have been generated in truly giant collisions, or by radioactivity in the rocks now on the maria surface? Much exploration will still be required to answer this question.

We have now had a brief survey of post-Apollo lunar science, and must leave the Moon to continue our survey outward through the universe.

REVIEW QUESTIONS

1. Why does the Moon exhibit phases?

2. How do astronomers determine the Moon's distance, size, and mass?

3. Does the Moon rotate?

4. Under what conditions do lunar eclipses occur?

5. Describe each of the following lunar surface features: (a) maria; (b) craters; (c) rilles; (d) mountains.

6. What is the current thinking on the origin of lunar craters? Lunar maria?

7

ATOMS AND LIGHT

How does the light from stars and galaxies tell us about the physical state of these celestial objects? In this chapter, two basic topics will be discussed: the nature of the light emitted by various materials; and the atomic nature of matter.

Chapter 4 dealt with the basic nature of light. There we learned how telescopes can be used to study the light received from distant bodies in the universe. This chapter takes up where that chapter left off.

THE SPECTRUM

When Newton showed that white light passing through a prism is spread out into a band of colors (Chapter 4), he initiated a long history of research that led up to our modern knowledge of the spectrum of emitting bodies. However, it took the invention of a device known as a **spectroscope** to help along the discoveries. Figure 7.1 shows a prism spectroscope. Light from a star, galaxy, terrestrial light source, or whatever is being studied passes through a slit, then through an achromatic lens arranged with the slit at its focus. The light beams are refracted by the lens, and emerge all parallel to the lens' optical axis and to each other. The light then passes through the prism, where it is refracted and dispersed into the spectrum. The light emerging from the prism is then examined by means of a telescope. The observer sees many images of the slit side by side. Roughly speaking, there is an image of the slit for each wavelength present in the initial light beam. If the source emits light of all wavelengths, then the slit images blend together and form a continuous band of color known as a **continuous spectrum**.

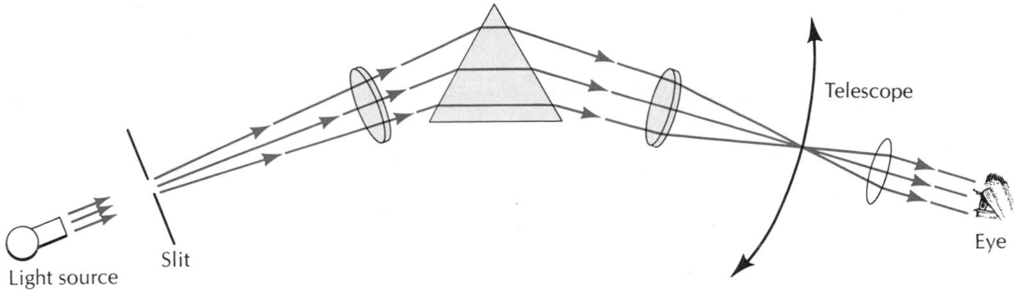

FIGURE 7.1 Internal workings of a spectroscope. The telescope turns about a pivot point located at the center of the prism to allow the viewer to examine all visible wavelengths.

When a spectroscope was first turned toward the Sun, it revealed that the Sun does not emit a continuous spectrum (Figure 7.2). Instead, the spectrum is crossed by a large number of dark lines: most of the colors or wavelengths are present in the solar spectrum, but some wavelengths appear to be absent. Where the color should be, there is a dark image of the slit. We know today that the radiation in these dark lines is not totally absent; it is greatly suppressed, though.

When a spectroscope is turned toward a cloud of the glowing vapor of some chemical element, a different type of spectrum is observed. Only a few colors are present in it. Each color is a distinct image of the slit, and is called a **spectral line**. One spectrum that is among the easiest to observe is that of sodium (Figure 7.3). A pinch of table salt in a flame colors the flame a brilliant yellow. Through a spectroscope, one sees two brilliant yellow spectrum lines very close together. If we measure the wavelength of the sodium lines by some means, we will find it to be roughly 6000 Ångstroms, with about a 5 Ångstrom separation between the two lines. A sodium-vapor lamp, such as is used for street lighting in large cities, appears yellow because of these two **emission** lines. Closer study of the sodium spectrum reveals many other fainter lines, as well.

The spectrum of hydrogen vapor (Figure 7.4) contains several emission lines: a brilliant red line at 6563 Ångstroms; then a blue-green line at 4860 Ångstroms; then a blue line at 4340 Ångstroms; and so on. As one looks from the red end of the hydrogen spectrum to the violet end, the lines crowd closer together, until they appear to coalesce at about 3650 Ångstroms.* These lines are referred to as the Balmer series of hydrogen (see p. 171).

---

*The wavelength at which the hydrogen lines appear to coalesce is an ultraviolet wavelength, which is invisible to the human eye. The phenomenon can easily be photographed, however.

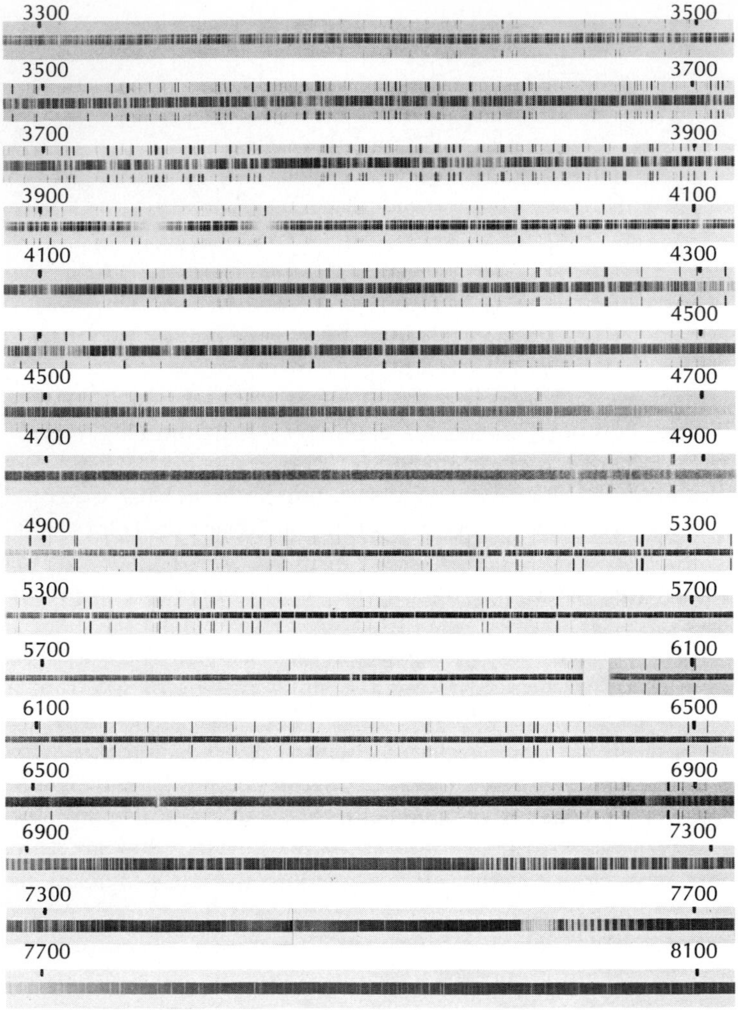

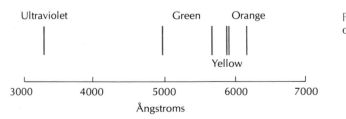

FIGURE 7.2 The spectrum of the Sun from 3300 Å to 8100 Å. The solar spectrum is flanked by an emission spectrum of iron, produced by an arc in the observatory, to provide a wavelength scale. Note that these are negative prints: the solar absorption lines appear white. (Hale Observatories.)

FIGURE 7.3 Spectrum of sodium.

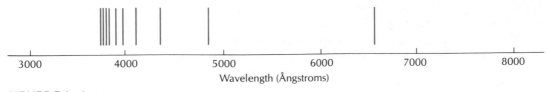

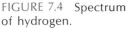

FIGURE 7.4 Spectrum of hydrogen.

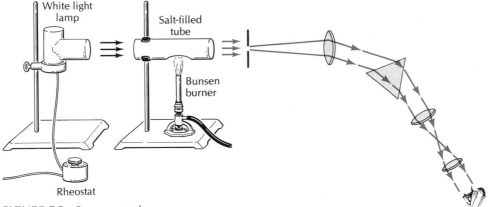

FIGURE 7.5 Set-up used to produce sodium absorption lines (see text).

Each element has its own characteristic pattern of spectrum lines, which, like a fingerprint, immediately identifies the element. As time progressed, laboratory work allowed scientists to study the spectra of most of the elements.

Interestingly, a study of the Sun's spectrum showed a pair of dark lines at the location where the yellow sodium lines should be. Scientists have found that each dark line in the solar spectrum corresponds to a bright line in the spectrum of a glowing laboratory gas. In this way, many of the elements found on Earth have been identified in the Sun.

Around 1860, Gustav Kirchoff tied all these observations together. First, he noted that a glowing solid, a liquid, or a very hot, dense gas always emits a continuous spectrum. A glowing gas that is not ultrahot or ultradense emits a bright line or emission spectrum. To explain the dark-line spectrum, let's look at an experiment that is easy to duplicate (Figure 7.5). Take a length of copper tubing about 2 centimeters in diameter and 10 centimeters long, cut a hole in the side of the tube, and line the tube with asbestos. Next, wet the asbestos, pour table salt in the tube, clamp it in a stand, and direct a bunsen-burner flame into the hole in the side of the tube. The tube will fill with glowing sodium vapor. If we look into the tube with a spectroscope, we will see the characteristic sodium spectrum.

Now, let's take a very bright lamp whose brightness can be varied by a rheostat, and shine the light through the tube toward the spectroscope slit. If the lamp is turned on low, we will see the bright sodium lines superimposed on the fainter continuous spectrum of the lamp's metallic filament. If we continue to turn up the brightness of the lamp, a point will be reached where the sodium lines are no longer brighter than the continuous spectrum; they become virtually invisible. What happens if we turn up the brightness of the lamp more? We will see dark lines at the location of the sodium lines, much like the dark lines of the solar spectrum.

Kirchoff generalized from many experiments like this one and arrived at the following law: a hot, glowing gas, placed in front of a hotter source of continuous spectrum, will absorb radiation of the same wavelengths that the gas would emit if viewed alone, and will produce a dark-line or **absorption spectrum**.

We have described three types of spectra: continuous, emission, and absorption (Figure 7.6), and have seen that emission and ab-

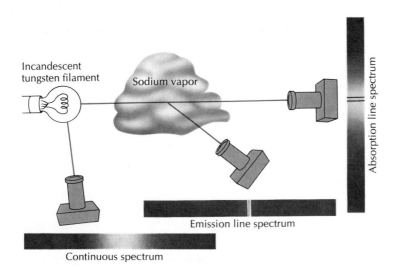

Incandescent tungsten filament

Sodium vapor

Absorption line spectrum

Emission line spectrum

Continuous spectrum

FIGURE 7.6 Origin of three types of spectra. (From *New Horizons in Astronomy* by J. C. Brandt and S. P. Maran. W. H. Freeman and Company. Copyright © 1972.)

sorption spectra tell us what elements are present in the gas. We now have a powerful tool for analyzing the light of all types of celestial objects in order to find out what chemical elements are present in them.

There is additional information in the spectrum of an object, too. A careful study will help us find out temperature, pressures, magnetic fields, and other things as well. In the rest of the chapter we will describe how to find some of these numbers.

THE CONTINUOUS SPECTRUM

Physicists talk about an idealized object, known as a **black body**, that absorbs all the light that falls on it: it does not reflect any light. Such an object will look black. Most things have color because they reflect only certain wavelengths from the white light that falls on them. For example, red ink appears red because it reflects red light. If the same ink were illuminated by a sodium-vapor light, it would appear black, because the sodium vapor emits no red for the ink to reflect.

A real black body is difficult to make, because almost all materials reflect some light. One trick that is used is to build a box with a small hole in it (Figure 7.7). Any photon that enters the hole bounces around inside the box until it is absorbed. The hole is a good approximation to the ideal black body. Let's carry out an experiment now. We will build a black-body box and fit it with a heating coil attached to a rheostat that allows us to control the temperature. Let's call it a furnace. We will study the light that comes out through the hole in the furnace with a **spectrophotometer** (Figure 7.8).

A spectrophotometer is just like the spectroscope described above, except that we remove the telescope used for viewing the

FIGURE 7.7 A small aperture of a box behaves like a black body. Once a photon enters the box, it will bounce around until it is absorbed. The chance it will find its way back out the hole is small.

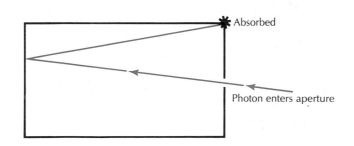

Absorbed

Photon enters aperture

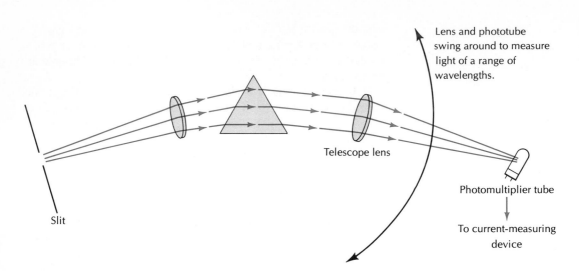

Lens and phototube swing around to measure light of a range of wavelengths.

Telescope lens

Photomultiplier tube

Slit

To current-measuring device

FIGURE 7.8 Spectrophotometer. The spectrophotometer is very much like a spectroscope, except a photomultiplier tube replaces the eye as the detector. The current from the photomultiplier is recorded as the lens and phototube are pivoted around the prism to view all wavelengths. The result is a graph of brightness versus wavelength.

spectrum, and replace it by a brightness-measuring device: a photometer or light meter. After the spectrum is dispersed by the prism, the light falls on a second slit, which lets one color at a time pass through to a lens, which is all that remains of the viewing telescope in the set-up. The eyepiece of the telescope has been replaced by a photomultiplier tube (Chapter 4). The light that passes through the lens is focused on the photomultiplier tube, where it is converted to a measurable current. Measuring that current allows the brightness to be inferred.

By swinging around the part of the spectrophotometer with the second slit and the photomultiplier tube, we can successively measure the brightness at each wavelength in the spectrum. A device of this type is used in many astronomical applications.

Now, we have the black-body furnace and spectrophotometer all set up; so let's begin our experiment. We turn on the electricity very low, and the furnace heats up until a faint red light is emitted at the exit hole. If we could measure the temperature in the furnace, we would find it to be about 3,000° C. If we continue to increase the electrical current flowing into the furnace and thereby raise the temperature, we will find the inside of the furnace to be first red-hot, then orange-hot, yellow-hot, white-hot, and finally blue-hot.

FIGURE 7.9 Black-body energy distributions for various temperatures: (a) 4,000° K and 6,000° K. (b) Hotter than 6,000° K. Note the 6,000° K curve on both halves of the figure for relative scale.

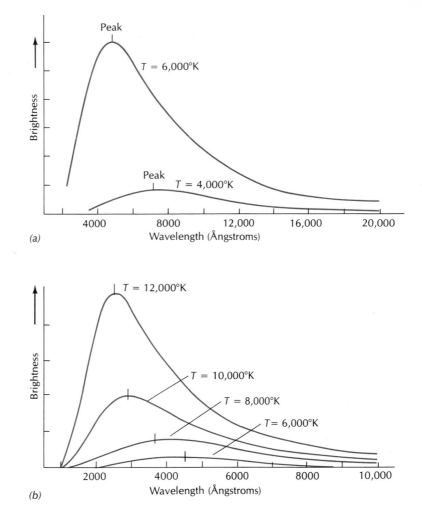

(a)

(b)

The brightness at visible wavelengths measured with the furnace at various temperatures is plotted in Figure 7.9. Notice that the red-hot furnace emits more energy in the red than at any other wavelength; the orange-hot furnace emits more energy in the orange; and so on. The exceptional case is that there is no such thing as green-hot: when the peak of the black-body energy curve is in the yellow-green, so much light of other colors is present that we get the sensation of white from the black body. The temperatures of the various cases are indicated on the curves in Figure 7.9. Of course, you can tell from these temperatures that the experiment we have described is not really possible. All known substances are gaseous at temperatures like 6,000° C, which is the temperature of the solar surface.

Temperature

Since we have talked about temperature in the last few paragraphs, it is useful to stop and discuss the concept briefly.

The old familiar Fahrenheit temperature scale is used in everyday life in the U.S. The scale is defined by two fixed points: the **ice point** and the **steam point**; that is, by definition, water freezes to ice at 32° F and boils to become steam at 212° F. Another common temperature scale is the centigrade scale, with ice point and steam point at 0° C and 100° C, respectively. (See Figure 7.10 for a comparison.) It takes a little arithmetic to convert from one scale to the other (see Appendix C).

A device for measuring temperature, the thermometer, consists of a reservoir of colored alcohol or mercury attached to a thin glass tube. All substances expand as their temperature increases and contract as their temperature decreases. Thus the volume of liquid in the reservoir changes with temperature, and a column of the liquid in the thin glass tube changes length to compensate. A thermometer scale is added to the tube by noting the position of the column in an ice bath and in boiling water (Figure 7.11).

This is not terribly new information. We use temperature and thermometers all the time; so we have a rough, intuitive feeling for what temperature means. But actually this intuitive feeling is based on the effect of temperature on our senses. We know what 95° F feels like: in the humid eastern U.S., it is uncomfortable. In this chapter we want to develop a physical idea of what temperature means. We will talk more about this later.

There is a limit to the lowest temperature that can be achieved. We will discuss the theoretical reasons for this later, too. The lowest temperature is 273.2° below centigrade zero; for our purposes −273° C is close enough. In many physical problems, it is convenient to use the Kelvin temperature scale, which makes mathematical equations for physical laws simple. Zero on the Kelvin scale is **absolute zero**: −273° C. The ice point is then 273° K and the steam point is 373° K. The human body temperature is 37° C or 310° K, for example.

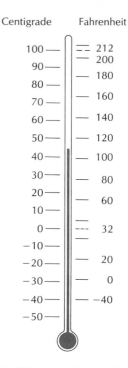

FIGURE 7.10 Comparison of temperature scales.

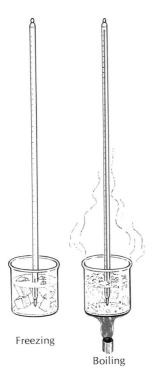

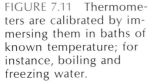

Freezing

Boiling

FIGURE 7.11 Thermometers are calibrated by immersing them in baths of known temperature; for instance, boiling and freezing water.

The Kelvin temperature scale will be used to describe temperatures of black bodies. The curves in Figure 7.9 are labeled with °K. Then why did we talk about °C in the text? Because at temperatures of 3,000° C or higher, the 273° difference between the scales is too small to make any difference for our discussion.

Radiation Laws

Several attempts were made during the nineteenth century to develop a theory that would explain the shape of the black-body curve: that is, the curve that results when we plot the intensity of the light emitted by a black body against the wavelength of the light. These attempts all failed. Each was based on the assumption that light behaves like a wave. Then Max Planck put together a successful theory by starting with the assumption that light behaves like a stream of particles. This theory was the origin of the modern photon concept that Einstein proved correct with his explanation of the photoelectric effect (Chapter 4). The mathematical expression which allows us to calculate the intensity of radiation at each wavelength for a given temperature is known today as Planck's law.

Two other radiation laws, which were known before Planck's time, can be derived from Planck's law. The earlier derivation of these laws was based on thermodynamic arguments. The laws are as follows:

(1) *The black-body intensity curve has a maximum value at some wavelength.* That is, for any temperature there is one wavelength where a black body emits more radiation than at any other wavelength. As the temperature of the black body is raised, the wavelength of the maximum shifts to shorter wavelengths (Figure 7.12). That's basically why the color of the black body appears to change as the temperature is changed.

(2) *The total light energy emitted from a black body equals the area under the black-body curve.* In other words, it is the sum of all the intensities at all wavelengths. The total energy is proportional to the fourth power ($T^4 = T \times T \times T \times T$) of the Kelvin temperature of the black body. We will return to this point in Chapters 8 and 12, when we use the T^4 law to compare the energy output of the Sun and stars.

In this section we have discussed the characteristics of the light emitted by a black body. A very hot, dense gas, such as is found in the interior of a star, emits a continuous spectrum like that of a black body. The solar spectrum, for example, has a shape roughly

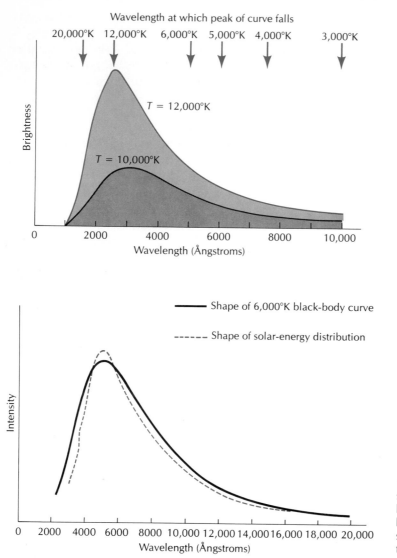

Wavelength at which peak of curve falls

20,000°K   12,000°K   6,000°K   5,000°K   4,000°K   3,000°K

$T = 12,000°K$

$T = 10,000°K$

Brightness

Wavelength (Ångstroms)

FIGURE 7.12   Properties of black-body energy-distribution curves. The total energy emitted by each square centimeter of the surface of a black body is related to the area inside the curve, which increases as T^4. The wavelength of the peak of the curve depends on temperature.

——— Shape of 6,000°K black-body curve

- - - - - Shape of solar-energy distribution

Intensity

Wavelength (Ångstroms)

FIGURE 7.13   Comparison of shape of 6,000° K black-body energy-distribution curve and the observed solar energy distribution.

like that of a 6,000° K black body (Figure 7.13). One fascinating fact, which is doubtless not coincidence, is that the human eye is most sensitive to the wavelength at the peak of the Sun's emission curve. We are designed to make maximum use of sunlight.

We wish next to look at the whole problem of line spectra and discuss what we can learn about celestial objects by a study of their spectra. We must look, therefore, at the atomic nature of matter.

A great deal of intellectual history is tied up in the knowledge that the chemical elements are made up of atoms. Like the history of gravitational astronomy, the history of the atomic nature of matter has its heroes and its periods of rapid development. It would carry us too far afield to review this historical development here. Instead, we will briefly summarize the modern picture of the atom.

There are two particles in the atomic nucleus, the neutron and the proton (Chapter 5). The number of protons in the nucleus determines its electric charge and ultimately its chemical nature. The number of protons and of neutrons together determine the mass of the nucleus. Atoms are tiny things, an Ångstrom or two across. The nucleus, on the other hand, is between 10^4 and 10^5 times smaller, depending on the element. The nucleus is an unbelievably small, dense unit of matter.

Orbiting the nucleus of the atom are the negatively charged electrons. In the undisturbed atom, there is one orbiting electron for each proton in the nucleus. Since the electron and proton charges are equal and opposite, the atom is electrically neutral.

Let's take a closer look at the simplest of all atoms, the hydrogen atom.

Hydrogen

Hydrogen is the simplest atom (Figure 7.14). Its nucleus consists of one particle, a proton. Orbiting the proton is one electron. The electron cannot move in just any orbit, but is restricted to certain allowed orbits. How are the allowed orbits determined?

Planck and Einstein showed that light can behave both like a wave and like a particle. In 1924 Louis de Broglie turned the tables and hypothesized that the electron (and other particles), which usually behaves like a particle, might also behave like a wave. The wavelength of the particle would be inversely proportional to its momentum (the product of its mass and its velocity). Soon after de Broglie's hypothesis, the wave nature of electrons was proved by an experiment in which electrons were demonstrated to show interference effects, which only waves can show. Planck's theory of the photon and the demonstration of the wave nature of electrons finally led to a modern theory of atomic-level phenomena known as **quantum mechanics**.

Picture the wavelike electron orbiting the nucleus in a hydrogen atom. The only orbits that the electron is allowed to occupy are

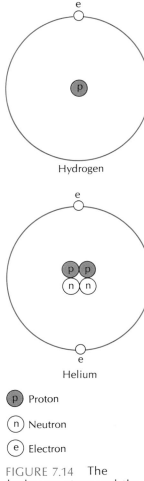

FIGURE 7.14 The hydrogen atom and the helium atom. The nuclei of these atoms are actually only 1/10,000th the size of the electron orbits.

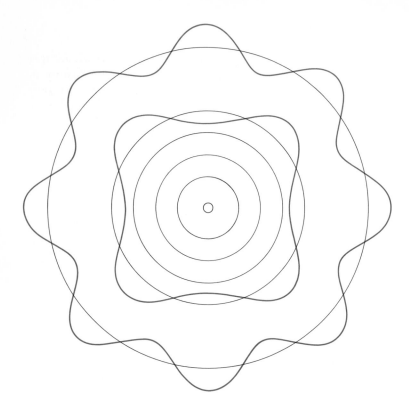

FIGURE 7.15 The al-
lowed orbits in an atom
are the orbits whose cir-
cumferences are an inte-
gral number of wave-
lengths of the wavelike
electron orbiting the nu-
cleus.

those with a circumference that is an exact multiple of the electron's
wavelength (Figure 7.15). The inner, smallest orbit is one wavelength in circumference; the next larger orbit is two wavelengths in
circumference; and so on. According to the theory, if the circumference were not an **integral** number of wavelengths, the electron
would destructively interfere with itself on subsequent circuits of
the orbit. Like the snake swallowing its own tail, the electron would
"eat" itself up. The total energy an electron has is different in each
orbit. The electron has more energy if it moves in a larger orbit.
Figure 7.16 shows the allowed orbits of a hypothetical atom.

In 1913, Neils Bohr developed this picture of the hydrogen atom.
He hypothesized, in addition, that electrons can jump from one
energy level to another. If the electron is in a high-energy orbit, it
can spontaneously jump to a lower orbit and emit a photon. The
energy of the emitted photon is equal to the difference between the
energy the electron had in the high-energy orbit and its energy in
the low-energy orbit. Since the electron moves only in certain
specific orbits, which have certain specific energies, only certain

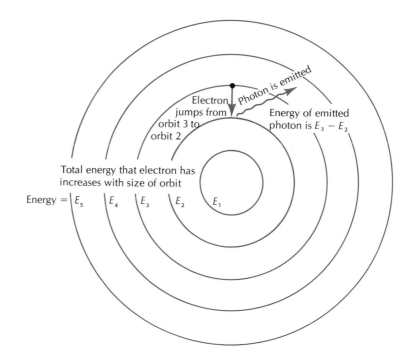

FIGURE 7.16 Allowed orbits in a typical atom, showing a photon being emitted by an electron jumping from a higher-energy orbit to a lower-energy orbit.

specific energy differences between orbits can exist; so only photons with certain specific energies can be emitted. Since the frequency is related to the energy, this also means that only photons of certain specific frequencies are emitted. Finally, this all tells us that only specific wavelengths of light can be emitted by the hydrogen atom. Each spectral line corresponds to a jump from a high orbit to a low orbit.

If a hydrogen atom has its electron in its third energy level, and the electron jumps to level 2, it will emit a photon with a 6560 Ångstrom wavelength (Figure 7.17). This jump is the origin of the red line in the hydrogen spectrum known as Hα. Electrons jumping from level 4 to level 2 produce the 4860 Ångstrom blue-green spectrum line, and so on. At higher energies, the hydrogen atom's orbits crowd together—or we can say that the energy levels crowd together. Thus electrons jumping from very high levels to level 2 produce spectrum lines which crowd together toward the blue (3650 Ångstrom) part of the spectrum. The series of lines produced by jumps down to level 2 is called the **Balmer series** of hydrogen. Electrons jumping from levels 2, 3, 4, etc., to level 1 produce a series of spectrum lines in the ultraviolet starting with a strong line at 1216 Å and crowding together toward 912 Å. This series of lines

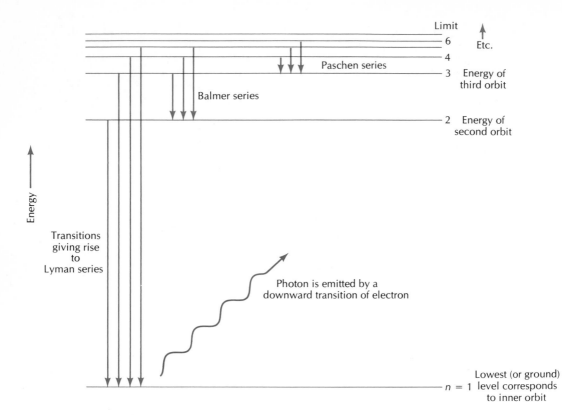

FIGURE 7.17 Energy-level diagram for hydrogen. In this type of diagram, the total energy of an electron in each allowed orbit is indicated. We show the transitions that lead to various series of spectral lines in hydrogen.

is known as the **Lyman series**. Figure 7.17 shows the origin of other series of lines in hydrogen, as well.

Of course, in any gas there are a vast number of hydrogen atoms, with electrons in all different energy levels. So each possible downward jump or transition of an electron is taking place in one atom or another. Thus the gas as a whole emits all the different spectrum lines.

More complex atoms, with more electrons, have a different set of energy levels and emit different photons. The basic mechanisms are the same, however.

More Complex Atoms

More complex atoms have more than one orbiting electron (Figure 7.18). Helium, the next most complex atom has two electrons, for instance. In undisturbed helium, the two electrons occupy the same orbit. These two electrons form the K-shell of the atom. The elements lithium through neon add electrons to the next shell, which is full when it has eight electrons.

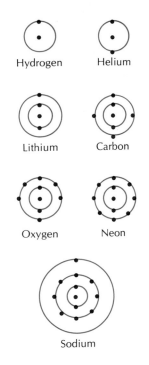

FIGURE 7.18 Structure of various atoms.

Hydrogen   Helium

Lithium   Carbon

Oxygen   Neon

Sodium

Lithium and sodium have one electron outside a full orbit of electrons. These elements, for that reason, are chemically similar, and have spectra which bear similarities to the hydrogen spectrum. Other elements with one electron outside full orbits are potassium and rubidium. If you look at a **periodic table** (see Appendix D), you see that these four elements (Li, Na, K, Rb) fall in one column. They are chemically similar and have similar spectra.

Beryllium, magnesium, calcium, etc., all have two electrons outside full shells and are chemically similar. Helium does not fit with this group, because it has two electrons in a shell that will only hold two electrons, whereas beryllium, etc., have two electrons in shells that hold more than two electrons.

Chlorine has seven electrons in an orbit that will hold eight electrons. It is chemically similar to fluorine, bromine, and iodine, which have the same kind of structure.

When sodium and chlorine combine to produce a molecule of table salt, the single electron of the sodium moves over to fill the single vacancy in the orbit of chlorine. The sodium atom thus becomes positively charged and the chlorine atom negatively charged; since opposite charges attract, the two atoms "stick" to-

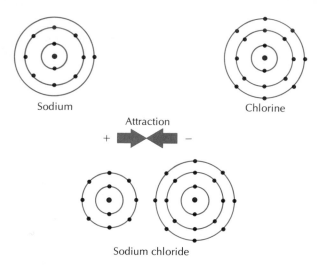

Sodium

Chlorine

Attraction

+ ◄► −

Sodium chloride

gether. This attachment is called an **ionic bond** (Figure 7.19) and is one example of a way in which compounds are formed.

We do not need to say anything more about this subject here. We have now talked about atoms and their emission spectra. How are absorption spectra formed?

Absorption Spectra

Absorption spectra occur when photons from a hot source of continuum radiation pass through a cooler gas. If a photon with the correct wavelength bumps into a hydrogen atom, it will give up its energy to the orbiting electron and cause it to jump to a higher energy level. If the electron is originally in level 2 and a photon with a 6563 Ångstrom wavelength comes along, for instance, the electron will absorb the photon and jump up to level 3. The photon that is absorbed and thus causes an electron to jump up from level a to level b must have the same wavelength as the photon that would be emitted by the electron if it jumped down from level b to level a. The wavelengths of absorption and emission spectrum lines are identical.

Excitation and Ionization

Leave a hydrogen atom alone, undisturbed, and it will spontaneously emit photons until its electron is in level 1. It can then emit no more light, for the electron is in its lowest energy level. A typical electron will remain in a high level for a tiny fraction (10^{-6}) of a

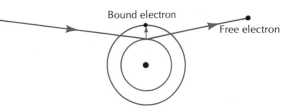

Bound electron

Free electron

FIGURE 7.20 A free electron can interact with an atom and jostle a bound electron from one state to another. If the bound electron jumps up to a higher energy level, the free electron loses energy to it; if the bound electron jumps down to a lower energy level, then the free electron gains energy from it.

second before it jumps to a lower level. Thus an atom will radiate its energy away in a very short time. What happens to **excite** the electron back to higher energy levels?

Atoms do not exist alone in space, but are part of a gas containing other atoms and photons. If two atoms collide, the energy of one atom's motion can bump an electron to a higher energy level (Figure 7.20). This process is known as **collisional excitation**. In addition, the photons which are usually around in the gas continually excite electrons to higher levels by being absorbed, in a process often called **photoexcitation**. Inside a star, the number of collisions and absorptions is large; so the atoms have their electrons in all possible energy states, and will emit all possible spectral lines.

The fraction of atoms that have their electrons in the various excited levels depends on the temperature. If the temperature were absolute zero, all electrons would be in their lowest possible levels. At increasingly higher temperatures, more and more electrons are excited to higher levels. Consider two spectrum lines that come from jumps which start on two different levels. Their relative strength depends on how many of the atoms have electrons on one or the other of the two levels, and is therefore related to the temperature (see Figure 7.21). Thus comparing the strengths of selected pairs of spectrum lines helps us find out the temperature of the source.

In any atom, there is a highest possible energy level that an electron can be in and still be bound to the atom. If the electron is

FIGURE 7.21 Spectral-line strengths help us find temperature. In a cool gas (a), most hydrogen atoms have their electrons in level $n = 1$; so they cannot absorb a 6563 Å photon, which can only be absorbed when an electron in level 2 jumps to level 3. In a hotter gas (b), more atoms have electrons excited to the $n = 2$ level and can absorb 6563 Å photons. Thus the strength of the 6563 Å line increases in strength with increasing temperature. However (c), as the gas gets very hot, more atoms have electrons in even higher levels; so, at very high temperatures, the 6563 Å line gets weaker again. Lines from other atoms can help decide whether a weak 6563 Å line means cool temperatures or very hot temperatures.

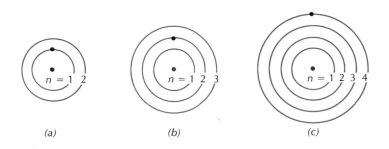

$n = 1$ 2

(a)

$n = 1$ 2 3

(b)

$n = 1$ 2 3 4

(c)

given more energy than it needs to jump to that orbit, it will be torn from the atom. The atom is then said to be **ionized**: it is an **ion**, (Figure 7.22). If helium, which normally has two electrons, is ionized, its one remaining electron produces a spectrum very similar to the spectrum of hydrogen.

If the electron of a hydrogen atom is in an energy level near the highest possible level and then drops down to level 1, it radiates a photon with a wavelength slightly longer than 912 Ångstroms. A photon with a wavelength shorter than 912 Ångstroms has enough energy to remove an electron from any level in a hydrogen atom if it is absorbed. Ionized hydrogen has no electrons and therefore emits no spectrum lines.

Incidentally, an atom that has many electrons, such as iron, can be ionized more than once. In the corona of the Sun, we can find iron atoms that have lost as many as 14 of their 26 electrons. In general, the hotter a gas is, the more highly ionized are the atoms. The spectrum of each ion is distinct; so we can find out the degree of ionization of a gas—a clue to its temperature—from the spectrum.

What actually happens in a gas as temperature changes? Studies of gasses in containers revealed a relationship between pressure, volume, and temperature known as the **equation of state**. For instance, if we heat a gas within a container with fixed volume, the pressure increases in proportion to the absolute temperature.

The properties of gasses were explained by a theory known as the kinetic theory of gasses. The energy of motion or **kinetic energy** of any body, including an atom, is proportional to the body's mass and the square of its velocity. According to the kinetic theory, the average kinetic energy of atoms in a gas is proportional to the temperature of the gas: the higher the temperature, the higher the kinetic energy, and vice versa. Basically, as temperature is increased, the speed of the atoms increases.

At absolute zero atoms cease to move about. Since there is nothing slower than a dead stop, no further energy can be removed from the system. Thus absolute zero is the lowest temperature possible. Interestingly though, all is not still inside the atoms at absolute zero. Physicists have found that a minimum-energy state exists in atoms and molecules. Theory says that there is no means by which this minimum energy can be removed from the molecules. It cannot be decreased to zero. Thus, at absolute zero, the internal energy is at its lowest possible value.

In a gas the atoms and molecules constantly collide with one another. Have you ever watched a tiny dust particle illuminated by a shaft of sunlight from a window? It dances around in a zigzag

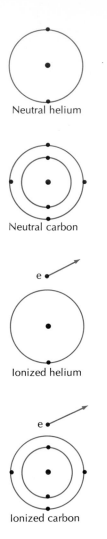

Neutral helium

Neutral carbon

Ionized helium

Ionized carbon

FIGURE 7.22 Ionized atoms.

path through the air. This effect, known as Brownian motion, is due to the air molecules, which are constantly colliding with the dust speck.

In a high-temperature gas, like the Sun's atmosphere, collisions between atoms can be very energetic. The resultant bump can excite an electron to a higher energy level or even tear it from the atom in a collisional excitation or collisional ionization process.

A gas exerts pressure on the wall of a box because of the large number of collisions between molecules and the wall. As the temperature increases, the molecules move faster, and the collisions give the wall harder bumps. The harder bumps appear as increased pressure.

In this section, we have given a quick look at the relationship between the temperature of a gas and the motion of the molecules in the gas. We will close the chapter with a description of one further important effect.

THE DOPPLER EFFECT

Have you ever listened to an ambulance racing by, with its siren blaring? As the vehicle passes, the pitch of the siren seems to lower slightly. The rapidly retreating siren appears to emit a lower sound than the approaching siren. This effect is an example of the **Doppler effect** (Figure 7.23).

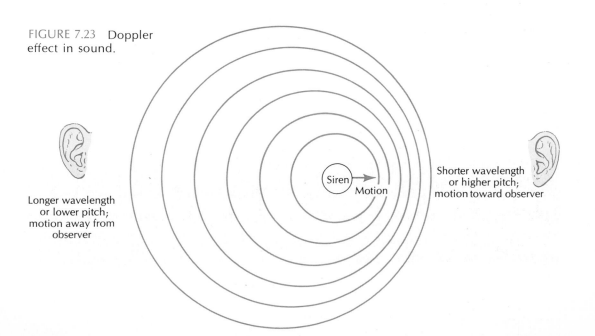

FIGURE 7.23 Doppler effect in sound.

Longer wavelength or lower pitch; motion away from observer

Siren → Motion

Shorter wavelength or higher pitch; motion toward observer

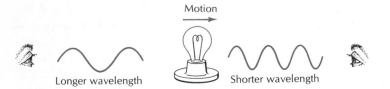

Motion

Longer wavelength   Shorter wavelength

The Doppler effect can be understood roughly as follows. When a wave approaches you from a source that is not moving, you can measure a certain wavelength or frequency for the wave. Now start the source moving toward you. A person who is moving along with the source would find that the frequency of the wave remains constant. However, since you are standing still, you would find that the frequency had increased. Since the source is moving toward you, you will encounter wave crests more often. The greater the velocity of the source, the greater the increase of the frequency. Similarly, if the source of the wave is moving away from you, the frequency is decreased by an amount that depends on the velocity of the source.

The last paragraph dealt with waves in general, for the Doppler principle applies to all kinds of waves. If you are out in the ocean in a speed boat, you will crash into wave crests more often if you are traveling toward the direction from which the waves come than if you travel in the same direction as the waves. In fact, it is conceivable that you might run your boat at the same velocity as the waves and stay in a trough, in which case the frequency of the waves appears to become zero.

The Doppler effect also applies to light waves (Figure 7.24). If you look at the spectrum of a star that is moving away from you, all its spectrum lines are shifted slightly toward the red end of the spectrum (Figure 7.25). The star's velocity of recession can be

FIGURE 7.25 Doppler-shifted spectrum lines from the star Arcturus. The apparent change in motion of the star is due to the Earth's motion around the Sun. (Hale Observatories.)

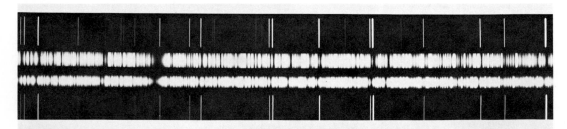

calculated from the amount of the shift.* If the star is approaching, its spectrum lines are shifted toward the blue end of the spectrum.

The Doppler effect will be very useful to us in our studies of the motions of celestial bodies. When we study the universe as a whole, for instance, we will find that the spectra of galaxies show red-shifted spectral lines. Distant galaxies all move away from us.

In this chapter, we have given a brief description of several facts about the physical nature of matter that will be needed in our later chapters. Let's now look at the Sun.

---

*The velocity of an object can be calculated from the wavelength shift of its spectral lines by the formula

$$\frac{\text{speed of motion}}{\text{speed of light}} = \frac{\text{wavelength shift}}{\text{wavelength at rest}}.$$

We should note also that the explanation of the Doppler effect given here is not precisely correct for light, as Einstein showed in his theory of special relativity. However, the explanation is correct in general, if not in certain details.

REVIEW QUESTIONS

1.   Describe or define: (a) continuous spectrum; (b) spectral line; (c) emission line; (d) black body; (e) spectrophotometer; (f) excitation; (g) ionization.

2.   How does the emission of a black body change with increasing temperature?

3.   What do we mean by temperature?

4.   Describe the structure of an atom.

5.   What physical processes lead to atoms having unique line spectra?

6.   What is the Doppler effect?

8

THE SUN

W hat do we know about the Sun? How do we carry out solar research? These questions will be answered in the following pages.

Let's start by introducing the Sun. We know that the Sun is 150,000,000 km from the Earth. Since its angular diameter is about $\frac{1}{2}°$, its true diameter is about 1,400,000 km, or about 109 times the Earth's diameter. Its mass, calculated from our planet's orbital motions, is approximately 2×10^{33} gm, or $\frac{1}{3}$ of a million Earth masses; so it has an average density of 1.4 gm/cm^3. The acceleration of gravity at its surface is 28 times that of the Earth.

With these numbers in hand to give a rough idea of the Sun's size, let's go on to a description of what the Sun looks like to an observer without expensive equipment.

THE VISIBLE SUN

Looking directly at the Sun through a telescope is very dangerous. The energy of sunlight entering the large aperture of a telescope, then focused by an eyepiece onto a piece of paper, is concentrated into a small image that can actually cause the paper to catch fire and burn. Of course, this energy would be very destructive if it got into your eye. Smoked glass is available to cover the eyepiece of the telescope and absorb some of the energy before it gets into your eye. Unfortunately, such glass sometimes cracks in the intense heat, allowing the eye to be exposed suddenly to the full intensity of the Sun's light. Partially reflecting mirrors can also be used to cover the aperture of the telescope; these reflect away most of the sunlight and prevent it from entering the telescope in the first place. These devices are safe, but can be expensive. Fortunately, there is a simple solution for the observer who wants an inexpensive way to look at the Sun. If you remove the eyepiece from the telescope, you can use

FIGURE 8.1 The McMath Solar Telescope at Kitt Peak. The mirror at the top of the tower at left shines sunlight down the sloping leg of the instrument to the underground observing room. (Kitt Peak National Observatory. Copyright © 1975 by the Association of Universities for Research in Astronomy, Inc.)

the objective lens (or mirror) to focus an image of the Sun onto a sheet of white paper, where you can examine it safely. A typical setup is shown in Figure 8.3. If we make observations of the Sun in this way, what will we see?

One thing that we will see is **sunspots** (see Figures 8.4, 8.5, and 8.6). When Galileo first turned his telescope toward the Sun, sometime around 1611, he discovered these dark spots on the Sun. He noted that sunspots change their position on the solar surface, and that long-lived spots returned to their original position on the Sun in about a month. His observations led him to conclude that the spots are clouds in the solar atmosphere, and that their motion is due to the rotation of the Sun on its axis. If we examine the image of a sunspot projected by a moderate-sized lens, we see that it consists of two parts: a dark central **umbra**, surrounded by a lighter **penumbra**, which is in turn fainter than the surrounding

FIGURE 8.2 The solar tower telescope at Sacramento Peak Observatory near Alamogordo, N.M. The tower is 136 feet high. (Sacramento Peak Observatory, Air Force Cambridge Research Laboratories.)

To sun

Telescope

Screen to prevent direct sunlight from reaching projected image

White paper

Solar image

FIGURE 8.3 Setup to observe the sun by projecting the image on a screen. Always use an expendable eyepiece, because it will get very hot and could break. *Never* look directly at the sun through the telescope; you could blind yourself.

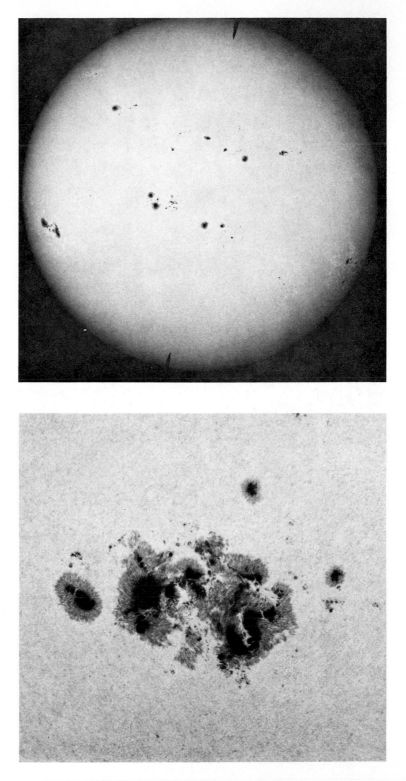

FIGURE 8.4 Whole-disk photograph of the Sun on July 13, 1937, near sunspot maximum. The Zurich sunspot number was 188. (Hale Observatories.)

FIGURE 8.5 Large sunspot group of May 17, 1951, photographed from the ground. (Hale Observatories.)

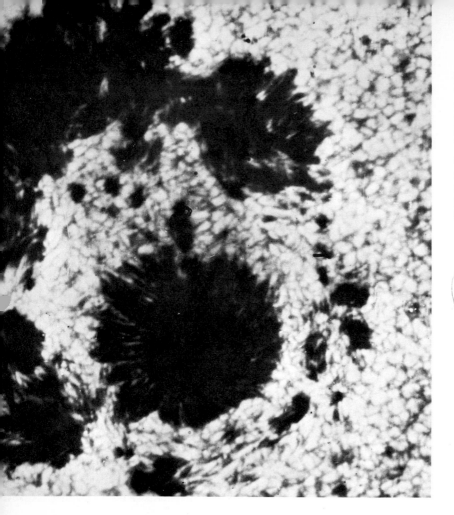

FIGURE 8.6 Closeup view of a sunspot taken with a 12-inch refracting telescope carried by a balloon to 80,000 feet. The telescope was above the portion of the atmosphere that causes the most severe seeing defects. (National Aeronautics and Space Administration.)

undisturbed solar surface. The modern view of the nature of sunspots will be described later.

The visible surface of the Sun has come to be called the **photosphere**, the sphere of light. If we examine the projected image of the photosphere carefully, we will find that it is not a smooth, uniform surface, but is, in fact, covered with a granular pattern, known, appropriately, as **granulation** (Figure 8.7). When observed with the best available solar telescopes, each granule is seen to be a bright area, shaped like an irregular polygon, and bordered by a narrow, darker lane. Granules are difficult to observe clearly, for they are just about the smallest objects that can be observed through the blurring atmosphere of the Earth. The best photographs of granules are taken at instants of unusual steadiness in the atmosphere, or from telescopes borne high above the densest parts

FIGURE 8.7 Solar granu-
lations. (Hale Observato-
ries.)

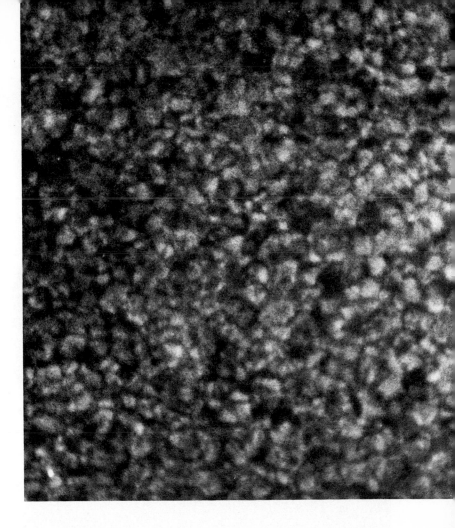

of the atmosphere in balloons. It is not possible to predict when
moments of unusual steadiness will occur and to then snap a picture; the moments come and go too fleetingly. Instead, astronomers
take many pictures of the Sun each minute, and pick out the very
best frames for study.

The photosphere is not uniformly bright. It is not too difficult to
see that the brightness of the image (which we often call the **solar
disk**) decreases from center to edge, or limb. This phenomenon is
known as **limb-darkening**. If you look very closely at the darkened limb, you will see occasional brighter areas which are known
as **faculae**.

Sunspots, solar rotation, granulation, limb-darkening, and
faculae are all phenomena of the visible Sun that can be discovered
by anyone with a small telescope and some care and patience.

FIGURE 8.8 Total eclipse of the Sun of June 8, 1918. Photographed from Green River, Wyoming. (Hale Observatories.)

Very rarely, when the Moon passes in front of the Sun and blocks out the intense photospheric light, we are treated to one of nature's truly beautiful phenomena, the total solar eclipse (Figure 8.8). At such times, we see that the Sun is surrounded by an extended halo of gas known as the **corona**, for it crowns the Sun. Just before and just after the eclipse is exactly total, we also can see a thin, fiery red layer of the Sun just beyond the edge of the Moon. This layer is known as the **chromosphere** because of its spectacular color (Figure 8.9).

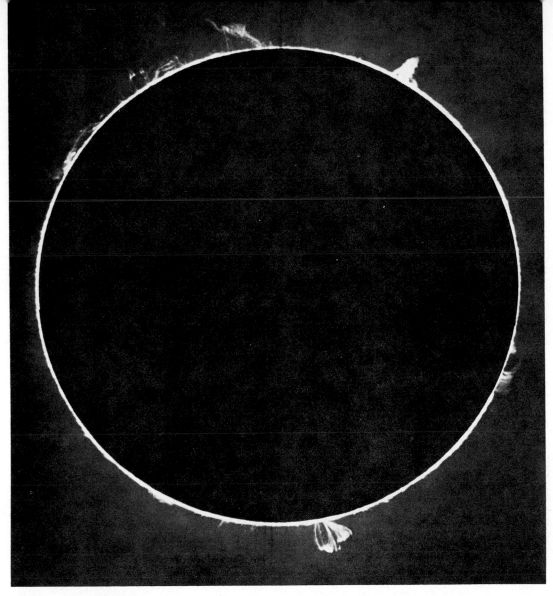

FIGURE 8.9 Artificial eclipse. The entire edge of the Sun is photographed in the violet spectral line of ionized calcium, while the solar disk is occulted by a metal disk inside the telescope. The ring of radiation around the occulting disk is the chromosphere. (Hale Observatories.)

SOLAR ROTATION

When Galileo announced his discovery than the Sun rotates in about a month, not everyone accepted his interpretation of the observations. A Jesuit, Christopher Scheiner, insisted that the sunspots were objects between the Sun and the Earth that were seen silhouetted against the bright solar disk. He believed that the apparent motions of sunspots were due to their orbital motions around the Sun. Scheiner's careful studies of the motions revealed a

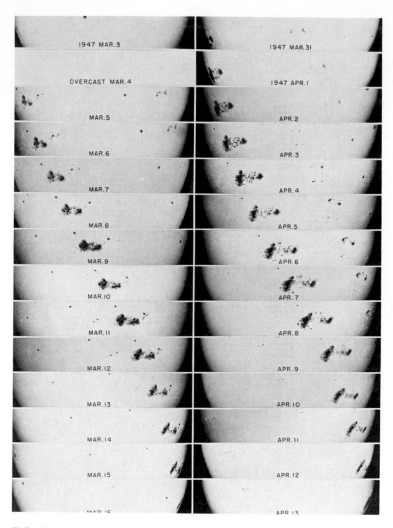

FIGURE 8.10 Solar rotation shown by sunspots. (Hale Observatories.)

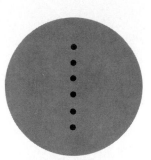

Hypothetical alignment of sunspots

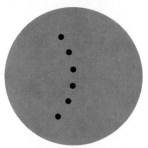

One rotation later

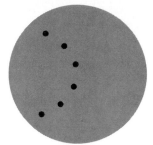

Two rotations later

most curious fact: the period of time it takes a spot to travel around the Sun at a high solar latitude is longer than the comparable time for a spot near the equator. We know today that Galileo was correct in his interpretation of the motion of sunspots as due to rotation of the Sun (Figure 8.10). However, Scheiner's observations of the motions of sunspots led to a discovery about solar rotation that is of great importance: the Sun does not rotate like a solid globe. Instead,

FIGURE 8.11 Differential rotation of the Sun. Since the Sun rotates most rapidly at its equator, high-latitude sunspots tend to lag behind equatorial sunspots after several rotations.

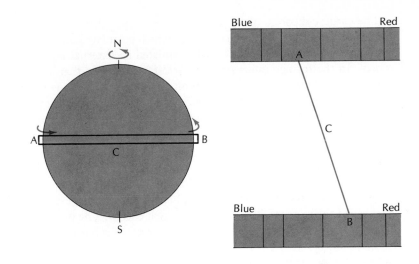

FIGURE 8.12 Solar rotation from the Doppler effect. (From *The Rotation of the Sun* by Robert Howard. Copyright © 1975 by Scientific American, Inc. All rights reserved.)

Table 8.1 Sidereal Rotation Period of the Sun

Solar Latitude (N and S)	Rotation Period (days)
0°	26.0
10	26.1
20	26.4
30	27.1
40	28.3
50	29.9
60	32.0
70	34.2
80	36.0

the material at the equator makes one complete rotation around the Sun's axis in less time than the material at higher latitudes. The Sun rotates faster at the equator than at the pole (Figure 8.11).

The easiest way to extend and verify the solar-rotation observations is by using spectroscopic measurements of Doppler shifts (Figure 8.12). If we set the slit of a spectroscope at the two extremes of the Sun's equator on the limbs, we will notice different Doppler shifts. One limb will be approaching us and the other receding from us, relative to the motion of the center of the disk. The relative velocities of approach or recession can be converted into a rotation period, since we know the Sun's circumference. Spectroscopic observations of this sort show that the Sun rotates once each 26 days or so at the equator. Sunspots have a rotation period of 25 days. The difference between the periods is probably due to a real motion of sunspots through the solar material.

Spectroscopic observations at different latitudes verify the fact that the Sun rotates **differentially**; that is, the rotation period increases with increasing latitude. Rotation periods are listed in Table 8.1.

SOLAR POWER OUTPUT

Why does the Sun shine? This is a question that scientists have asked for centuries, and which was not satisfactorily answered until the 1930s.

The earliest theories of the Sun's energy output date from the 1840s and fall into two categories: those which postulated an internal energy source; and those which postulated an external energy source. In the former category of theories was the idea that solar power comes from a chemical reaction; that is, the Sun is viewed essentially as a massive lump of burning matter. The problem with this theory arose when it was calculated that, even if the Sun were a solid mass of coal, it could not burn at its present rate of power output for more than 6,000 years. By 1840, it was known that civilization on Earth had existed for more than this short time. In the second category of theories was the idea that a great influx of meteors strikes the Sun from space, heating the solar gases by converting the meteors' kinetic energy to heat energy. The problem with this theory also became clear when exact calculations were carried out. Each year an amount of material equal to one-30,000,000th of the solar mass would have to fall into the Sun to maintain its energy output. An increase in the Sun's mass of this magnitude would cause an increase in the speed with which the Earth orbits the Sun. Roughly speaking, an increase of one-30,000,000th in the solar mass would speed up the Earth's revolution by two-30,000,000ths. Since the year is 31,558,000 seconds, this speed-up would shorten each successive year by 2 seconds, an amount that would have been detectable in the mid-nineteenth century.

A more acceptable theory was put forward in 1853 by Helmholtz, who hypothesized that the Sun is slowly contracting. Such a contraction would compress the solar gasses and heat them. The rate at which the Sun is observed to radiate heat requires that its diameter shrink by 75 meters each year, a completely imperceptible rate of contraction. If the Sun had begun as a gas cloud larger than the solar system, it would have taken roughly 18,000,000 years for it to contract to its present size, assuming that its radiation rate remained constant throughout that time. Helmholtz's theory remained a viable alternative until almost 1900.

At about the turn of the present century, radioactivity was discovered (see Chapter 5). Using this important discovery, scientists were able to date the rocks of the Earth's crust and establish the ages of the fossil remains of ancient life forms. The Earth was shown to be at least 3.5 billion years old and probably more like 4.5 billion years old. The swamps which produced the coal fields of Pennsylvania were found to have thrived about 300 million years ago. With these discoveries, the seemingly ample solar lifetime of the Helmholtz theory became inadequate by a large margin.

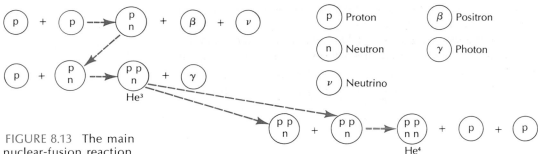

FIGURE 8.13 The main nuclear-fusion reaction that is thought to be responsible for the solar energy. Several other reactions contribute small amounts of energy.

However, the discovery of radioactivity led to the study of the atomic nucleus, and ultimately to the discovery of the processes which do in fact provide the necessary energy to keep the Sun shining for billions of years. The process involved is known as **nuclear fusion**.

We know now that the temperature at the Sun's center is very high, in excess of 15,000,000° K. At this temperature, some atomic nuclei have sufficient energy to fuse together at a collision, resulting in the formation of a heavier nucleus. By a complicated series of reactions, known as the proton-proton chain, four hydrogen nuclei fuse together to form a helium nucleus (Figure 8.13). In the process 0.7% of the mass of the four hydrogens is converted into an amount of energy given by Einstein's famous $E = mc^2$ formula. This formula says that if a mass m is converted to energy, the amount of energy created, E, is equal to m times the speed of light squared. If all the hydrogen in the Sun underwent fusion to helium, the energy created would be sufficient to keep the Sun shining at its present rate for 100 billion years, an entirely adequate length of time. However, the Sun cannot use all its hydrogen without changing its internal structure. Major changes in the Sun's structure will begin to occur after it is about 10 billion years old, according to present theories. If we assume the Sun is about as old as the Earth, roughly 4 to 5 billion years old, then it has used up half its lifetime in its present form. But we need not panic; it still has a long way to go. If we are not careful, we will use up the Earth's energy resources long before the Sun's energy store is exhausted.

In this section we have discussed the solar power output and have stated that nuclear fusion is the only process known that can provide the tremendous solar energy output for billions of years. In the next section, we will describe one aspect of our understanding of the internal structure of the Sun.

We should say in passing what the total solar energy output is, for the record. The Sun radiates energy at the rate of 4×10^{26} watts. To put this in context, in 1972 the U.S. produced 1.7×10^{12} kilowatt hours of electrical power. The Sun radiates that much power in around 10^{-8} seconds!

SOLAR STRUCTURE

In the last section, we mentioned the high temperature at the center of the Sun and how it contributes to nuclear-energy generation. How do we know the temperature deep inside the Sun? We see only the outer surface of the Sun; so how do we find out about its interior?

The answer to these questions is both simple and complex at the same time. The simple answer is that the physics of matter under the conditions we expect inside the Sun is fairly well understood. Furthermore, the equilibrium state of the interior of the Sun can be expressed by a set of mathematical equations that can then be solved on modern computers relatively easily. The complex part of the answer arises for two reasons: first, there are a large number of

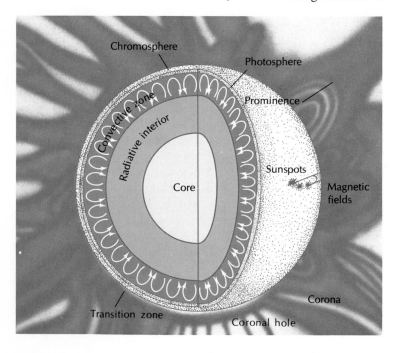

FIGURE 8.14 Solar interior in cross-section. (From *The Sun* by E. N. Parker. Copyright © 1975 by Scientific American, Inc. All rights reserved.)

detailed processes going on inside the Sun which must be specified in order to proceed with the solution of these equations; second, the Sun has aged. In the core of the Sun a fraction of the hydrogen has already been converted to helium by nuclear reactions, thus changing the composition of the material. This change in composition, in turn, alters the internal structure somewhat. Thus, to be precise we must begin with the newborn or initial Sun and study how its

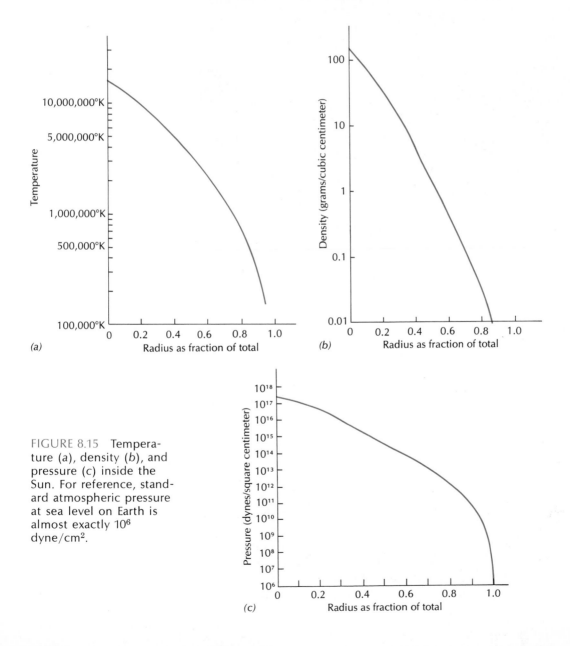

FIGURE 8.15 Temperature (a), density (b), and pressure (c) inside the Sun. For reference, standard atmospheric pressure at sea level on Earth is almost exactly 10^6 dyne/cm².

FIGURE 8.16 The tank of perchloroethylene used to capture solar neutrinos. (Courtesy of Raymond Davis, Jr., Brookhaven National Laboratory.)

structure changes with time to arrive at a description or **model** of the interior of the present-day Sun.

We cannot probe the great depths of the solar interior by any direct means; so how can we test our theories? One answer to this question leads to a discomforting problem. The nuclear reactions in the stellar core produce, along with energy, a most peculiar particle: the **neutrino**. The neutrino is an electrically neutral particle with no mass. Once a neutrino is created, it is difficult to destroy. Most of the neutrinos created by the solar-core nuclear reactions actually pass through the solar interior and escape into space. If we could measure the number of neutrinos escaping from the Sun and arriving at the Earth, we could say something about the nuclear reactions in the Sun and test our theories about its core.

Scientists have set up a neutrino detector. It is in a deep mine, surrounded by a jacket of water to keep out the effects of cosmic rays (Figure 8.16). The detector is a tank containing thousands of gallons of dry-cleaning fluid—perchloroethelene, C_2Cl_4. Neutrinos will react with the chlorine nucleus and produce an argon nucleus, which is then torn from the molecule. By collecting the argon from the tank periodically, the experimenters can measure the number of reactions. The number of neutrinos passing through the tank can be inferred from the number of reactions.

The discomforting problem is simply that no neutrinos have yet been detected, whereas, if our model of the Sun is correct, the experiment should have found some. Something is wrong. We may not know the correct rates for the nuclear reactions inside the Sun. We may have left some important factor out of our solar model. Or maybe our understanding of the physics of matter under conditions inside the core of the Sun is not as good as we think.

Before we can be comfortable with our understanding of the Sun's interior, and for that matter with our understanding of the interior structure of any star, we must solve the mystery of the missing neutrinos. Needless to say, this is a hot research topic these days.

One process of special relevance to the solar interior is the transport of solar energy from its core to the surface. We will look at the process now.

Energy Transport in the Sun

The gasses deep inside the Sun are fairly transparent; so the energy released by the nuclear reactions, which is largely in the form of energetic photons, can flow outward toward the surface. In the process, each photon undergoes many absorptions, re-emissions, and scatterings, and is degraded into several less-energetic photons. However, when the radiation has traveled about 86% of the way to the surface, it reaches a point where the opacity of the gas increases markedly. This increase occurs when the temperature has fallen low enough that not all the atoms are completely ionized. Some of the heavier nuclei pick up electrons and become partially ionized. When this happens, radiation no longer flows easily, and if something drastic did not happen, the radiation would back up behind the dam of opaque material. But something drastic does happen: something called **turbulent convection**. In the simplest terms, turbulent convection can be understood as follows. Bubbles of gas are heated at the bottom of a convective layer and become buoyant (Figure 8.17). The bubble then rises rapidly up to the photosphere, where it radiates away its heat, loses its buoyancy, and sinks back into the atmosphere, mixing with other sinking material. The granules observed on the photosphere are the tops of fountain-like cells of rising and falling material. The material rises in the center

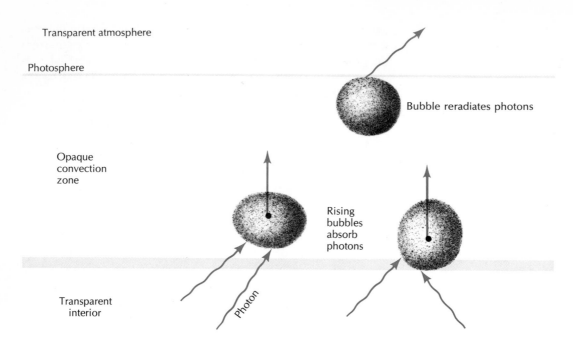

Transparent atmosphere

Photosphere

Bubble reradiates photons

Opaque
convection
zone

Rising
bubbles
absorb
photons

Photon

Transparent
interior

FIGURE 8.17 Bubbles of gas in the solar convection zone. Bubbles absorb photons at the bottom of the zone, then rise, carrying the energy across the opaque layer. At the photosphere, the photons are reradiated into the Sun's transparent atmosphere.

of each granule and sinks back at the edges. The bottom of the convection zone is where heavy nuclei become partially ionized. Much higher in the layer, the partial ionization of hydrogen and the existence of the **H⁻ ion** contributes to the convection. Frequently the whole layer is referred to as the **hydrogen convection zone**, although just **convection zone** is really more correct.

Actually, one of the major unsolved problems in physics is to find a theory of turbulent convection. Because of the lack of a theory, it is not possible to calculate an exact model of the convection zone, and approximate methods must suffice. The existence of the convection zone affects profoundly even those outer layers of the Sun which are again transparent enough for energy transport to occur by radiation.

What defines the top of the convection zone? The answer is related to the existence of the H⁻ ion we mentioned above. This ion is a hydrogen atom with a second orbital electron loosely attached (Figure 8.18). Since this second electron is loosely attached, any photon passing an H⁻ ion is likely to tear the electron from the ion and be absorbed. As a result, gas containing a small fraction of H⁻ ions is very opaque compared to a gas containing only neutral hydrogen. At the top of the convection zone, H⁻ is relatively

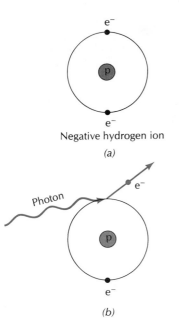

Negative hydrogen ion

(a)

(b)

FIGURE 8.18 The negative hydrogen ion *(a)* is a hydrogen atom with an extra electron. The second electron is loosely bound, and is easily torn off the ion *(b)* when it absorbs a photon.

plentiful, but higher up in the photospheric layer, conditions change: H⁻ begins to disappear; the gas becomes relatively more transparent; and photons can again flow through the gas. This region where H⁻ begins to decrease with height defines the top of the convection zone. The layers above the convection zone are called collectively the **solar atmosphere**.

In this section we have dealt mainly with the problem of convective motions in the outermost 14% of the Sun's interior. We have discussed the fact that convective motions are required to carry the tremendous energy through the opaque gasses of the convection zone.

THE PHOTOSPHERE

Above the convective zone is the layer described earlier where the very opaque H⁻ ion becomes relatively less abundant than in the convective zone, and where the gas therefore becomes more transparent again. In this layer, known as the photosphere, energy is transported outward by a flow of photons rather than by a convective flow of gas. The photosphere is not a very thick layer of the atmosphere, as we shall see. The H⁻ abundance drops from an appreciable value to almost zero within a layer about 200 km deep, and the solar gas goes from very opaque to very transparent in this 200-km distance. That is why the Sun appears to have a sharp edge. If the opacity of the gas were to decrease slowly throughout a depth of thousands of kilometers, the Sun would have a fuzzy edge instead.

If we make a graph of the brightness of the Sun as a function of wavelength, the curve looks like a black-body curve with a few bumps (Figure 7.13). The black-body curve that is closest to the shape of the Sun's energy distribution is the curve corresponding to 6,000° K.

Superimposed on the continuous spectrum of the Sun are numerous dark spectrum lines (Figure 8.19). These are caused by the many chemical elements in the solar atmosphere (Chapter 7). The visible part of the solar spectrum contains tens of thousands of lines, created by about 65 different elements that are also found on Earth. The lines in the yellow region of the spectrum caused by sodium were among the first to be identified with a terrestrial material.

In the late nineteenth century, another line in the yellow portion of the solar spectrum was a puzzle, because it did not correspond to

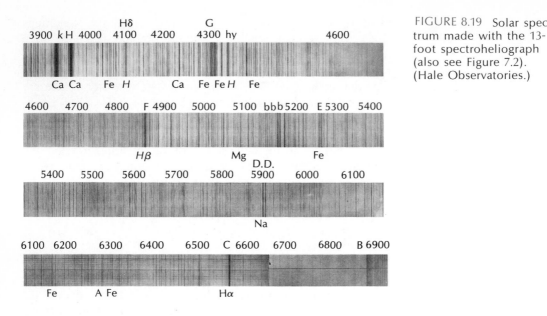

FIGURE 8.19 Solar spectrum made with the 13-foot spectroheliograph (also see Figure 7.2). (Hale Observatories.)

any known terrestrial element's spectrum. Scientists named it helium from the Greek word *helios*, which means "sun." Roughly a quarter-century later, helium was found on Earth.

The temperature of the solar gasses steadily declines from 15,000,000° K at the center to about 6,000° K at a level that we can call the visible surface, according to calculated models of the Sun. Above the visible surface, the temperature continues to fall until it reaches a value of about 4,500° K at a height of about 500 km. Near the limb of the Sun we look into the photosphere at an oblique angle and, because of the opacity of the gas, do not see as deeply into the photosphere as we do when looking at the disk center (Figure 8.20). Since we don't see as deeply, we see cooler, higher gasses, which radiate less light than the 6,000° K gasses we see at disk center. Thus the limb of the Sun is fainter than the center of the disk, and we see limb-darkening as described earlier. Limb-darkening verifies that the temperature decreases with height in the photosphere.

The fact that the solar temperature decreases outward from the visible surface also allows one to give a rough explanation of why the solar spectrum is an absorption spectrum. The continuous part of the spectrum is produced by the hot, ionized gas below the visible surface, but the absorption lines are produced in the cooler, overlying gasses.

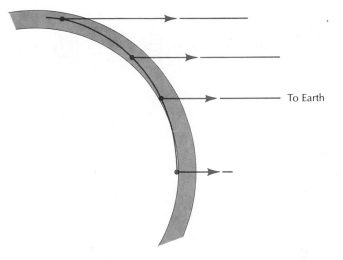

FIGURE 8.20 Cause of limb-darkening. The dots show the effective depth from which radiation arises. Near the limb we see radiation from shallower, cooler layers than we do near the center of the disk.

To Earth

In this section, we have described why the Sun, though gaseous, has a sharp edge; we have discussed limb-darkening and why the Sun has an absorption spectrum. From the discussion of the convection zone, we know the explanation of granulation. To continue the explanation of the phenomena of the visible Sun, we next talk about sunspots.

SUNSPOTS AND
THE SUNSPOT CYCLE

As described earlier, a sunspot is a dark area on the photosphere which consists of two parts: a dark central umbra, and a lighter penumbra surrounding the umbra. The umbra is not black, but only appears dark by contrast to the very bright photosphere. A large sunspot group actually radiates enough light that, if the Sun were to stop radiating except for the spot group, it would seem like bright twilight on the earth. The umbra of a sunspot is about 1,500° K cooler than the surrounding photosphere; that is, its temperature is about 4,500° K.

It is believed that the sunspot's temperature and pressure structure is influenced in a complicated manner by the presence of strong magnetic fields, which were first discovered by G. E. Hale in 1908. It turns out that, if an atom finds itself in a strong magnetic field, its energy levels are split into two or more separate levels, shifted slightly from their normal position (Figure 8.21). The splitting of

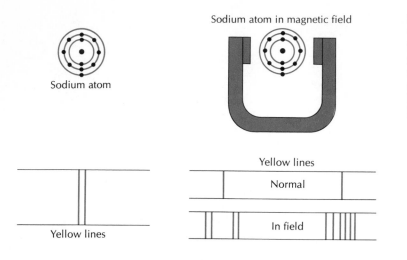

Sodium atom

Sodium atom in magnetic field

Yellow lines

Yellow lines

Normal

In field

Yellow lines

FIGURE 8.21 The spectrum of an atom is changed if the atom radiates in a region of strong magnetic field. For example, each of the two yellow lines from sodium is broken up into several lines.

the energy levels in turn causes the spectrum lines to be split into several components, each of which is polarized. The theory of this magnetically induced splitting was worked out at the end of the nineteenth century by the Dutch physicist P. Zeeman, and his theory can be used to calculate the strength of the magnetic fields that produce an observed effect on the spectrum. Hale, working at Mount Wilson in California, found the Zeeman effect in the spectra of sunspots, and inferred the existence of intense magnetic fields in the spots.

Sunspot Cycle

When the Sun is observed day after day and year after year, an interesting discovery can be made. The number of spots on the disk changes, in a cyclical fashion (Figure 8.22). This variation was first announced by a German amateur astronomer, Heinrich Schwabe, in 1851. Sunspots and groups of sunspots are born on the disk of the Sun, usually grow in size, then slowly disappear. On a short-term basis, the number of spots fluctuates, as old spots disappear and new spots appear. However, during a longer time-period, the number steadily increases to a maximum value, then declines to a minimum. The difference between the numbers at minimum and maximum is much larger than the range of variation during the short-term fluctuations. The length of time from one maximum number of sunspots to the next, on the average, is 11 years, though it varies from a short $7\frac{1}{2}$ years to a long 16 years. We often speak of the

FIGURE 8.22 (*Opposite*) Sunspot numbers, showing (a) daily numbers for 1968, (b) monthly mean numbers, and (c) annual mean numbers.

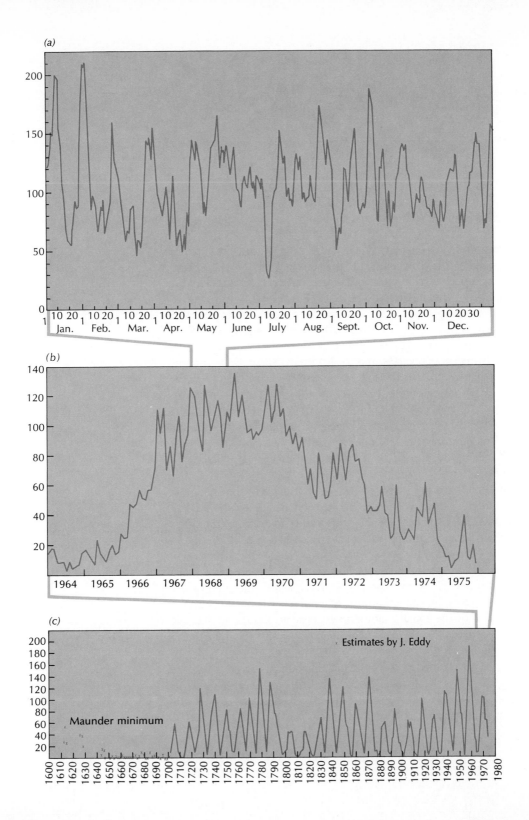

(a)

(b)

(c)

Maunder minimum

Estimates by J. Eddy

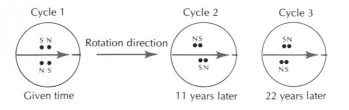

FIGURE 8.23 Magnetic polarity of sunspots from cycle to cycle.

Cycle 1   Cycle 2   Cycle 3

Rotation direction

Given time   11 years later   22 years later

11-year sunspot cycle; however, it is better described as a 22-year cycle.

In any cycle, if one looks at a large pair of sunspots in one hemisphere of the Sun, the leading spot (that is, the spot which leads a group as the Sun rotates) has one magnetic polarity and the following spot has the opposite polarity (Figure 8.23). In the other hemisphere, the polarity of leading and following spots is reversed. Furthermore, during that cycle all leading spots in each hemisphere have the same polarity. In the next cycle, the polarity of the leading and following spots reverses, and the leader has the polarity that the follower had in the previous cycle. In each successive cycle, the magnetic polarities of the leading and following spots alternate. Therefore, if we consider both the number of sunspots and the magnetic fields, the cycle takes 22 years—covering two maxima and minima of numbers and two polarity shifts—before it comes back to the starting values of both sunspot number and polarity of leading spot.

Another interesting aspect of the sunspot cycle is the change in the latitude of sunspots (Figure 8.24). First of all, sunspots never occur outside a band between 45° north and 45° south of the Sun's equator. At the beginning of a new sunspot cycle, at minimum number of sunspots, the first spots of the cycle usually occur at 35° to 40° latitude, north and south. As the cycle progresses toward maximum, new spots are born successively nearer the equator. At the end of the cycle, near minimum again, the last spots are within a few degrees of the equator. Sometimes, near minimum, we can see both a spot from the old cycle near the equator and a spot from the new cycle at a higher latitude.

There is increasingly good evidence that sunspots were very rare during the years 1645 to 1715. We will talk more about this in Chapter 11, when we discuss solar-terrestrial interactions. It turns out that winters in Europe were unusually cold during that time; so there may be a relationship between sunspots and weather.

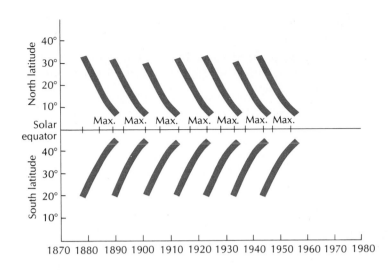

FIGURE 8.24 The average latitude at which sunspots occur on the Sun (see text).

In this section we have talked about the magnetic fields of sunspots and the sunspot cycle. Except for some very rare phenomena, the sunspot cycle is the only visible manifestation of what we call **solar activity**. In the rest of this chapter, we will discuss the phenomena one observes when special observing tools are used to isolate the radiation of a particular species of ion, such as neutral hydrogen. We will discuss the other layers of the solar atmosphere, the chromosphere, and corona, and will describe other manifestations of the solar activity cycle.

MONOCHROMATIC SOLAR IMAGES

Most of the solar phenomena that we have discussed so far can be seen by viewing the image of the solar disk that is projected by a small telescope or binoculars. Now we proceed to phenomena that can only be seen by using specialized equipment.

The key piece of special equipment is an optical filter that isolates the radiation from one specific ion, for instance, neutral hydrogen. Essentially, the filter isolates the radiation within one line in the solar spectrum. Even though the absorption lines appear black, there is nevertheless enough light in the darkest part of the line to form an image of the Sun. Two filter devices which are in wide use are the spectroheliograph invented by G. E. Hale and the French solar astronomer H. Deslandres, and the interference filter perfected by another Frenchman, B. Lyot. These instruments allow us to look at a narrow band of wavelengths emitted by the Sun, and are often called **monochromators** (*mono* = one, *chroma* = color). The monochromatic images of the Sun are called **spectrohelio-grams** or **filtergrams**.

THE CHROMOSPHERE

Without doubt, the most common monochromatic images of the Sun used by solar astronomers are images in the red line of neutral hydrogen—called hydrogen-alpha or Hα—and images in the strong, violet lines of singly ionized calcium (Ca II)—called the H and K lines of Ca II.

If we compare ordinary light, Hα, and Ca II images of the Sun, we can easily discover one interesting fact. Sunspots, which are obvious in ordinary light, are much less obvious in Hα; they appear to be partly covered over by bright material. In Ca II, the sunspots are even less visible, the covering area being more pronounced than in Hα. We conclude that the material radiating in Hα and in the Ca II lines lies above the photosphere. In fact, this radiation comes from the layer of the Sun known as the chromosphere, which was mentioned earlier as being seen as a bright, thin, red layer beyond the line of the Moon during a solar eclipse. We know now that the red color is from Hα, which is one of the strongest emissions from the chromosphere.

Detailed study of the chromosphere shows it to be about 5,000 km thick, though its upper boundary is poorly defined. Detailed studies also reveal a very puzzling feature of the chromosphere: it is *hotter* than the photosphere. We pointed out above that the temperature of the photosphere falls to 4,500° K about 500 km above the visible surface of the Sun. Incredibly, above this "temperature minimum" the temperature again rises with increasing height, going from 4,500° K at 500 km to around 50,000° K at the top of the chromosphere (Figure 8.25). One of the fundamental questions the solar astronomer tries to answer is: why is the chromosphere hotter than the photosphere? This increased temperature cannot be explained by the upward flow of photons. It is believed that the source of the heating ultimately is "noise" generated by the turbulent convection in the convection zone. But how the noise energy propagates upward and how it finally gets deposited in the chromosphere remains a mystery.

Several phenomena can be noted on careful study of the mono-chromatic solar images (Figures 8.26, 8.27, and 8.28). In Hα, and

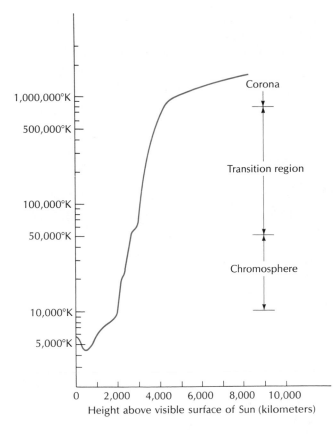

FIGURE 8.25 Temperature versus height in the outer solar atmosphere. In the transition region, the temperature can increase hundreds of thousands of degrees in a few kilometers.

Corona

Transition region

Chromosphere

1,000,000°K

500,000°K

100,000°K

50,000°K

10,000°K

5,000°K

0   2,000   4,000   6,000   8,000   10,000

Height above visible surface of Sun (kilometers)

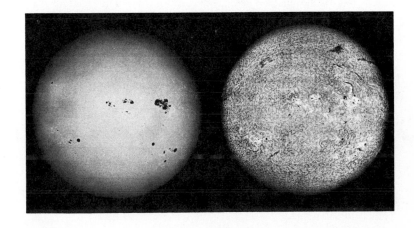

FIGURE 8.26 Comparison of an Hα spectroheliogram and a whitelight picture (left) of the Sun. (Hale Observatories.)

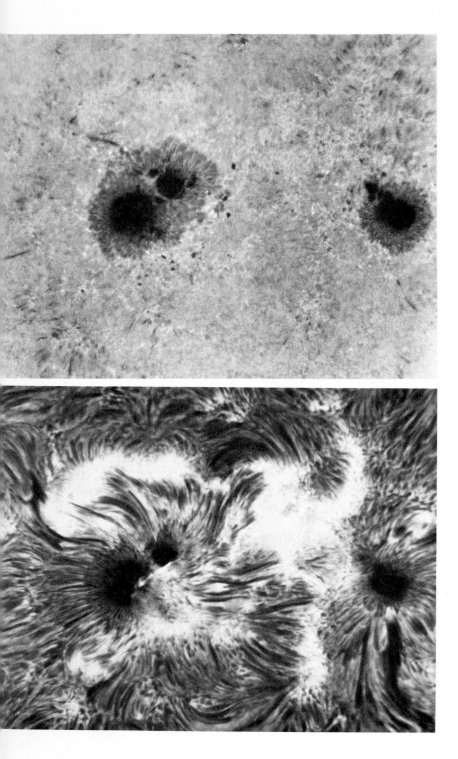

FIGURE 8.27 White-light
picture of a sunspot
group and a beautifully
detailed Hα spectrohelio-
gram of the same group.
(Hale Observatories.)

FIGURE 8.28 Comparison of (a) white-light photograph, (b) Hα spectroheliogram, and (c) ionized calcium spectroheliogram of the sun. (d) An enlarged Hα spectroheliogram. (Hale Observatories.)

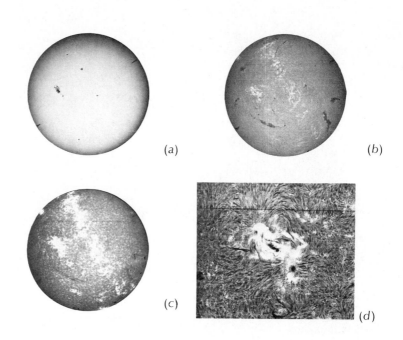

(a)

(b)

(c)

(d)

especially in Ca II, the locations where sunspots occur on the photosphere are marked by bright material in the chromosphere. Furthermore, bright material is scattered along bands parallel to the equator at the latitude where sunspots are common at the time of the observations. Bright material may mark an area without sunspots, though usually the spot would have been there and would have disappeared a solar rotation or so earlier. These regions of bright material are called **plages** or **active regions**.

In Hα, dark **filaments** can be seen scattered across the solar disk. We shall see later that these filaments are one aspect of a phenomenon known as solar **prominences**.

If one observes the limb of the Sun in a Ca II spectroheliogram, jets of gas can be seen rising in the chromosphere. These jets are called **spicules** (Figures 8.29 and 8.30). They are about 10,000 km long and exist for a few minutes. Spicules do not cover the solar disk uniformly, but outline a network pattern (Figure 8.31). This **chromospheric network**—sometimes called **supergranulation**— consists of cells about 30,000 km in diameter.

At the top of the chromosphere the temperature climbs abruptly, increasing from 50,000° K to 200,000° K in slightly more than 10 km. The increase then tapers off, and climbs more slowly to a

FIGURE 8.29 Spicules near the limb of the sun. The spicules are the dark, grass-clump-like objects. (Hale Observatories.)

FIGURE 8.30 Spicules silhouetted against the darker corona. (Sacramento Peak Observatory, Air Force Geophysics Laboratory.)

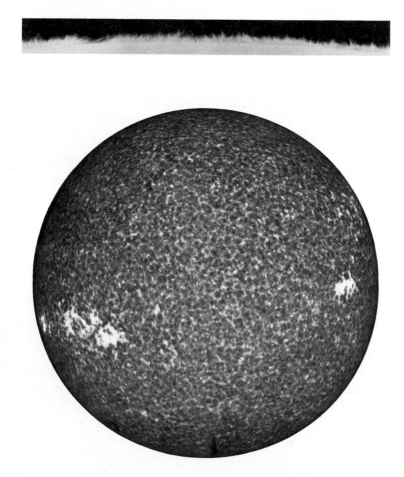

FIGURE 8.31 Calcium spectroheliogram. Note that the bright emission forms a network pattern over the sun. (Sacramento Peak Observatory, Air Force Cambridge Research Laboratories.)

value of about 2,000,000° K in the next few thousand kilometers. This region of rapid temperature rise is called the chromosphere-corona **transition region**. Above the transition region is the corona.

In this section we have described very briefly the chromosphere and some of the fine details of its structure. This now brings us to the corona.

THE CORONA

It has already been mentioned that the corona can be seen as a tenuous crown surrounding the Sun during a total eclipse (Figure 8.32). At non-eclipse times, it is difficult to observe the corona, in part because of the brightness of the sky. There are a few mountaintop observatories, with very clear, hazeless skies, where a specially designed telescope, called a coronagraph, can be used to produce artificial eclipses and observe the corona at almost any time (Figure 8.33).

FIGURE 8.32 Coronal photograph from the 1970 eclipse expedition of the High Altitude Observatory. (Courtesy High Altitude Observatory, National Center for Atmospheric Research, under sponsorship of the National Science Foundation.)

FIGURE 8.33 Photograph made by the white-light coronograph on board Skylab on June 10, 1973. The bubble of material in the corona is moving out from the Sun with great speed. (National Aeronautics and Space Administration.)

The physical state of the corona was first revealed by the Swedish physicist, B. Edlén. In 1942 he explained the coronal emission-line spectrum, which was first observed at the eclipse of 1869, but remained a mystery for the intervening decades. He showed that the spectrum lines emitted by the corona come from iron and calcium atoms that have lost many electrons (9 to 13 for the visible spectrum), and are due to transitions that become less infrequent at extremely low gas densities. The corona, according to Edlén, would have to have a temperature of at least 1,000,000° K and a density 1,000,000 times less than that of the photosphere. Studies of the conditions in the corona in recent years show that its temperature is close to 2,000,000° K throughout its volume; that is, it is *roughly* isothermal.

The over-all structure of the corona changes with solar activity. During sunspot minimum, the corona is most extended at the solar equator and least extended at the poles. During sunspot maximum, the corona is nearly circular in extent and more intense than the "minimum" corona (Figure 8.34).

Filaments, seen dark against the solar disk in Hα spectroheliograms, are actually flamelike clouds of gas extending up into the corona. When seen at the limb, we call the phenomenon **prominences** (Figure 8.35). The Hα pictures show prominences that extend hundreds of thousands of kilometers above the limb. Prominences can take several forms: **quiescent** prominences (Figure

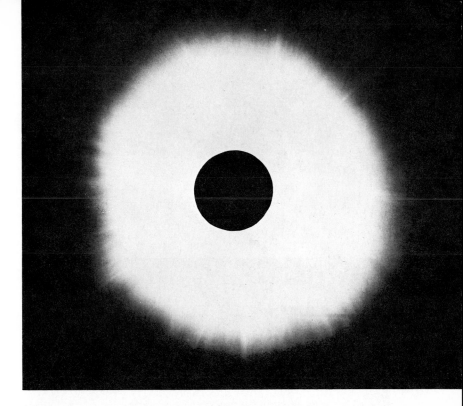

FIGURE 8.34 The corona at sunspot maximum. (Yerkes Observatory.)

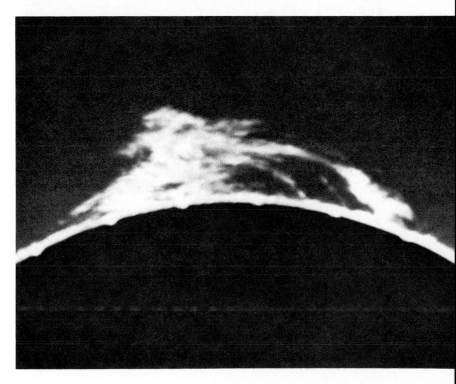

FIGURE 8.35 Solar prominence, over 200,000 kilometers high. Photographed in violet light of calcium. (Hale Observatories.)

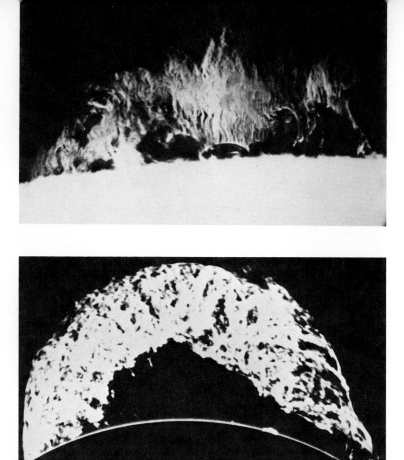

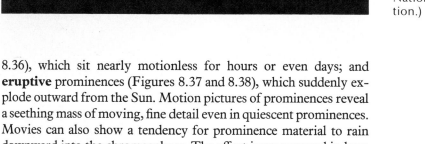

FIGURE 8.36 An Hα photograph of a 50,000-kilometer-high quiescent prominence. The vertical filaments in the prominence are magnetic field structures. (Sacramento Peak Observatory, Air Force Cambridge Research Laboratories.)

FIGURE 8.37 One of the most spectacular eruptive prominences ever observed. This giant erupted rapidly outward from the Sun. (Courtesy High Altitude Observatory, National Center for Atmospheric Research, under sponsorship of the National Science Foundation.)

8.36), which sit nearly motionless for hours or even days; and **eruptive** prominences (Figures 8.37 and 8.38), which suddenly explode outward from the Sun. Motion pictures of prominences reveal a seething mass of moving, fine detail even in quiescent prominences. Movies can also show a tendency for prominence material to rain downward into the chromosphere. The effect is pronounced in loop prominences (Figure 8.39).

The corona is not a stable layer, but is constantly expanding away from the Sun and being replenished from below. The expansion of the corona results in a flow of material out through the solar system which has been dubbed the **solar wind**. At the Earth, the wind blows at the supersonic velocity of 400 km/sec, but consists of only a few particles per cubic centimeter. As the solar wind

FIGURE 8.38 Skylab 4 photograph of the Sun in the extreme ultraviolet light of ionized helium. The prominence is erupting rapidly away from the Sun, following a flare at its base. (National Aeronautics and Space Administration.)

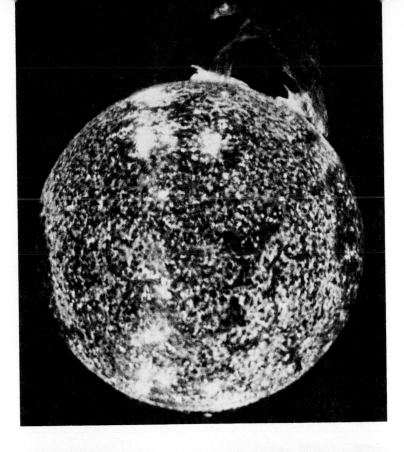

FIGURE 8.39 Loop prominences at the solar limb. (Sacramento Peak Observatory, Air Force Cambridge Research Laboratories.)

blows by the Earth, it interacts in a complicated manner with the Earth's magnetic field to create the magnetosphere. In Chapter 11 we will look at the terrestrial effects of the solar wind.

The last two sections have described briefly what has been discovered about the chromosphere and corona from the observations on the ground. Let us next look briefly at what can be discovered using spaceborne instruments.

THE SUN FROM SPACE

Since World War II there has been a continuing effort to study the Sun from above the atmosphere. The hot chromosphere and corona emit radiation in the far ultraviolet and X-ray regions of the spectrum, where the Earth's atmosphere—fortunately for us living on the ground—is completely opaque. To see these radiations, we must build instruments and launch them into space on satellites. The results of the many satellite studies have added much new information to our knowledge of the Sun.

The temperature rise in the chromosphere-corona transition region described above was first studied in detail by means of far-ultraviolet observations. The corona has been shown to be cooler over the Sun's poles than over the equator. Cooler areas elsewhere in the corona, known as coronal holes, have been discovered. A complete description of discoveries about the Sun made from space would fill many volumes; so we must be satisfied with this brief mention of the subject here.

FIGURE 8.40 Skylab picture taken as the final crew left the orbiting laboratory for the last time. (National Aeronautics and Space Administration.)

FIGURE 8.41 Ultraviolet view of the corona by the Orbiting Solar Observatory, OSO-7. (Laboratory for Solar Physics and Astrophysics, Goddard Space Flight Center, NASA.)

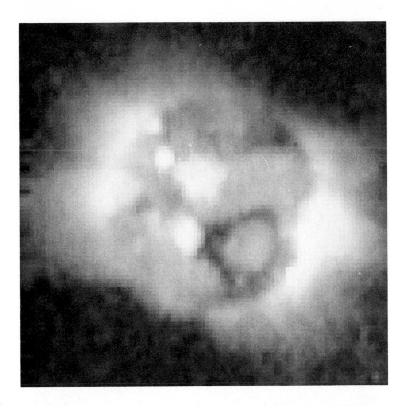

FIGURE 8.42 Corona in X-rays. This picture was taken on June 12, 1973, from Skylab. Note the numerous bright points and the dark "coronal hole." A tiny point flare is in progress in the coronal hole. (Solar Physics Group, American Science and Engineering.)

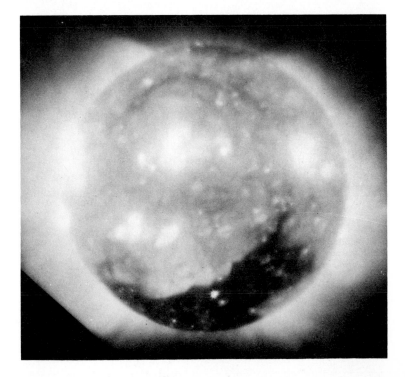

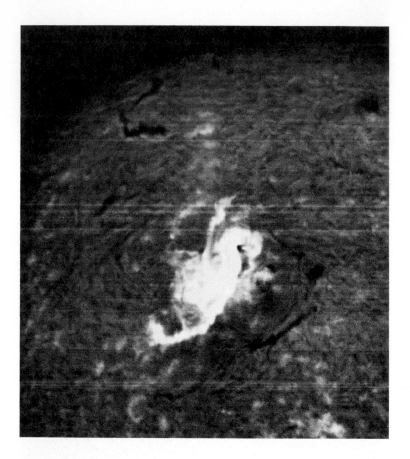

FIGURE 8.43 Solar flare in Hα, July 16, 1959. (Hale Observatories.)

FLARES

One of the most spectacular events on the Sun is the solar flare (Figure 8.43). A flare is a fantastically violent explosion in the Sun's atmosphere that evaporates a large volume of the chromosphere and leaves the corona filled with a cloud of superhot material. When a flare occurs, filaments and prominences erupt; material surges outward from the flare site; and blast waves travel great distances over the solar surface.

When seen in an Hα spectroheliogram, a flare appears as a localized brightening within a plage, which then spreads rapidly. The brightening, which eventually may cover a large fraction of the plage area, takes only a few minutes to reach maximum. The fading away of the Hα brightening may take from 20 minutes to several hours.

FIGURE 8.44　Great flare of August 7, 1972. (Hale Observatories.)

The flare is accompanied by the emission of large quantities of X-ray, ultraviolet, and radio emission, as well as by charged particles, such as protons and electrons, with great energies. When the ultraviolet and X-rays reach the Earth's ionosphere, they increase the amount of ionization. The effect on radio communications can be severe, with intercontinental transmissions being disrupted. The charged particles reach the upper atmosphere in the Earth's polar region and cause spectacular aurora displays.

The energy released in very large flares is truly immense. In August 1972, several flares occurred which were among the largest of that solar cycle (Figure 8.44). It is estimated that each of these great flares released an amount of energy equal to the United States' current power consumption continued for 100,000,000 years.

We have talked about the many faces of solar activity: sunspots, active regions, and flares. In the final section of the chapter, we will talk about a theoretical explanation that tries to tie many solar phenomena together.

THE THEORY OF SOLAR ACTIVITY

Beginning in 1959, a theory of solar activity has been developed and refined by H. W. Babcock and R. B. Leighton. The basic elements of the theory are the Sun's magnetic field and its interaction with convective gas motions and differential rotation in the photosphere and chromosphere. Needless to say, the theory is very complex; so we will give only a thumbnail sketch here.

At sunspot minimum, the magnetic field of the Sun is rather simple, roughly like the field of a bar magnet. The field is frozen into the ionized gas of the solar atmosphere. That is, as the gas moves about, the magnetic field moves with it. We know that the gas in the photosphere and in the convective zone is very turbulent. This turbulent motion, together with the Sun's differential rotation, tends to tangle up the magnetic field and destroy its nice, simple pattern. In the process, areas of extra-strong field are sometimes created. These strong field areas in turn lead to sunspots and plages.

If you look at some of the Hα spectroheliograms, (Figures 8.26, 8.27, and 8.28), you will notice a salt-and-pepper effect of tiny bright and dark features. Near sunspots, the salt-and-pepper pattern takes on a very swirled appearance. The tiny bright and dark features, like iron filings sprinkled over a paper covering a bar magnet, indicate the presence of a magnetic field and show the field to be jumbled near sunspots.

The problem that is not yet fully understood is the fact that the strong magnetic fields in sunspots somehow cool the gas, and make it much cooler than the photosphere. At the same time, the presence of strong fields in the plage act to heat the plage. Perhaps it's like an air conditioner in some manner: the field acts to remove heat from the sunspot and dump it in the plage.

As the field gets more and more mixed up, the areas of especially strong field become more numerous, and the number of sunspots becomes greater. Thus the activity level on the Sun increases.

Apparently, a point is reached when the field is as mixed up as possible. This point would be sunspot maximum. Slow motions in the chromosphere now tend to unjumble the field, and differential rotation now acts in a way to unwind the field. The number of sunspots begins to decline. Finally the field returns to its simple starting point, and another sunspot minimum occurs.

Our simple verbal description of this complex theory cannot do justice to it. The theory does succeed in predicting many of the observed characteristics of the sunspot cycle and solar activity.

One final phenomenon related to the jumbled fields near plages is the solar flare. The magnetic field can be viewed as consisting of so-called "lines of force" which act in many ways like rubber bands. In the twisting and tangling of the field, some of the lines of force get stretched greatly. Occasionally something happens to release the energy stored in the stretched field lines, and some material (mostly electrons, as it turns out) is propelled into the chromosphere with great velocity. The "paper wad" of electrons hitting the chromosphere heats it explosively, and we have a flare.

We have now completed our brief survey of the Sun and solar activity, and will move on to look at the rest of the solar system. In Chapter 11, we will return to both solar and terrestrial phenomena, and look at how happenings on the Sun may affect the Earth.

REVIEW QUESTIONS

1. Describe: (a) sunspots; (b) granulation; (c) limb darkening; (d) filament; (e) spicule; (f) prominence.

2. What is meant by differential rotation?

3. What produces the solar power output?

4. What is the so-called "neutrino" problem?

5. Describe three layers of the solar atmosphere.

6. Describe several aspects of solar activity.

9

THE PLANETS

It seems fair to say that, if space exploration brought a revolution to lunar science, then its effect on planetary science has been nothing short of a holocaust. Many of the ideas we held before the early 1960s have been utterly destroyed, and a new, exciting picture of the planets has emerged. Furthermore, we have yet to make close-up studies of the planets beyond mighty Jupiter. All the credit is not due to space exploration alone. Radio astronomy and its natural outgrowth, radar astronomy, have contributed greatly to our new understanding, too.

In this chapter, we will take a planet-by-planet tour of the solar system, working from the Sun outward. Since we know so much more about our home Earth than we do about any other planet, it is treated separately in Chapters 5 and 11. Before we look at the individual planets, let's talk briefly about their dimensions.

PLANETARY DIMENSIONS

The methods used to measure planetary dimensions are similar for all planets; so let's look at them briefly together.

The sizes of the planets can be measured by combining many observations made from Earth. To begin with, the scale of the solar system in astronomical units was established by the pioneering work of Kepler and others (Chapter 3). The Earth's distance from the Sun, the astronomical unit, was translated into kilometers by radar studies of Venus (Chapter 5). Combining these numbers allows us to calculate the distances to all the planets in kilometers. With our telescopes, we can measure the apparent angular sizes of the planets as seen from Earth. Knowing both the distances of the planets from Earth in kilometers and the angular sizes, we can calculate the linear sizes of the planets. Proceeding in this way, we

Table 9.1 Planetary Dimensions

	Diameter (kilometers)	Mass (Earth = 1)	Density (gm/cm³)	Sidereal Rotation Period*
Mercury	4,880	0.055	5.4	59 days
Venus	12,104	0.815	5.2	-243 days
Earth	12,756	1.0	5.5	$23^h\ 56^m\ 04^s$
Mars	6,787	0.108	3.9	$24^h\ 37^m\ 23^s$
Jupiter	142,800	317.9	1.3	$9^h\ 50^m\ 30^s$
Saturn	120,000	95.2	0.7	$10^h\ 14^m$
Uranus	51,800	14.6	1.2	-11^h
Neptune	49,500	17.2	1.7	16^h
Pluto	6,000	?	?	6 days

*The minus sign on a rotation period means that the planet rotates in the retrograde direction.

can calculate the diameter in kilometers of each planet with great accuracy (see Appendix A), except for distant Pluto, which presents special problems.

The masses of Earth, Mars, Jupiter, Saturn, Uranus, and Neptune can be calculated by measuring the motions of their natural moons (see Chapter 6). Mercury and Venus presented a more difficult problem, since they have no moons. Originally, their masses were calculated by studying their gravitational effects on one another and on the Earth. Today, however, both the U.S. and the Soviet Union have flown spacecraft to the vicinity of these planets. The gravitational pull of each planet on the spacecraft has been used to find improved values for their masses. Distant Pluto again presents a special problem: it has no moon, and no spacecraft has flown past it. Thus its mass is uncertain.

Table 9.1 is a summary of the best available information on size, mass, and density for the planets. The density follows immediately from the size and mass, of course.

The planetary diameters and densities point out a fact of fundamental importance in the study of the solar system. Leaving out Pluto, the planets fall into two distinct groups: the small, dense planets near the Sun; and the large, much less dense planets far from the Sun. We call Mercury, Venus, Earth, and Mars the **terrestrial** (or Earth-like) planets, and Jupiter, Saturn, Uranus, and neptune the **Jovian** (or Jupiter-like) planets. This division is one of the many clues about the origin and early history of the solar system. As we explore the planets one by one, we will discover more about the common properties of the two groups of planets. Pluto is often classed with the terrestrial planets.

Mercury is the most elusive of the "naked eye" planets. Since it is never more than 28° from the Sun, it is usually hidden in the glare of the sunrise or sunset. However, on occasion, when it is in a favorable position, as far from the Sun as it can get, the planet can be seen as a point of light in the sky, somewhat brighter than the brightest stars.

An unusually beautiful configuration occurs on rare occasions when Mercury, Venus, and the crescent moon are all seen together in, say, the evening sky. Out in the countryside, away from the polluted air of large city life, the three celestial objects hang in the sky with the red glow of sunset still illuminating the horizon. Venus is a brilliant, starlike object; Mercury is slightly fainter than Venus. The thin crescent moon nestles the earthlit Moon in its arms. The only sound is the chirping of crickets and the occasional screech of a nighthawk winging about for insects. It's a lovely stage for the quiet contemplation of the universe.

Through an earthbound telescope, Mercury is not terribly interesting. It is possible to see that the planet exhibits phases like Venus or the Moon. However, only fuzzy, dark surface features can be discerned.

The consensus of planetary observers prior to 1965 was that Mercury keeps one face constantly toward the Sun, just as the Moon keeps one face toward the Earth. This conclusion was reached by tracking the indistinct surface features as the planet moved. Since Mercury revolves around the Sun once each 88 days, its rotation period was assumed to be 88 days.

In 1965, powerful radar signals were sent toward Mercury from the giant Arecibo radio-observatory dish (Figure 9.1). The return signal from the planet contained a band of wavelengths, due to the Doppler shifts introduced by the rotating planet, which led to the deduction that Mercury rotates once every 59 ± 5 days, which meant that its true rotation rate was very likely to be between 54 and 64 days.

The measured period of 59 ± 5 days led astronomers to speculate that the planet might actually rotate in 58.65 days, so that it rotates three times as it revolves twice around the Sun. Studies of the planet from the fly-by Mariner spacecraft have verified that the rotation rate is, in fact, 58.65 days. There can be little doubt that this rotation rate results from a tidal interaction between the planet and the Sun.

The motion of the Sun in Mercury's sky is interesting. To a well-protected astronaut standing on the planet, the Sun would

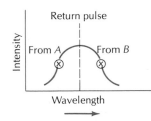

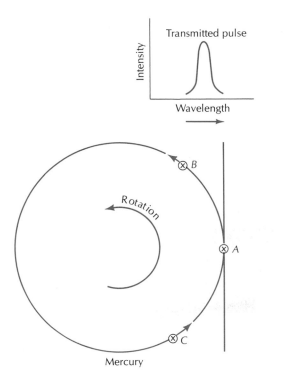

FIGURE 9.1 The rotation of Mercury broadens a radar pulse reflected from the planet.

appear to move from east to west across the sky at a slow rate. When the Sun rises, it spends 88 days above the horizon. Then it sets and spends 88 days below the horizon. The solar day on Mercury is therefore 176 days. Because the planet moves in an elliptical orbit, and its orbital speed changes considerably between perihelion and aphelion, the Sun does not appear to move across the sky at a uniform speed. When the planet passes perihelion, the Sun appears to execute a retrograde loop in the sky (see Figure 9.2).

Because the day is so long on Mercury, the sunlit surface of the planet becomes very hot. At the subsolar point—that is, the point where an observer would see the Sun directly overhead—the temperature reaches 700° Kelvin, hot enough to melt lead. Like the Moon, Mercury has little or no atmosphere. Thus, when the Sun sets, the temperature at Mercury's surface drops rapidly. At the antisolar point (or midnight point) on the planet, the temperature is 600° cooler than at the subsolar point.

The Mariner 10 spacecraft swept to within a few hundred kilometers of the planet's surface in 1974, and sent to Earth a magnificent series of close-up photographs. The cratered landscape re-

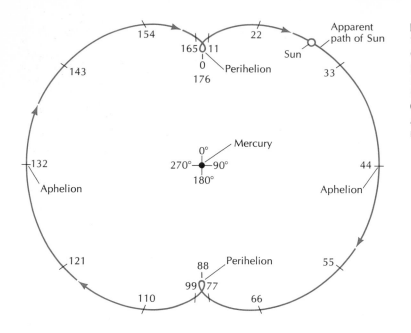

vealed in the pictures is so startlingly like the lunar surface that only an expert eye can tell the two apart. Figures 9.3, 9.4, and 9.5 show examples of the Mercurian terrain. It is easy to identify craters with central peaks, craters with rays, and craters within craters within craters.

A careful comparison between Mercury and the Moon does reveal some significant differences, however. Mercury has far fewer craters that are 20 to 50 kilometers in diameter than the Moon does, and there are areas of Mercury that are relatively crater-free, even in the heavily cratered regions. The history of Mercury's surface clearly is different from the history of the Moon's surface.

Why does Mercury look different from the Moon? There are several reasons. Mercury has a larger gravity than the Moon. Material ejected from crater-producing meteor impacts on Mercury will be spread over a much smaller area, since the material cannot fly as far in Mercury's larger gravity. Thus material ejected from craters will cover up less old terrain on Mercury than on the Moon. Most ancient craters on the Moon have been buried under material that was ejected when recent impact basins, such as Imbrium, were formed, whereas the most ancient craters on Mercury are still

FIGURE 9.3 Photograph of Mercury made by the Mariner 10 spacecraft at a distance of 200,000 kilometers from the planet on March 29, 1974. (National Aeronautics and Space Administration.)

FIGURE 9.4 Mariner 10 photograph taken when the spacecraft was 78,000 kilometers from Mercury. Note the line of cliffs near the planet's limb. A data-collection problem can be seen extending across the picture a fourth of the way down from the top. (National Aeronautics and Space Administration.)

FIGURE 9.5 Mariner 10 photograph of a 290-by-220-kilometer area of the Planet Mercury shows a heavily cratered area with many low hills. Note the long narrow valley near the top of the picture. The crater from which it extends is 80 kilometers across. (National Aeronautics and Space Administration.)

visible. Some of these old craters rival the Moon's largest craters in size. Incidentally, Mercury does have a giant ringed plain like the Imbrium basin: it is called Caloris Basin.

The Interior of Mercury

Mercury has virtually the same average density as the Earth. However, the planet is much smaller than the Earth, and the pressure on the material in its deep interior is less than that in the Earth. We must conclude, then, that the material in Mercury's core is less dense than the material in the Earth's core. How can this be true, if Mercury has the same *average* density as the Earth? The answer is that Mercury's iron core is relatively larger than the Earth's, so that relatively more of Mercury's interior is high-density material. Figure 9.6 shows the relative sizes of the cores of Mercury and Earth.

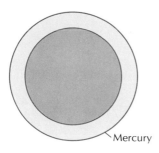

Mercury

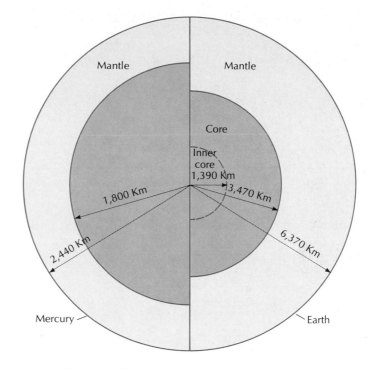

Mantle   Mantle

Core

Inner
core
1,390 Km

1,800 Km   3,470 Km

6,370 Km

2,440 Km

Mercury   Earth

FIGURE 9.6 Comparison of the relative scales of the interior structures of Mercury and Earth, showing the relatively larger size of the iron core of Mercury. Mercury is much smaller than Earth, of course. (From *Mercury* by Bruce C. Murray. Copyright © 1975 by Scientific American, Inc. All rights reserved.)

VENUS

Venus is the third brightest object in the sky, exceeded ordinarily only by the Sun and the Moon. We have to say ordinarily, because on occasion a temporary phenomenon, such as a comet, may be brighter. Because of its brightness and the fact that the planet may be as much as 47° from the Sun, Venus is as obvious as Mercury is elusive.

To the Greeks, Venus was Phosphorus the morning star or Hesperus the evening star, depending on whether it was east of the Sun or west of the Sun. Venus' synodic period is 584 days. The synodic period of an **inferior** planet like Mercury or Venus (that is, a planet that is nearer the Sun than the Earth is) is the time it takes for the planet to go from some position in the sky and some starting phase, through all possible positions and phases, and back to the starting position and phase again. The planet will reach its greatest brilliance in the evening sky every 584 days. Thus we should look for the planet near the point where the Sun has just set in February 1977, October 1978, May 1980, December 1981, and every 584 days thereafter.

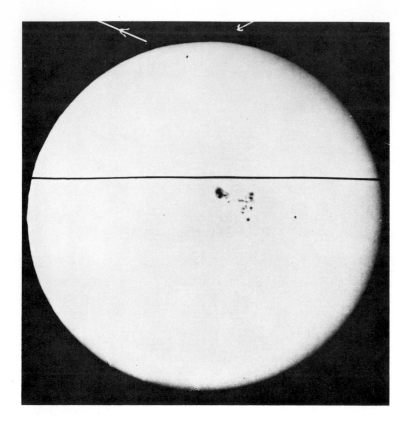

Venus, like Mercury, goes through a complete series of phases. When either planet is at "new" phase, it lies roughly between the Earth and the Sun. However, since their orbits are inclined to the ecliptic, they usually appear to pass above or below the solar disk. On rare occasions one of the planets may pass directly between the Earth and the Sun, and its tiny disk will be seen silhouetted against the Sun (Figure 9.7). The passage of a planet across the Sun is called a **transit**. The telescope can tell us little about the surface of Venus: it has a dense atmosphere which perpetually hides the surface from view (Figure 9.8).

We had two strong bits of evidence that Venus is shrouded in a dense atmosphere before close-up space observations proved the point. First, its **albedo** is high. The albedo of a planet is a measure of its reflectivity: it is the fraction of light falling on it that is reflected back to space. The albedo or airless bodies like the moon and Mercury is about 6%; the albedo of the Earth is 40%; but that of Venus is 75%. That exceptionally high reflectivity indicates

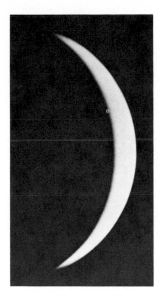

FIGURE 9.8 Venus at crescent phase, photographed by the great Hale reflector. Virtually no detail can be seen in the cloud cover on the planet. (Hale Observatories.)

something bright and shiny over the surface, and dense clouds are the best guess.

The second bit of evidence that Venus has a dense armosphere can be seen when the planet is at a very thin crescent phase. The "horns" of the crescent extend beyond the directly illuminated hemisphere, because of light scattered in the dense atmosphere.

Information about surface conditions on Venus began to be established in the 1960s with radar and radio-wavelength studies of the planet. These long-wavelength radiations penetrate the atmosphere and allow us to see the planet's surface. The Doppler effect in radar signals bounced off the planet indicated that it has a very slow rotation rate. It rotates once every 243 days, in the retrograde direction; that is, opposite to most other rotations and revolutions in the solar system.

The truly remarkable feature of the planet's rotation lies in the fact that whenever it approaches as close as possible to the Earth, Venus presents the same face to the Earth. How the planet's rotation could by synchronized with its motion relative to the Earth remains something of a mystery.

The radar signals bounced off Venus are reflected by its solid surface. Using complex analysis procedures on the radar pulse returned to Earth, scientists can map the terrain of the planet (Figure 9.9). The surface has a very shallow relief; for instance, it is very smooth compared to the Moon. Among the features discerned are several clearly craterlike formations. Venus thus seems to join many of the other solar-system bodies in being strewn with craters.

Venus' surface emits radio radiation of a kind known as thermal radiation. Basically this radio emission is the long-wavelength end of the planet's black-body emission curve. All warm bodies emit some radio radiation, including you and I. The total intensity of Venus' radio emission can be measured, and the planet's surface temperature can be deduced from the result. The temperature inferred for the surface of the planet shocked astronomers. It turned out to be above 600° Kelvin (which is above 620° F). Then the Soviet Union soft-landed several Venera probes on the planet, and established that the temperature is actually close to 750° K (890° F). The planet is hot, indeed. The Venera probes also showed the surface atmospheric pressure to be high, about 90 times greater than atmospheric pressure at the Earth's surface.

As if the high temperature and pressure were not bad enough, spectroscopic studies of the planet have revealed features in its absorption spectrum due to corrosive substances such as sulphuric, hydrochloric, and hydrofluoric acid. As Carl Sagan has pointed out, the surface of Venus comes close to the biblical description of hell.

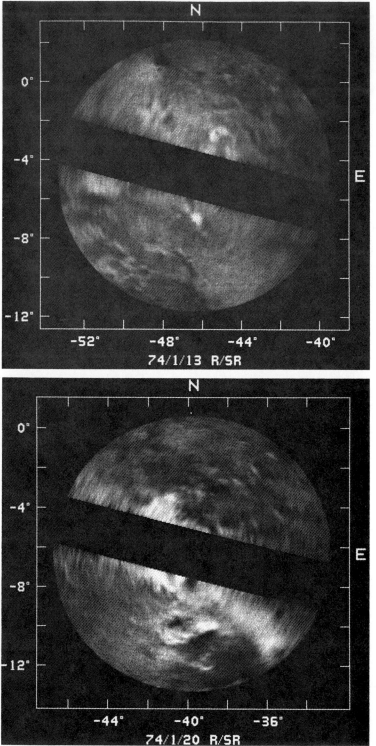

FIGURE 9.9 Two radar images of circular areas of the surface of Venus near the center of the planet's apparent disk. Note several features that appear to be craters. (Courtesy R. M. Goldstein, Jet Propulsion Laboratory.)

FIGURE 9.10 Venus photographed in ultraviolet light in 1962. Seen in the ultraviolet, some details can be discerned in the planet's cloud cover. (Lick Observatory photograph.)

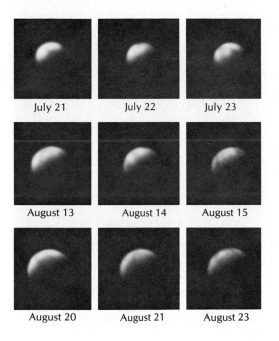

July 21	July 22	July 23
August 13	August 14	August 15
August 20	August 21	August 23

The main constituent of the dense atmosphere of the planet is carbon dioxide. The poisonous gas carbon monoxide has also been identified, but the oxygen so important to life on Earth has not been found. The presence of water vapor also has been inferred from observations by the Soviet Venera probes. The total amount of water is very small.

In February 1974, the Mariner 10 spacecraft sped within 5,800 kilometers of Venus. Photographs were made by cameras fitted with special filters and films that recorded the ultraviolet from the planet; these revealed a wealth of fine details in the cloud cover (Figures 9.10, 9.11, and 9.12). We knew that, as seen from the Earth, the planet was featureless in ordinary white light, but showed some detail when viewed in the ultraviolet. That is why ultraviolet cameras were flown on Mariner 10.

Series of photographs made at regular time-intervals by Mariner 10 show cloud motions, indicating windspeeds up to 100 meters /second (200 miles/hour) in the planet's upper atmosphere. At that high rate, a cloud would circle the planet in slightly more than four days. This apparent rapid atmospheric motion is strange compared to the planet's slow (243-day) rotation period. Reconciling these two rates of motion will require much additional study of Venus.

FIGURE 9.11 Venus photographed by Mariner 10 on February 9, 1974. The pattern of black dots on the image was placed there by the TV system which made the picture. (National Aeronautics and Space Administration.)

FIGURE 9.12 Venus from 700,000 kilometers. (National Aeronautics and Space Administration.)

FIGURE 9.13 Energy balance of a rotating, atmosphereless, black-body planet. The surface will reach a temperature such that the total energy of black-body radiation that the planet emits is equal to the total solar energy it absorbs.

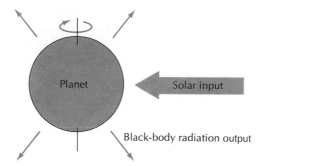

Planet

Solar input

Black-body radiation output

Sun

We have pictured a hot planet veiled with a dense, cloudy atmosphere that consists mainly of carbon dioxide with small amounts of corrosive acids. To close this section, we will look at two theoretical problems: Why is Venus as hot as it is? Why is Venus' atmosphere so unlike that of its twin sister, the Earth?

If Venus were a perfectly smooth black body, without an atmosphere, it would absorb all the sunlight that fell on it and warm up until it reached a temperature where it radiated away to space the same amount of energy it absorbed (Figure 9.13). The planet would then be in an equilibrium state, and its temperature would remain constant thereafter. To calculate the temperature Venus should have when it is in an equilibrium state, we need to know the total solar energy that reaches the top of the planet's atmosphere, and the rate at which it would radiate energy if it were a black body. The temperature turns out to be about 350° K. If we allow for the fact that the planet reflects 75% of the solar energy that reaches it, the equilibrium temperature will be even lower (around 235° K). How can these theoretical temperatures be made to jibe with the observed 750° K temperature of the planet's surface?

The answer lies in the planet's dense atmosphere and in an effect known as the **greenhouse effect**. A tightly closed automobile left out in direct sunlight gets very hot on the inside because of the greenhouse effect. Sunlight enters through window glass and is absorbed by the upholstery and other material inside the car, heating it up. The material reradiates the energy, mostly as long-wavelength infrared. However, the windows are nearly opaque to infrared, and most of the heat is trapped inside the car. More light enters through the window to heat the interior more. The temperature inside the car rises until enough infrared energy is radiated away by the car's interior so that as much energy leaks out through the glass as enters. An equilibrium is then reached.

If we replace the car's interior by Venus' surface, and the windows by Venus' atmosphere, the above argument extends to the planet. A tiny amount of sunlight filters to the planet's surface and heats it. The infrared that the surface radiates cannot escape: the carbon-dioxide atmosphere is opaque to infrared. Therefore the temperature rises until an equilibrium is established by convective motion in the atmosphere, which carries heat away from the surface and up to the top of the atmosphere.

The carbon dioxide in the Earth's atmosphere is removed by three processes: photosynthesis; absorption in the water of the oceans; and chemical reactions with the surface rocks. Four obvious sources add carbon dioxide to the atmosphere: respiration by animals; the decay of organic matter; volcanoes; and combustion (e.g., the use of great quantities of fuel by human beings). A delicate balance between all but the last of these processes has kept the carbon-dioxide content of the Earth's atmosphere low. However, industrial output of carbon dioxide increases the amount of the gas in the atmosphere.

Vast amounts of carbon dioxide are dissolved in the ocean, and equally vast amounts are tied up in the planet's chalk sediments, such as the white cliffs of Dover. Chalk, or calcium carbonate ($CaCO_3$), is created by innumerable tiny, shelled animals dying and falling to the ocean floors. The animals used the carbon dioxide dissolved in the oceans as raw material for their shells.

If all the carbon dioxide tied up in these two reservoirs were released to the atmosphere, the Earth might look more like Venus. This suggests that the existence of life and of large quantities of liquid water are responsible in part for the differences between the atmospheres of the two planets.

Venus is almost identical in size to the Earth. Its equatorial diameter is 12,104 kilometers; the Earth's is 12,756 kilometers. The densities of the two planets are also very similar. Venus' mean density of 5.2 gm/cm^3 compares with Earth's 5.5 gm/cm^3. It is safe to assume, therefore, that the interior structure of Venus is similar to the Earth's.

MARS

Mars has inspired more fanciful speculation than any other planet in the solar system. At the end of the nineteenth century, a few scientists were convinced that the planet was the abode of an advanced civilization which eked out a frugal existence on the cold, dry surface.

In 1877 the Italian astronomer G. Schiaparelli announced the discovery of *canali* on Mars. These were a series of fine linear markings crisscrossing the planet's surface. The Italian word *canali,* which means channels, was mistranslated by American astronomers of the time as canals.

The leader of the "life on Mars" school was Percival Lowell, scion of a socially prominent Boston family. Lowell founded an observatory near Flagstaff, Arizona, in 1874 which continues in operation today and bears his name.

Lowell's early studies at his observatory helped establish that the planet has a thin dry atmosphere; as he put it, "thinner at least by half than the air upon the summit of the Himalayas." Sunlight reflected from the planet's surface passes twice through the Martian atmosphere before it reaches Earth. Thus the atmospheric gasses can absorb their characteristic spectrum lines and make their presence known to astronomers on Earth. Since some of the gasses in Mars' atmosphere are the same as those in the Earth's atmosphere, we must observe Mars when it is moving toward or away from the Earth, so that spectrum lines due to Martian atmospheric gas are Doppler-shifted away from lines formed in the Earth's atmosphere (Figure 9.14).

The Earth-based observations of Mars revealed polar caps which grow and shrink with the Martian seasons. Mars' polar axis is tipped 24° to its orbit, much like the Earth's $23\frac{1}{2}$° tilt; so its seasons are very similar to our own. The Martian year of 687 days is almost twice our year; so its seasons are proportionately longer.

As one polar cap shrinks with the coming of spring in that hemisphere, the dark surface markings become much more distinct, then fade again as winter approaches. These darker markings appear gray-green against the ochre-red of the rest of the planet.

In Lowell's vision of Mars, advanced beings had dug canals to carry the sparse water from the melting polar caps to irrigate their crops grown in the dark areas on the planet. The narrow canals would show up because of broad bands of vegetation along their banks.

One of the more careful observers of the same period, E. E. Barnard, studied Mars using better equipment than was available to Lowell and could not see the canals at all. Barnard and many other people concluded that the canals were a psychological trick of the eye, which perceives lines between faint details. The canals could not be photographed. Turbulence in the atmosphere during the extended exposures would smear out such fine details.

Interest in the canals of Mars continued right up through the early 1960s as the U.S. prepared to send Mariner spacecraft to the

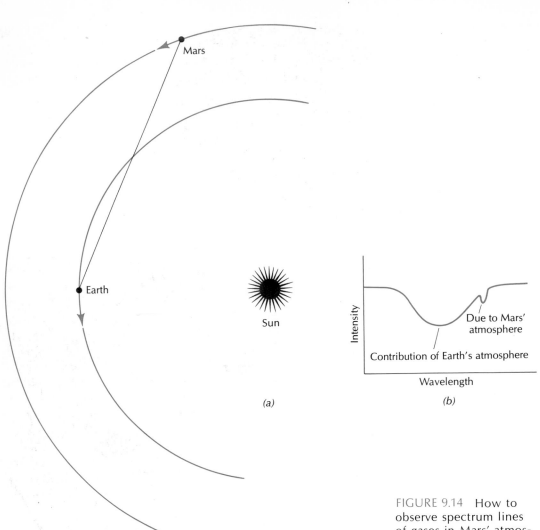

(a)

(b)

FIGURE 9.14 How to observe spectrum lines of gases in Mars' atmosphere that also occur in Earth's atmosphere. We observe Mars when it is moving rapidly relative to the earth and its spectrum lines are Doppler-shifted (a). A plot of the intensity of a spectrum line (b) shows a strong line from Earth's atmosphere and a smaller, Doppler-shifted contribution from Mars' atmosphere.

mysterious planet. The early Mariner spacecraft, which photographed the planet in 1965, 1969, and 1971, showed a crater-pocked surface not unlike the Moon's, but no canals. The idea that the canals were a trick of the eye seems to have been correct. Mars appeared to be a dead planet.

In 1971, Mariner 9 went into orbit around the planet, ultimately mapped its entire surface, and surveyed its atmosphere. Viking also analyzed the planet's atmosphere while sitting on the surface.

FIGURE 9.15 Mars as seen from Earth, taken with a 0.2-meter (8-inch) reflector. (Courtesy J. C. LoPresto, Edinboro State College, Edinboro, Pennsylvania.)

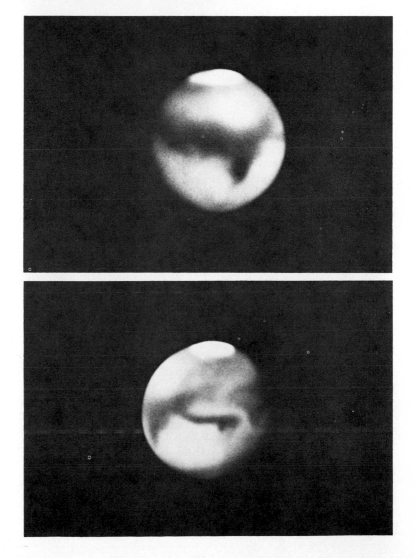

FIGURE 9.16 Among the best Earth-based photographs of Mars, made with a 1-meter (36-inch) reflector. (Lick Observatory photograph.)

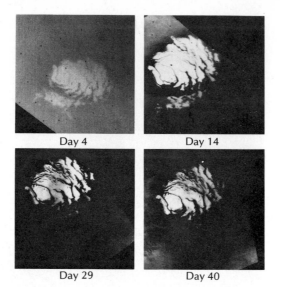

Day 4   Day 14

Day 29   Day 40

FIGURE 9.17 Four views of the southern polar cap of Mars, taken by Mariner 9. The time period covered is the southern summer on Mars, when the polar cap is melting rapidly. (National Aeronautics and Space Administration.)

Atmosphere and Water on Mars

The Martian atmosphere is very thin, its surface pressure being less than 1% of the Earth's. The major atmospheric constituent is carbon dioxide. In addition, traces of argon, ozone, oxygen, carbon monoxide, and hydrogen have been found. Water vapor makes up a tiny percentage of the atmosphere.

The polar caps are composed primarily of water ice and frozen carbon dioxide (Figure 9.17). As the caps shrink in spring, they release carbon-dioxide gas directly into the atmosphere. No liquid flows on Mars' surface today. The Viking spacecraft that soft-landed on the surface of Mars in 1976 provided clear evidence that in midsummer only water ice remains in the Martian polar caps. This ice probably never melts, but what does melt will immediately evaporate into the thin atmosphere. Viking photographs of several areas of Mars' surface taken at different times of the Martian day show fog lying in low spots on the surface in early morning hours. Subsurface water ice is being melted by the rising Sun, and the vapor immediately condenses into fog.

The Mariner-9 pictures of the planet revealed a truly astonishing fact. The surface of the planet near its equator is marked with many long, sinuous channels (Figure 9.18). Unlike the lunar rilles, these channels give the impression of having been cut by the flow of a

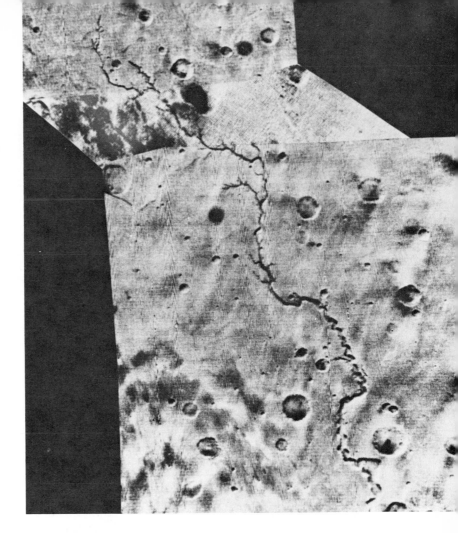

FIGURE 9.18 Mariner-9 photograph of Mars showing a sinuous rille which convinced scientists that water flowed on the planet in the past. (National Aeronautics and Space Administration.)

very fluid substance. Most workers feel the most likely substance was water. The channels look for all the world like dry river beds. The big question is, where is the water now? In addition, Viking photographed many teardrop-shaped landforms around craters (Figure 9.19). The appearance suggests strongly that these shapes were formed by a flood of water.

If Mars had a somewhat denser atmosphere in the past, the larger greenhouse effect would have warmed the surface. If it also received more sunlight, the climate could have been much warmer than it is now. Scientists speculate that both of these factors may have been important in the very distant past, perhaps as much as a billion years ago. If the speculation is true and Mars was much

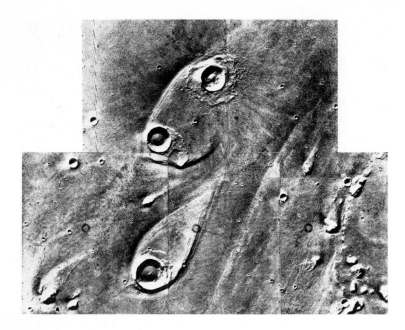

FIGURE 9.19 Mosaic of Viking photographs of Mars showing teardrop-shaped landforms associated with craters and other raised features. Scientists feel that the picture shows clear evidence of winds. (National Aeronautics and Space Administration.)

warmer, the water now tied up in the polar caps and in permafrost may have been liquid, thus providing water for Martian rivers. We must conclude that Mars is locked in an ice age today.

Surface Features of Mars

The first surface features discovered on Mars were craters. Figure 9.20, which shows a small portion of the planet's surface photographed on July 14, 1965, by Mariner 4, is typical of the early photographs. Craters are visible in most of the other photographs in this section.

One does not have to study these photographs very carefully to discover that the Martian craters look substantially different from those on Mercury on the Moon. The Martian craters have been smoothed by billions of years of weathering by dust-carrying winds at the planet's surface.

One of the most interesting craters on Mars was photographed by the Viking I orbiter. The crater, called Yuty, is surrounded by a flower-petal-shaped flow pattern (Figure 9.21). Scientists feel that the impact of the meteor that caused the crater melted the subsurface permafrost layer, and splashed a semi-liquid mass of material from the crater.

FIGURE 9.20 Mariner-4 picture of Mars taken in July 1965 shows a cratered area of the planet's surface. (National Aeronautics and Space Administration.)

FIGURE 9.21 Viking-I photograph of a Martian crater. The flower-petal-like flow pattern surrounding the crater indicates that the impact may have ejected wet material from the crater. (National Aeronautics and Space Administration.)

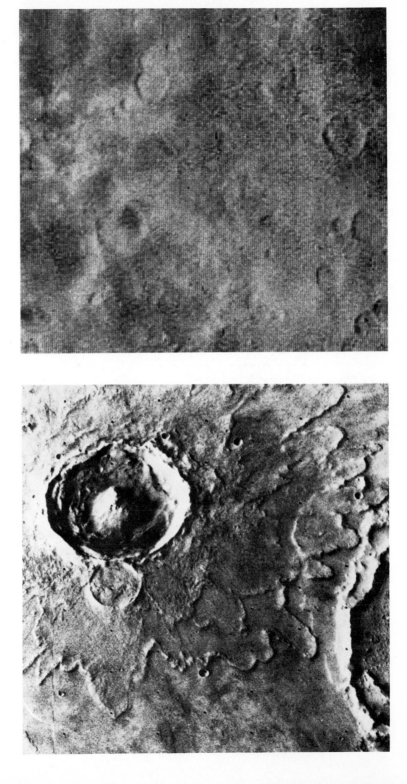

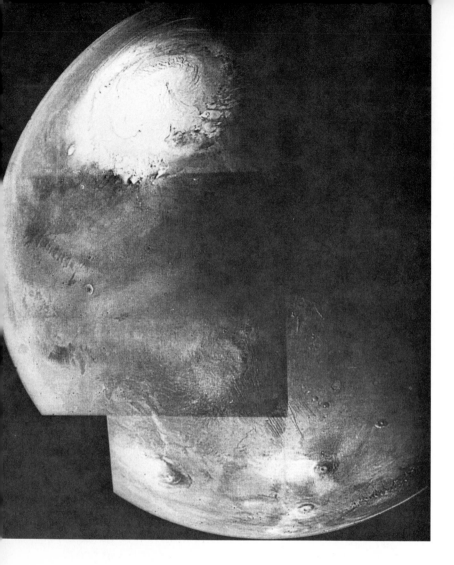

FIGURE 9.22 Composite picture of Mars made by Mariner 9 as it approached the planet. Note the polar cap and the large volcanos. (National Aeronautics and Space Administration.)

Mariner 9 revealed two types of features which indicate that Mars had been geologically very active in its remote past (Figure 9.22). There are several enormous volcanic cones scattered across the planet's surface, and a gigantic canyon system that would dwarf the Grand Canyon.

The largest of the Martian volcanos in Olympus Mons (Figure 9.23). This conical mountain is 600 kilometers in diameter, and stands 25 kilometers higher than the plain from which it rises. To put these numbers in perspective, the volcanic cone would almost cover the state of Nevada. Mount Everest, whose summit is nearly 9 kilometers above sea level, actually rises from a base that is almost 3 kilometers above sea level. Thus the mountain proper is 6

FIGURE 9.23 The Martian volcano Olympus Mons. (National Aeronautics and Space Administration.)

kilometers tall. Olympus Mons towers over four times higher than Everest, from base to summit. According to volcanologists, it would have required 100 million years to build Olympus Mons if the cone grew at the same rate as terrestrial volcanic cones.

What an incredible sight the mountain must be from the surrounding plain! The base of the mountain is ringed by two-kilometer-high jagged cliffs, then the massive lava cone rises another 23 kilometers. Perhaps some day we will see this mountain on our home television sets, as astronauts at Mars Base I pan their camera around the landscape.

Near Mars' equator is an extensive system of canyons known as Marine Valley, which covers an area 2,700 kilometers long and 500 kilometers wide (roughly the area of Alaska). Figure 9.24 shows the Tithonian Chasm, one small part of Marine Valley. The cliffs flanking the chasm are several kilometers high.

Some of the pictures look like one idea of the Earth's ancient ocean beds (or the Great Rift Valley in Africa) just after two continents fractured and began to drift apart. However, experts feel that we are not seeing the results of ancient continental drift on Mars. Rather, the canyons are the result of a long chain of erosion processes which began when volcanic eruptions tapped subsurface

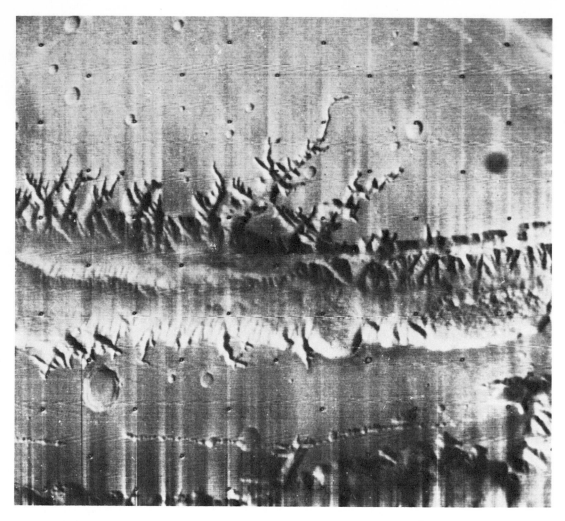

magma chambers, causing the overlaying surface to slump as the lava flowed out onto Mars' surface.

FIGURE 9.24 A Martian canyon, part of the Marine Valley. (National Aeronautics and Space Administration.)

This brief summary of a few of the types of features on Mars' surface gives a picture of a planet that must have been geologically active in the distant past. Ancient river beds indicate that the planet was much warmer long ago, so that liquid water could have flowed across its surface. Today what water exists on Mars is frozen permanently in the ice caps and in the surface soil.

Before we leave Mars, it is interesting to take a look at its two moons, and talk about one of the most unbelievable literary coincidences of all time.

FIGURE 9.25 Viking-I Lander photographs of the rocky surface of Mars. The pitted rocks on the surface are thought to be volcanic in origin. (National Aeronautics and Space Administration.)

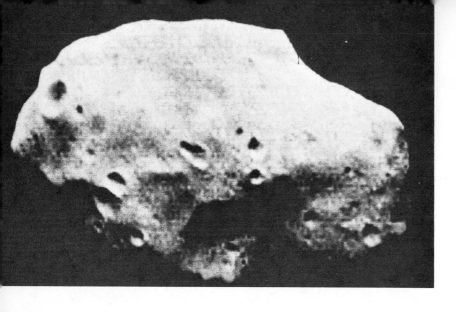

FIGURE 9.26   Mariner-9 photograph of Mars' inner moon, Phobos, taken at a distance of about 5,000 kilometers. (National Aeronautics and Space Administration.)

The Moons of Mars

According to Swift in his *Gulliver's Travels* (written in 1726), the astronomers of the island of Balnibarbi discovered that Mars has two moons, "whereof the innermost is distant from the center of the primary planet exactly three of its diameters, and the outermost five; the former revolves in the space of ten hours, and the latter in twenty one and a half."

In 1877, Asaph Hall at the Naval Observatory in Washington, D.C., discovered that Mars *does* have two moons, which were subsequently named Phobos and Deimos. Phobos, the inner of the two, is 9,000 kilometers (1.3 diameters) from Mars' surface and revolves around the planet in 7.65 hours (Figures 9.26 and 9.27). Deimos is 23,000 kilometers (3.4 diameters) from the planet and revolves in 30.24 hours (Figure 9.28). Swift's guess was amazingly close to the true values.

Phobos actually revolves around Mars faster than the planet rotates on its axis, so that it alone rises in the west and sets in the east: its rising and setting is determined by its orbital motion rather than by the planet's rotation. While Phobos is in the Martian sky, it goes through more than one set of phases.

Mars' two moons are tiny. Phobos is about 25 kilometers in diameter, and Deimos is about 15 kilometers. When it is directly overhead on Mars, Phobos would appear to be roughly 15 arc minutes in diameter, about half the size of our Moon as seen from Earth. Thus it would be seen as an elongated, irregular shape in the sky. Mariner 9 photographed both of Mars' moons.

Let's now push onward and take a look at the next planet in order of distance from the Sun, Jupiter.

FIGURE 9.27   A second view of Phobos, taken from the Viking-I orbiter. (National Aeronautics and Space Administration.)

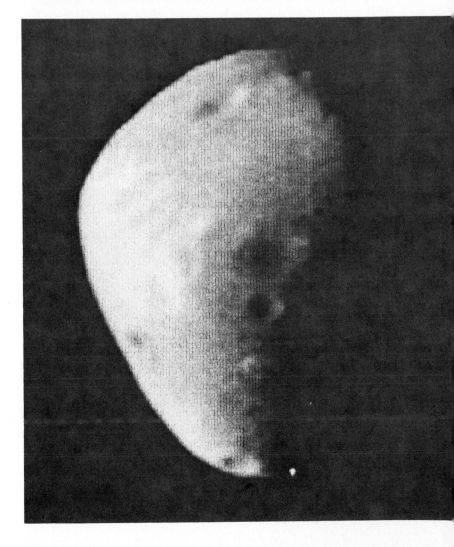

FIGURE 9.28   Mars' outer moon Deimos, photographed by Mariner 9. (National Aeronautics and Space Administration.)

JUPITER

We now come to a fascinatingly different region of the solar system, where we encounter the giant planets. They probably do not have rocky surfaces like the four inner planets, but consist largely of the light elements hydrogen and helium in various states. Thus their densities are low.

Mighty Jupiter is the most massive planet in the solar system, containing more than 70% of all the mass outside the Sun: it is 300 times more massive than the Earth. Even so, Jupiter has only a thousandth of the mass of the Sun.

The star with the smallest known mass is a nearby, faint object whose mass is 0.07 times the mass of the Sun. It is not clear to astronomers how small the mass of a star can be, but it is probably true that if an object has much less than 0.07 times the mass of the Sun, it will look more like a planet than a star. Thus Jupiter would have to be almost 70 times more massive than it is to qualify as a star.

Despite the fact that Jupiter could not make it as a star, it probably does have an internal heat source. As calculated from terrestrial observations, the temperature of the planet is about 150° K. However, the temperature for a black body at Jupiter's distance that is absorbing and reradiating sunlight should be 100° K. The most widely accepted explanation for the elevated temperature is that the interior of the planet is hot and radiates its own heat away, in addition to absorbing and reradiating sunlight.

Although Jupiter has a large mass, it is still much less dense than the Earth, because of its enormous size: 1,300 times the volume of the Earth. The Jovian density is only 1.3 gm/cm³, compared to the Earth's 5.5 gm/cm³.

Through even a small telescope, Jupiter is a lovely sight (Figure 9.29). The planet's disk can be discerned with a pair of 7-power binoculars, but no details can be seen on the surface. However, if the binoculars are steadied against a building or on a tripod, one can see the four bright moons, first discovered by Galileo.

With a small telescope (say, around 4-inch or 10-centimeter aperture) and an eyepiece that will give moderate (50-power) magnification, one can see that the planet's disk is crossed with narrow, parallel bright and dark bands. Historically, the bright bands have been called **zones,** and the dark bands have been called **belts.** A careful study of the color of the planet shows the zones to be white to pale yellow, with an occasional tinge of blue, and the bands to be reddish brown.

FIGURE 9.29 Jupiter as seen from Earth. This photo was taken with a 0.4-meter (16-inch) reflecting telescope. (Courtesy J. C. LoPresto, Edinboro State College, Edinboro, Pennsylvania.)

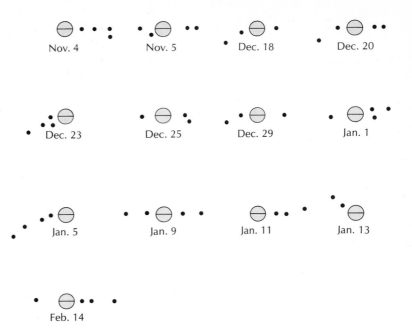

FIGURE 9.30 Drawings of Jupiter and its moons made with a 7.5-centimeter (3-inch) telescope by the author. All drawings were made about 8:30 A.M. in the winter of 1952–1953.

Nov. 4   Nov. 5   Dec. 18   Dec. 20

Dec. 23   Dec. 25   Dec. 29   Jan. 1

Jan. 5   Jan. 9   Jan. 11   Jan. 13

Feb. 14

Our small telescope shows the Galilean satellites of Jupiter easily. These four satellites revolve around the planet in 1.8 days, 3.6 days, 7.2 days, and 16.7 days. It can be fun to see the moons move around the planet. The motion of the innermost, rapidly moving Galilean satellite is obvious after a few hours. It is relatively easy to draw the planet and its moons one or two times each night and watch the motions (Figure 9.30). Sometimes the shadow of one of the moons can be seen on the planet.

Galileo discovered Jupiter's four brightest moons in 1610. Almost three centuries later, in 1892, a fifth moon was discovered by E. E. Barnard, using the 36-inch (0.9-meter) Lick Observatory refractor. This faint moon, closer to the planet than the Galilean moons, completes one circuit of its orbit in roughly 12 hours. Seven more moons were discovered between 1904 and 1951, bringing the total number to 12. Seth B. Nicholson, working at Mount Wilson Observatory, discovered four of these faint moons, to equal Galileo. Moons number thirteen and fourteen were discovered in 1974 and 1975, using long-exposure photographs made with the Schmidt telescope at Palomar Observatory.

It is not too difficult to tell, even with our small telescope, that Jupiter is not spherical. The planet is flattened at the poles and

FIGURE 9.31 Jupiter, showing the great red spot, the satellite Ganymede, and its shadow, taken with the 5-meter (200-inch) reflector. (Hale Observatories.)

bulges at the equator just like the Earth. However, for Jupiter the effect is immediately obvious without careful measurement. The planet's equatorial radius is 71,400 kilometers, but its polar radius is 67,000 kilometers. Jupiter is oblate because of its very rapid rotation. The planet's rotation period has been measured by astronomers watching features on its surface move around the planet. In round numbers, it rotates once every ten hours.

A fascinating feature of Jupiter's disk is the great red spot: an elongated spot, south of the planet's equator, which was first discovered near the end of the seventeenth century (Figure 9.31). At times the spot has a prominent reddish-brown color, but it does fade in color on occasion.

The visible features on Jupiter's disk are the tops of clouds in a deep atmosphere. When astronomers turned their spectroscopes on the planet, they identified molecular hydrogen (H_2), helium, ammonia (NH_3), methane (CH_4), and water vapor: a rather poisonous brew. These gasses are colorless, however; so other substances must be present to give the cloud bands their characteristic colors. One possibility is the molecule ammonium hydrosulfide (NH_4SH).

FIGURE 9.32 Photograph of Jupiter by the Pioneer 11 spacecraft. Note the great red spot, and the smaller dark and bright spots near it. (National Aeronautics and Space Administration.)

When the Pioneer 10 and Pioneer 11 spacecraft flew by Jupiter in 1973 and 1974, they made many pictures of the planet, which were sent back to Earth (Figure 9.32 and Plate IV). These close-up pictures revealed the disk of the planet in detail never seen before. Several infrared pictures were especially revealing. They showed that the zones are much colder than the planet's average temperature, and the belts are much warmer.

The clouds in the close-up Pioneer photographs show a distinct mottled appearance, suggesting the presence of convective cells, where bubbles of gas rise from deep in the planet's interior (Figure 9.33). The current idea is that gas rises in the zones and cools, causing clouds to condense. The gas then moves north or south and sinks in the belts, where the clouds disperse. The rapid rotation of the planet causes the cloudy and clear regions to spread out into parallel bands. The cold-zone clouds are composed of crystals of water ice and ammonia, according to experts.

The great red spot appears to be an immense, long-lived hurricane in the Jovian atmosphere. Why it persists if this theory is true remains something of a mystery. A smaller red spot was discovered

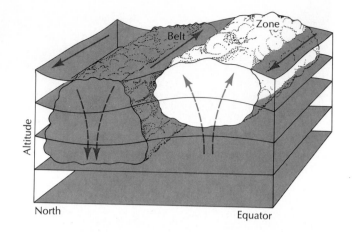

by earth-based telescopes in 1972 and was photographed by Pioneer 10 as the spacecraft swung past the planet. When Pioneer 11 reached Jupiter, the spot had vanished. Thus this small spot persisted for about two years.

Fortunately, hurricanes on Earth do not persist for such great lengths of time. However, these terrestrial storms are known to take their strength over the oceans and disperse rapidly once they reach land. If the Earth were uniformly covered with water, there is no telling how long a hurricane might last.

With this brief look at the telescopic appearance of Jupiter and the meteorological conditions which account for them, let's now turn to the interior structure of the planet.

Jupiter's Interior

Figure 9.34 illustrates today's best idea of the interior structure of Jupiter, as inferred from studies of the planet's gravitational field by Pioneer spacecraft.

The planet's core is suspected to be a small rocky ball containing materials like those found in the Earth. The remainder of the planet is primarily hydrogen, in two distinct layers. The inner layer is composed of liquid hydrogen that is subjected to the tremendous pressures of the overlying layer. It is believed that under the conditions found in this inner layer, hydrogen acts like a metal. Surrounding the layer of liquid metallic hydrogen is a layer of liquid molecular hydrogen.

Calculations indicate that the metallic layer is 46,000 kilometers thick, and has a temperature which decreases from 30,000° K to

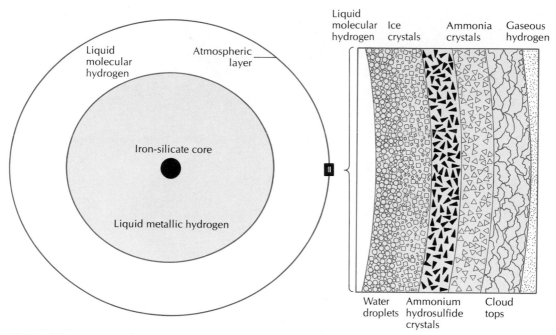

Liquid molecular hydrogen

Atmospheric layer

Iron-silicate core

Liquid metallic hydrogen

Liquid molecular hydrogen

Ice crystals

Ammonia crystals

Gaseous hydrogen

Water droplets

Ammonium hydrosulfide crystals

Cloud tops

FIGURE 9.34 Internal structure of Jupiter. (From *Jupiter* by John H. Wolfe. Copyright © 1975 by Scientific American, Inc. All rights reserved.)

10,000° K from its inner edge to its outer edge, whereas the liquid molecular layer is about 25,000 kilometers thick and has a temperature which falls from 10,000° K at its inside to the cold temperature of the atmosphere, which it blends into rather gradually.

In the next chapter, we will discuss the entire question of the origin of the solar system, and explain why the terrestrial planets and Jovian planets differ so. Jupiter's magnetic field will be discussed in Chapter 11.

SATURN

We have now reached the outskirts of the solar system, where all our information has been gathered from Earth-based telescopes and an occasional balloon flight above our atmosphere. Since we have much less information, there is much less to say about the remaining planets.

Saturn is the second largest planet in the solar system. Its diameter is 85% of Jupiter's, but its mass is only 30% of that planet's; thus Saturn is much less dense than Jupiter. In fact, Saturn's density of 0.7 gm/cm³ makes it even less dense than water: it would float.

FIGURE 9.35 Saturn
photographed with the
2.5-meter (100-inch) telescope on Mount Wilson,
California. (Hale Obser-
vatories.)

FIGURE 9.36 Saturn
photographed with an
0.2-meter (8-inch) reflect
ing telescope. (Courtesy
J. C. LoPresto, Edinboro
State College, Edinboro,
Pennsylvania.)

Out at 9.5 astronomical units from the Sun, Saturn receives very little sunlight, and its temperature is only 97° K, or roughly −285° F. Like Jupiter, Saturn is warmer than it ought to be, indicating that it has an internal heat store.

Through a moderate-sized telescope, Saturn is a spectacular sight (Figures 9.35 and 9.36). The most amazing feature of the planet is its beautiful ring system (Figure 9.37). The rings' rotation can be measured from the Doppler shift of their spectrum lines (Figure 9.38). They do not rotate like a phonograph record: instead, the inner edge rotates faster than the outer edge. In fact, the rate at which each point in the ring rotates follows Kepler's laws for bodies orbiting the planet. It is believed that the rings are an immense number of tiny particles, each circling the planet on a circular orbit. Stars can be seen through the rings, indicating that there is a lot of space between the particles.

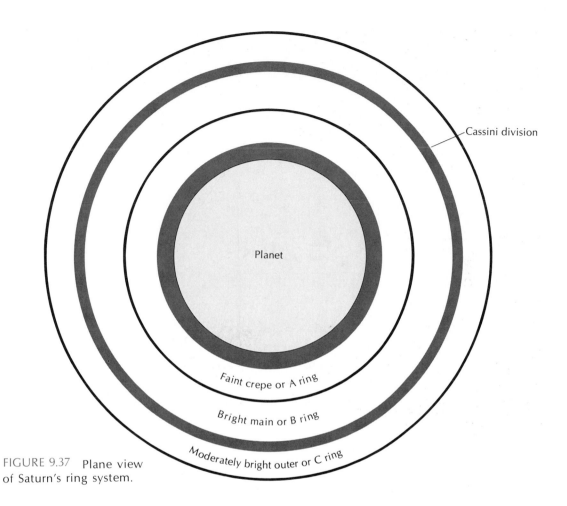

Cassini division

Planet

Faint crepe or A ring

Bright main or B ring

Moderately bright outer or C ring

FIGURE 9.37 Plane view of Saturn's ring system.

Saturn's orbit is tipped to the plane of the Earth's orbit, and as the planet moves in its 29-year circuit of the Sun, it is sometimes above (north of) the Earth's orbit and sometimes below (south of) it. As a result, the apparent tip of the planet's rings as seen from Earth changes (Figure 9.39). When Saturn crosses the plane of the Earth's orbit and its rings are seen edge on, they virtually disappear; so they must be very thin.

Saturn has ten natural moons, the largest of which, Titan, is the size of Mercury: 4,880 kilometers in diameter. Saturn's innermost moon, Mimas, orbits the planet in slightly under a day.

The motion of Mimas is interesting because it has an effect on Saturn's rings. Notice in Figure 9.37 that the rings are divided by a

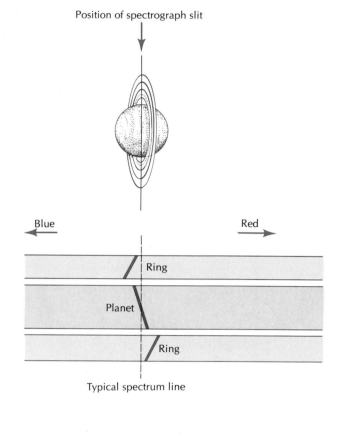

Position of spectrograph slit

Blue   Red

Ring

Planet

Ring

Typical spectrum line

FIGURE 9.38 Rotation of Saturn's rings as measured by Doppler shifts of spectrum lines. Note that the inner edge of the rings shows the greatest shift and therefore the greatest velocity.

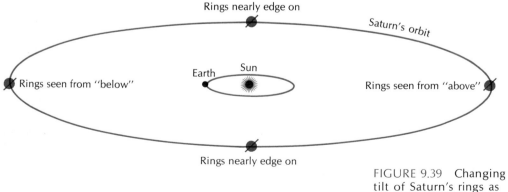

Rings nearly edge on

Saturn's orbit

Earth   Sun

Rings seen from "below"   Rings seen from "above"

Rings nearly edge on

FIGURE 9.39 Changing tilt of Saturn's rings as seen from the Earth.

broad gap known as Cassini's division. Particles moving in this gap would orbit Saturn in exactly one-half the period of Mimas. Very regular periodic perturbations by Mimas on these particles swept them from the division.

The surface of Saturn is crossed by cloud bands, not unlike those of Jupiter. The meteorology of Saturn's atmosphere is probably similar to Jupiter's.

The interior of Saturn is roughly similar to Jupiter's, but with a few differences. The rocky core is relatively larger, and is surrounded by an icy layer. The layer of metallic hydrogen is relatively thin, and the bulk of the planet's radius is liquid molecular hydrogen.

URANUS AND NEPTUNE

In the darkness of the outer reaches of the solar system, the planets Uranus and Neptune wend their icy cold ways. These two planets—with diameters of 51,000 kilometers and 49,500 kilometers; masses of 14.6 and 17.2 times the Earth's mass; densities of 1.2 and 1.7 gm/cm³—are very similar. They are also similar in being so far away that they are shrouded in mystery.

Through the telescope, Uranus appears as a virtually featureless, pale-green disk (Figure 9.40). The best photograph taken of the planet was made with Stratoscope II, a 36-inch (0.9-meter) optical telescope carried up to 80,000 feet by a giant balloon (Figure 9.41). Neptune is a tiny featureless disk even through the largest telescope (Figure 9.42).

Spectroscopically, both planets show strong absorption spectra due to methane. Apparently the upper atmospheres of these planets are very clear, and sunlight penetrates to a great depth. The long path the sunlight takes causes the absorption to be very strong. On the other hand, ammonia, which is strong in Jupiter's spectrum, is not found in the spectra of Uranus and Neptune. The exceedingly cold temperature of these planets probably freezes the ammonia out of the high atmosphere.

Astronomers predicted that it would be possible to see Uranus pass in front of (or **occult**) a faint star in the constellation Libra in early 1977. Occultations of stars by planets are important events. The way that the star dims as it passes behind the planet tells astronomers about the structure of the planet's atmosphere, and the length of the occultation allows astronomers to calculate an accurate diameter for the planet.

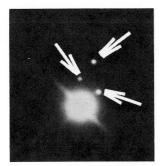

FIGURE 9.40 Uranus as seen from Earth, with arrows pointing out three of its moons. (Lick Observatory photograph.)

FIGURE 9.41 Uranus, taken by a meter-size telescope carried to an altitude of 25 kilometers on a balloon. The planet is virtually featureless. (National Aeronautics and Space Administration.)

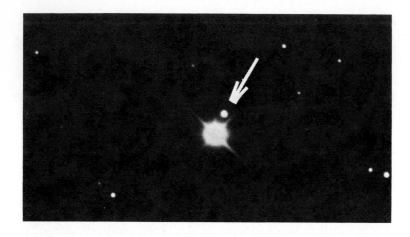

FIGURE 9.42 Neptune and one of its moons. (Lick Observatory photograph.)

On March 10, the long-awaited event was observed by several groups of astronomers. As Uranus approached the star on the sky, all the groups observed the star to disappear for several seconds, then reappear, then disappear four more times for about a second each. Later, after the star emerged from behind the planet, the disappearances and reappearances occurred again, but in a time-sequence that was exactly the reverse of the one observed before the occultation.

Astronomers have concluded that Uranus has a set of rings like Saturn's, but much fainter, narrower, and closer to the planet. The behavior of the light of the star was caused by its passing in turn behind rings and the gaps between the rings.

Uranus has the most extreme seasons of any planet in the solar system. Its rotation axis actually lies in the plane of its orbit. As the planet revolves around the Sun, there are times when one pole or the other points toward the Sun, and only one hemisphere receives sunlight.

Both Uranus and Neptune were unknown before the late eighteenth century. The story of their discovery is told in Chapter 3.

PLUTO

Mysterious Pluto is aptly named for the god of the underworld, for it orbits in perpetual darkness at the outer edge of the solar system.

Very little is known about Pluto (Figure 9.43). Its brightness varies regularly, with a period of a little over six days, probably because of the planet's rotation and an unevenly bright surface.

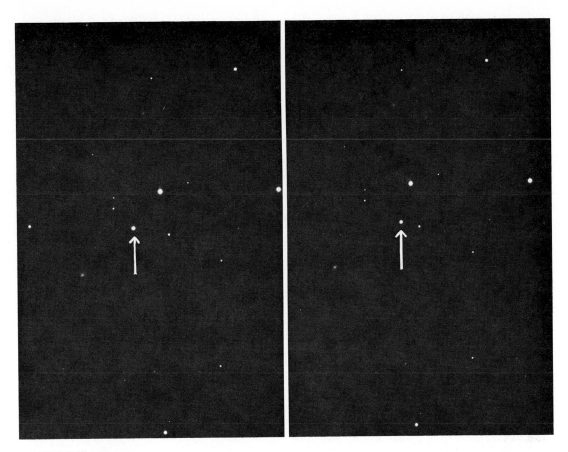

FIGURE 9.43 Pluto, showing its motion in one day. This photograph, made with the mighty 5-meter reflector on Palomar Mountain, shows the planet as a mere point of light. (Hale Observatories.)

Pluto seems to be roughly the same size as Mercury, but it is so far away that its angular size cannot be measured with certainty.

Recently, astronomers have studied infrared radiation from Pluto. The infrared spectrum of the planet strongly suggests that its surface is covered with frozen methane. Pluto has an equilibrium black-body temperature of 45° K, much lower than the 89° K freezing point of methane; so the result is not surprising. If Pluto is covered with ice its albedo may be relatively high, perhaps even as high as 50%. If this is true, the planet may be only around 3,000 kilometers in diameter.

We have now reviewed briefly what is known about each of the planets in the solar system. As we neared the outer edge of the system, the story for each succeeding planet got shorter. Our knowledge gets increasingly meager for the most distant planets.

Perhaps in a few decades we will have sent space probes to each planet, and will know as much about Pluto as we do about, say, Mars. There are no doubt some fascinating discoveries waiting to be made.

In the next chapter, we will first talk about some of the lesser bodies of the solar system, then try to put all the facts together and tell how the system may have been formed.

REVIEW QUESTIONS

1. List the planets in order of distance from the Sun.

2. Describe the gross characteristics of each planet: (a) visual appearance; (b) size; (c) atmosphere; (d) surface characteristics; (e) interior structure.

3. Describe the seasons on Uranus.

10

THE SOLAR SYSTEM

S o far in our discussion of the solar system, we have talked about the morphology—that is, structure and form—of the planets. There is some rhyme and reason behind these seemingly unconnected facts that tell us about the origin of the solar system. In this chapter we will try to tie the picture together. However, before we can give the complete story, we must look at a few more facts. We will therefore begin this chapter with a brief look at two subjects that are important: first, the regularities of the solar system; second, the lesser bodies of the system.

THE FORM OF THE SOLAR SYSTEM

There are several regularities of the solar system that are of vital importance for understanding its origin and evolution.

The Scale of the Solar System

The solar system is mostly empty. If the Earth and Venus were shrunk to the size of peas (0.5 centimeter, say), then their closest distance of approach to one another (roughly 40,000,000 kilometers) shrinks to a little over 16 meters (50 feet). On this scale, Pluto would orbit the 50-centimeter-diameter Sun at a distance of 2.4 kilometers (1.5 miles).

The shape of an elliptical planetary orbit is described by a quantity called its **eccentricity**. If the eccentricity is zero, the orbit is a circle. The eccentricity of the Earth's orbit is 0.017, which means that when the Earth is closest to the Sun, it is $0.017 \times 100\% = 1.7\%$ closer than average, and when it is farthest away, it is 1.7% farther than average. Table 10.1 summarizes information on the scale of the solar system. Recall that the synodic

Table 10.1 Scale of the Solar System

	Mean Distance from Sun (million km)	Revolution Period		Inclination of Orbit to Ecliptic
		Sidereal	Synodic	
Mercury	58	88 days	116 days	7°
Venus	108	225 days	584 days	3°
Earth	150	365¼ days	—	—
Mars	228	687 days	780 days	2°
Jupiter	778	12 years	399 days	1°
Saturn	1,427	29 years	378 days	2°
Uranus	2,870	84 years	370 days	1°
Neptune	4,497	165 years	368 days	2°
Pluto	5,900	248 years	367 days	17°

period of revolution of a planet is the period as seen from the moving Earth. For Mercury or Venus it is the interval, for instance, between two successive times when the planet is between the Earth and the Sun. For Mars, Jupiter, and the other Jovian planets, it is the interval between successive times when, say, the Earth is between the planet and the Sun.

The Regularities

The Solar System Is Flat. Most bodies in the solar system move in orbits that lie close to the plane of the ecliptic (Figure 10.1). Thus the majority of the mass of the system is confined to a very flat, pancake-like volume of space.

Motions Are in the Same Direction. Almost all of the orbital motions in the system are in the same direction. When seen from far above the north pole of the Earth, the planets revolve around the Sun in a *counterclockwise* direction.

The natural moons of the planets also revolve in orbits that lie near the plane of the ecliptic, and 26 out of 33 of them revolve in the counterclockwise direction as seen from above the north pole. All the planets except Venus and Uranus rotate in the same direction as the revolutions. Venus and Uranus rotate in the opposite direction. Pluto's direction of rotation is unknown.

The Titius-Bode Law. There is a regularity to the distance of the planets from the Sun that has come to be called the Titius-Bode law. It goes as follows: first, write down the numbers 0, 3, 6, 12, 24, . . . , on to 384. Each number except 0 and 3 is double the previous number in the sequence. Second, add 4 to each number in the

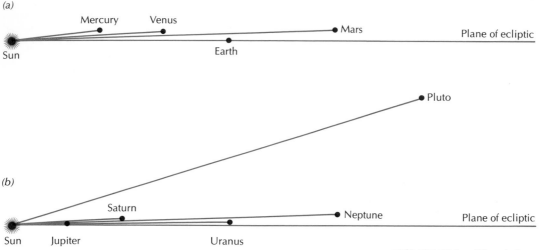

(a)

Mercury Venus · Mars Plane of ecliptic

Sun Earth

· Pluto

(b)

Saturn

Sun Jupiter Uranus · Neptune Plane of ecliptic

FIGURE 10.1 Tilt of the planes of the orbits of the planets are indicated schematically for (a) the inner solar system and (b) the outer solar system. If one ignores Pluto, the planets do not move far from the ecliptic plane.

sequence. Third, divide each sum by 10. The results are close to the distances of the planets from the Sun, in astronomical units, except for the gap at 2.8 a.u. Is this result a coincidence or a physically important relationship? We will answer this question later.

Terrestrial Versus Jovian Planets. In the last chapter we pointed out that the planets are divided into two classes. The four terrestrial planets are small, dense, and near the Sun, whereas the four Jovian planets are large, low-density, and at great distances from the Sun.

With this brief look at the regularities of the solar system, let's discuss some of the minor bodies of the system.

ASTEROIDS

The Titius-Bode Law was first brought to prominence in 1772 by J. E. Bode, who had found it inserted as a footnote by J. D. Titius in a book he had translated from the French. The law predicted the distances of the known planets from the Sun in astronomical units quite well, except for one mysterious gap. The law predicted a planet at 2.8 a.u., between the orbits of Mars and Jupiter. However, no planet was known there.

The law gained credibility in 1781, when Herschel stumbled upon Uranus. That planet's distance was found to fit perfectly into the scheme. So why was there no planet at 2.8 a.u.?

Astronomers of the day felt that there should be a planet in that gap, and began a search. On January 1, 1801, G. Piazzi, working at the observatory of the University of Palermo, noticed a faint star-like point of light moving across the celestial sphere. It was slightly fainter than the faintest star that can be seen with the naked eye. Piazzi realized the object might be a planet and began to plot its position so that he could calculate the size and shape of its orbit. Unfortunately, the object moved close to the Sun before he obtained enough observations, and it was lost in the glare of the daytime sky.

Piazzi computed a rough orbit based on his observations, and months later, when the object had moved back to the night sky, he predicted where it should be. The prediction was wrong. The object was lost. However, about this time one of the giants of mathematics came to the rescue. Carl Friedrich Gauss invented a new, superior mathematical scheme for calculating the size and shape of the orbit of a body in the solar system. Gauss applied the scheme to Piazzi's observations, and predicted the position of the lost body. It was quickly found near that location.

In the process of looking for the lost object, astronomers found yet another moving, starlike image, about as faint as the first. Its orbit, too, was calculated by Gauss' method. Both bodies moved in orbits in the gap between Mars and Jupiter. Both turned out to be tiny compared to the other planets: the first, now called Ceres, is 950 kilometers in diameter; the second, Pallas, is 560 kilometers in diameter. In the following years, more and more tiny planets were found between Mars and Jupiter. Today, almost 2,000 of these objects, which we call **asteroids** or minor planets, have well-determined orbits (Figures 10.2 and 10.3). Astronomers sought to find a missing planet and discovered thousands, most of which orbit the Sun in the 2.8 a.u. gap between Mars and Jupiter.

Over the years astronomers have often asked themselves whether the asteroids are the remains of a planetary catastrophe. Was there an ordinary planet at 2.8 a.u. which was torn apart by some unbelievably great explosion? Perhaps two planets collided. We will return to this point later, and mention some of the arguments for and against the idea. First we need to discuss the nature of these objects and other solar-system debris.

Figure 10.4 illustrates the distribution of asteroids in the solar system between Earth and Jupiter. The curve represents the number of asteroids in each 0.01 astronomical unit of distance out from the Sun. It is immediately obvious that the distribution is not uniform. In particular, there are a series of gaps in the distribution where few astroids move. These gaps are named for D. Kirkwood,

FIGURE 10.2 When a photograph is made of a region of the sky near the ecliptic and far from the Sun's position, the image of an asteroid can always be found. In this one-hour time-exposure, the images of two asteroids can be recognized, since they moved far enough during the exposure to appear as small elongated images. In this picture, the brighter of the two images is the asteroid Bellona. (Yerkes Observatory.)

FIGURE 10.3 A 1.5-hour exposure, taken with the great 1-meter (40-inch) refracting telescope at Yerkes Observatory, shows the trail of the asteroid Eros. The orbit of this asteroid brings it within 20 million kilometers of the earth on occasion. (Yerkes Observatory.)

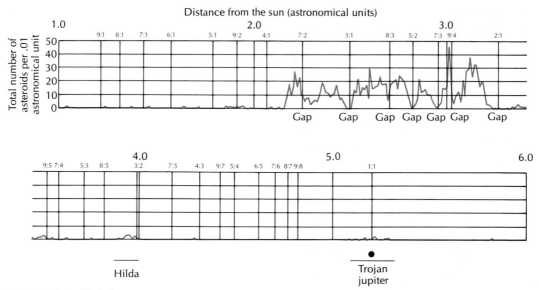

Distance from the sun (astronomical units)

FIGURE 10.4 Distribution of asteroids in the belt between Mars and Jupiter. Notice that a few asteroids fall between Earth and Mars. (From *The Nature of Asteroids* by C. R. Chapman. Copyright © 1974 by Scientific American, Inc. All rights reserved.)

who explained them in 1866. The gaps are labeled in the figure by ratios such as 3:1. This means, for instance, that a hypothetical asteroid moving in the 3:1 Kirkwood gap would orbit the Sun exactly three times as Jupiter orbits once. Every third revolution, the asteroid, Jupiter, and the Sun would line up, with the asteroid at precisely the same place in its orbit each time. The resulting periodic tug on the asteroid by Jupiter would ultimately change the asteroid's orbit and remove it from the gap. The same process worked to produce the gaps in Saturn's rings (Chapter 9), caused by that planet's moons.

A familiar example of how a periodic push has a large effect in the long run is the method we use to get a stuck car out of a snow bank or mud. A steady forward shove will not get the car unstuck nearly as effectively as rocking it back and forth: shove forward, let it roll back, shove forward a little farther, let it fall back, back and forth, faster and further, until finally the car is free.

Some asteroids move in very elliptical orbits (Figure 10.5). Icarus gets closer to the Sun than any other asteroid. It is aptly named for the legendary Greek lad who attached feathers to his arms with wax and learned to fly. But he flew too close to the Sun; the wax melted; and he fell to his death.

The Trojan asteroids travel in two clumps in Jupiter's orbit, one 60° ahead of the planet, and one 60° behind. It can be shown mathematically that bodies at the location of the Trojan asteroids have very stable orbits and will continue to lead or follow Jupiter.

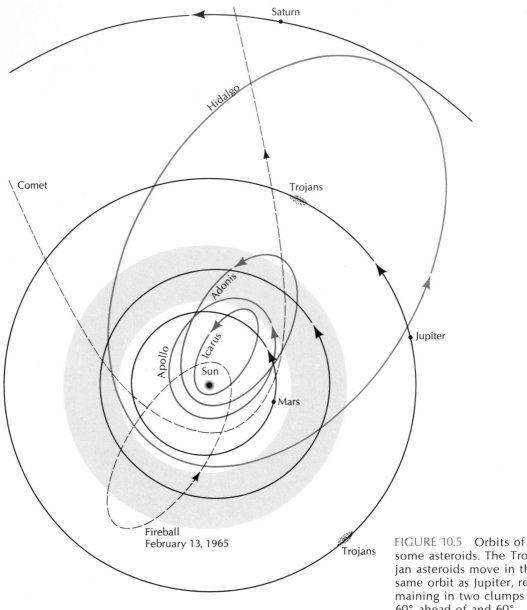

Saturn

Hidalgo

Comet

Trojans

Adonis

Apollo

Icarus

Sun

Mars

Jupiter

Fireball
February 13, 1965

Trojans

FIGURE 10.5 Orbits of
some asteroids. The Tro-
jan asteroids move in the
same orbit as Jupiter, remaining in two clumps
60° ahead of and 60°
behind the giant planet.
(From *The Smaller Bodies
of the Solar System* by
W. K. Hartmann. Copy-
right © 1975 by Scientific
American, Inc. All rights
reserved.)

How can we study the physical makeup of asteroids? They are solid bodies with no atmosphere; so their spectra show no characteristic spectrum lines. It turns out, though, that minerals reflect light of different wavelengths in a manner that is characteristic for each specific mineral. If an asteroid were a perfect mirror, its spectrum would look exactly like the Sun's spectrum. However, it might in reality reflect relatively more blue than red. The result would be a spectrum that looked roughly like the Sun's, but with the red relatively too faint. This sort of reflectivity variation has been measured for a great many asteroids.

There are detailed variations among the asteroids, but basically they fall into one of two groups. The asteroids of one group have relatively high reflectivities and a reddish color. Their reflectivities are characteristic of silicate minerals with a mixture of metals. The asteroids of the other group have low reflectivity and a neutral, grayish color. Their surface reflectivities are characteristic of carbonaceous chondrites. The latter are a type of stony meteorite with a high carbon content. The difference between the two groups, according to one theory, depends on how far from the Sun they were initially formed.

We have now briefly described asteroids, most of which fall in the 2.8 a.u. gap between Mars and Jupiter. They consist of two types: silicate-metal asteroids and carbonaceous asteroids. Let's next look at a related subject, meteoroids, meteors, and meteorites.

METEOROIDS, METEORS, AND METEORITES

If you go outside on a clear night away from city lights, the sky is fun to watch. You do not have to look for long to see a streak of light flash through one of the constellations. This is an example of a **shooting star** or **falling star**, more properly called a **meteor**. A meteor is not a star: it is a small particle of material from space which hits the earth's atmosphere at very high speed and vaporizes because of the heat generated by friction between the matter and the air.

The chunk of matter in space and in the atmosphere is called a **meteoroid.** When it reaches the ground, it becomes a **meteorite.** The term meteor refers only to the flash of light.

Most meteors are faint flashes which travel a few degrees across the sky; these are caused by sand-grain-sized meteoroids. On very rare occasions, a large enough chunk will hit the atmosphere that some of it survives to reach the ground. In such a case, the meteor would be extremely bright, possibly bright enough to be seen in daylight.

On August 19, 1972, a great fireball (an extremely bright meteor) was observed over the northwestern United States, in broad daylight. The meteoroid entered the atmosphere over Utah, traveled about 1,500 kilometers northward, and finally disappeared near Edmonton, Alberta. Over Montana, it produced sonic booms that were heard on the ground. Scientists believe that the meteoroid skipped out of the atmosphere after entering, like a stone skipped on water. This is fortunate, for it probably weighed as much as four tons, and could have caused considerable damage if it had reached the ground.

Meteor Showers

Sometimes an unusually large number of meteors are seen during the space of a few nights. An interesting feature of such occurences is that if you extend backwards the streaks the meteors make in the sky, the lines all meet in one small area. This area is called the **radiant** of the shower (Figure 10.6). Showers are frequently named for the constellation in which their radiant lies.

Meteor showers occur when the earth crosses a meteoroid stream, which is a large number of meteoroids all traveling in the same orbit. The Earth will cross the orbit at the same time each year; so meteor showers are annual events (see Table 10.2). If the meteoroids are uniformly spread out in their orbit, the shower is the same intensity each year.

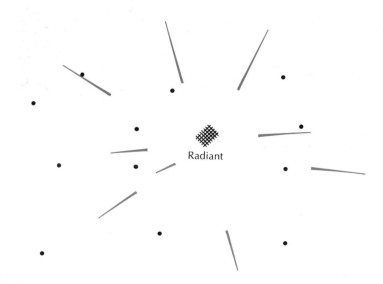

Radiant

FIGURE 10.6 Radiant of a meteor shower. The meteoroids are all moving toward us along parallel paths. Like railroad tracks converging at the horizon, these paths all appear to converge at the radiant.

PLATE I   The full impact of the fact that we inhabit a small, fragile
world was brought home dramatically in this Apollo 8 photograph of
Earth from lunar orbit. (National Aeronautics and Space Administration.)

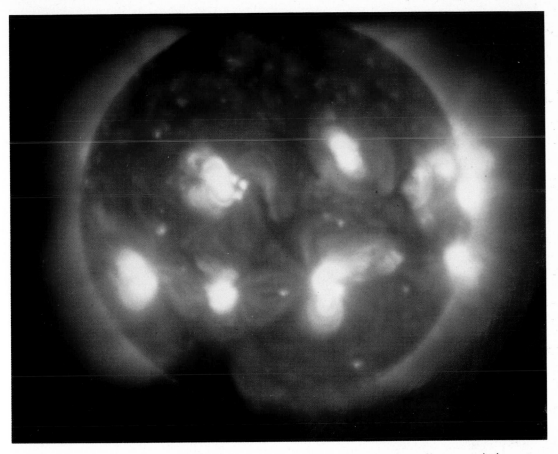

PLATE III   Photograph of X-ray emission from the sun made from
Skylab. Notice the bright points scattered over the solar surface.
(Solar Physics Group, American Science and Engineering, Cambridge,
Massachusetts.)

PLATE II   Comet Bennett as
photographed on April 16,
1970. The blue molecular-
emission tail and the yellow
reflected-sunlight tail can
be seen, albeit very subtly,
in this color photograph.
(J. C. Brandt, R. G. Roosen,
and S. B. Modali, Goddard
Space Flight Center.)

PLATE IV   Jupiter as seen from the Pioneer 11 spacecraft. Notice the great red spot. (National Aeronautics and Space Administration.)

PLATE V   The Crab Nebula. The white central region of the nebula emits a continuous synchrotron spectrum, whereas the red filaments emit predominantly the Hα line of hydrogen. (Hale Observatories. Copyright © by the California Institute of Technology and The Carnegie Institution of Washington. Reproduced by permission from the Hale Observatories.)

(a)

(b)

PLATE VII   Color picture of Mars made by the Viking I spacecraft as it approached the planet in June 1976. Near the terminator, Argyre basin can be seen below the center of the picture and the great Marine Valley above the center. (National Aeronautics and Space Administration.)

PLATE VI   (a) The Lagoon Nebula (M8). The emission from this nebula is dominated by the red Hα line of hydrogen. (b) The Trifid Nebula (M20). The blue part of the nebula is due to starlight reflected from a cloud of dust. (Courtesy, Astrophotography Group, E. Alt, Dr. E. Brodkorb, K. Rihm, J. Rusche, Neustadt-19, Maxburgsattel, Germany).

PLATE VIII   Color view of the Martian surface taken by the Viking I
lander. The sun is only 3° above the horizon, causing a great deal of
shadowing, and accentuating surface features, such as the shallow de-
pression near the center of the frame. (National Aeronautics and Space
Administration.)

Table 10.2 Selected Meteor Showers

Shower	When Visible (Best Night)	Radiant in Constellation
Quadrantids	Jan. 2–4 (Jan. 3)	Ursa Major
Lyrids	Apr. 20–22 (Apr. 21)	Lyra
Eta Aquarids	May 3–10 (May 5)	Aquarius
Delta Aquarids	July 24–Aug. 6 (July 28)	Aquarius
Perseids	Aug. 1–20 (Aug. 12)	Perseus
Orionids	Oct. 15–26 (Oct. 20)	Orion
Leonids	Nov. 11–20 (Nov. 15)	Leo
Geminids	Dec. 9–14 (Dec. 12)	Gemini

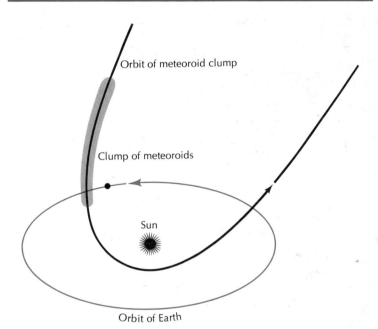

FIGURE 10.7 A meteoroid stream. A clump of meteoroids moves around the Sun in an elongated orbit. If the Earth passes through the clump, a meteor shower occurs.

In some meteoroid streams, many of the bodies are bunched up in a single clump which moves around the orbit (Figure 10.7). In the Leonid shower (named for the constellation Leo), the meteor clump takes roughly 33 years to orbit the Sun. In 1866 the Leonid shower produced 100,000 meteors per hour for a night or so. In subsequent years, the shower was very weak. In 1899 a large shower was again observed. We will discuss the connection between meteor streams and comets when we talk about the latter.

Not all meteors belong to streams. Those that do not are called sporadic meteors. These meteors are most common after midnight. The sector of the Earth on which local times are between midnight and dawn lies in the direction that the Earth travels in its orbit

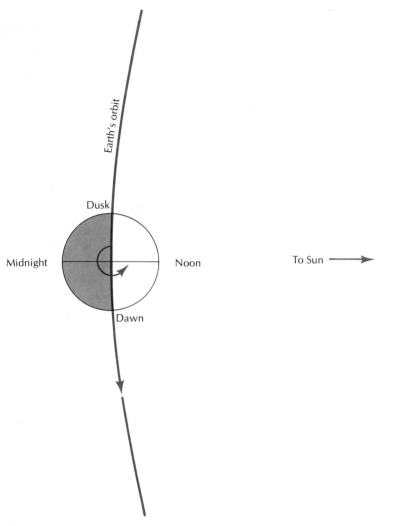

FIGURE 10.8 As the Earth moves in its orbit, the leading hemisphere, where local times range from midnight to dawn to noon, sweeps up meteoroid debris. The meteors which result can be seen in the dark midnight-to-dawn sector.

(Figure 10.8). That part of the Earth sweeps up meteoroids as it moves, causing many meteors. Meteors seen before midnight, on the other hand, had to catch up with the moving Earth, and are less common.

Meteorite Impacts on Earth

There seem to be objects in space of all sizes, from the large asteroids to the tiny meteoroids. It is clear from the cratered appearance of the Moon and some of the other planets that the solar system was filled with large fragments in its early history. It took some very large impacts to produce the innumerable craters and giant ringed basins we see on the other solar-system bodies.

In the Arizona desert there is a crater which is about 1.2 kilometers across and 200 meters deep (see Figure 6.44). Scientists estimate this crater was blasted out by the impact of a chunk of iron weighing a few million tons and traveling about 15 kilometers/second. The meteoroid would have been less than 50 meters in diameter: small compared to the size of the crater. We may hope there are not too many chunks lurking in space that could create a hole the size of the Arizona crater.

However, there have been some large meteorite falls in recent history. One particularly spectacular fall occurred near the Tunguska River in Siberia in 1908. The impact, which registered on seismographs all over our planet, toppled trees within a 30-kilometer radius of the impact point. Scientists who visited the remote site later found the trees, like pick-up sticks, stripped of bark and limbs, and all pointing away from the impact point. No debris from the original meteoroid has been found; so there has been much speculation about what it was. One guess is that it was a small comet. As we shall see, the solid parts of comets are mostly ices, which would vaporize on impact and leave little trace.

Meteorites

Meteorites that have been picked up from the ground and taken into the laboratory tell us a great deal about our solar system. Before the space age, they were our only direct access to the system outside Earth. Now we have studied samples of material returned from the Moon, too.

Meteorites are divided into three main classes according to their composition. The first class is the stony meteorites, which have compositions roughly like the crustal material of the Earth and the Moon (Figures 10.9 and 10.10). The remaining two classes, stony iron meteorites and iron meteorites, have very high concentrations of iron (Figures 10.11, 10.12, and 10.13). As the names imply, stony iron meteorites contain both silicate minerals and iron, whereas iron meteorites are almost pure iron.

It is interesting to compare the compositions of the meteorite types with the composition of the interior of the Earth at various depths. The stony meteorites are like the planet's crust, whereas the iron meteorites are like its core. The stony iron meteorites are similar to the iron-enriched material in the deeper mantle between the crust and the core. What does this mean?

FIGURE 10.9 A stony meteorite known as an ordinary chondrite. The name chondrite comes from small spherical inclusions, known as chondrules, that are found in the meteorite. (Smithsonian Institution.)

FIGURE 10.10 A slice of a stony meteorite showing chondrules and other inclusions. (Smithsonian Institution.)

FIGURE 10.11 A stony iron meteorite has been sliced to show the iron inclusions in the stone. (Smithsonian Institution.)

FIGURE 10.12 A nickel-iron meteorite. This 12.7-kilogram chunk, which is about 15 centimeters across, was found in Saskatchewan, Canada. (Courtesy H. H. Nininger, American Meteorite Museum.)

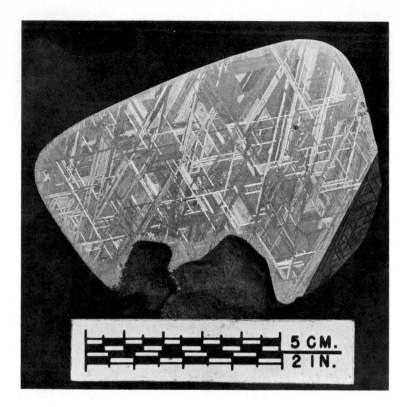

FIGURE 10.13 A nickel-iron meteorite has been cut, polished, and etched with nitric acid. The pattern markings, known as the Widmanstätten pattern, are characteristic of iron meteorites. (Smithsonian Institution.)

One idea is as follows. If we could heat up the entire Earth until it melted to a liquid, then stir it until it was thoroughly mixed, it would still end up looking like it does today when it resolidified. The process that changes the composition with depth is **gravitational differentiation**, which works in any quiet gas or liquid.* The heavy elements like iron settle to the Earth's core because the force of gravity on iron atoms is larger than on the lighter atoms. Iron is the most abundant heavy metal in the solar system, which is why the core is mostly iron. A large fraction of the rarer, heavier metals like uranium wind up in the core, too. Not all the heavy metals wind up in the core, of course. Some remain in the crust and mantle. The crust, which is made up of the lighter elements, does not experience as large a force from gravity as the iron, and it floats to the top like slag in a blast furnace.

---

*As we shall see in the next chapter, for instance, the Earth's atmosphere is gravitationally differentiated above 100 kilometers, where there are no longer any winds to keep it stirred up and mixed.

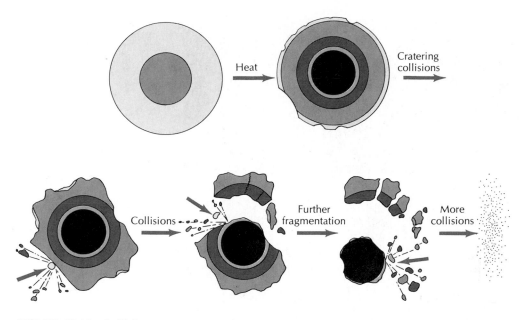

Heat

Cratering collisions

Collisions

Further fragmentation

More collisions

FIGURE 10.14 Collisions between asteroid-sized bodies, the larger of which have been gravitationally differentiated into an iron core and a stony crust, produced the different types of meteorites we see today. (From *The Nature of Asteroids* by C. R. Chapman. Copyright © 1974 by Scientific American, Inc. All rights reserved.)

According to some experts, the asteroid belt long ago contained many more bodies like the bigger asteroids. As these bodies solidified, they formed an iron core by gravitational differentiation. This means the bodies had to be fairly large. If they were too small, differentiation would not take place. In the early history of the solar system, when there were many large asteroids, collisions would occur frequently (Figure 10.14). In a collision between two asteroid-sized bodies, material would be thrown into space because of the bodies' low gravity. Many such collisions would grind the bodies apart, and the material thrown into space would have a range of compositions like the compositions at various depths in the asteroids. This material formed the meteoroids we study today.

With this brief discussion, we will leave meteorites and pass on to another fascinating type of object, comets.

COMETS

When an astronomer takes a photograph of a portion of the sky, many different types of celestial bodies in addition to stars will be recorded. On occasion, the astronomer may see a faint, featureless, fuzzy image on the photographic plate. The object whose image

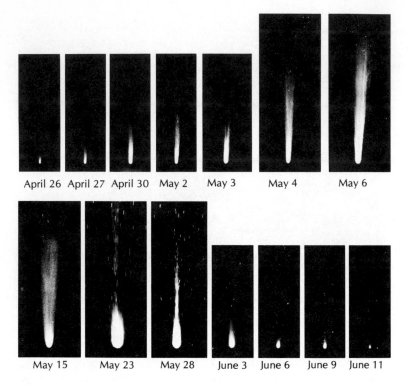

April 26 April 27 April 30 May 2 May 3 May 4 May 6

May 15 May 23 May 28 June 3 June 6 June 9 June 11

FIGURE 10.15 Halley's Comet in 1910. (Hale Observatories.)

was photographed could be many things: a distant galaxy; a distant cloud of gas in our own Galaxy; or a comet. To find out, the astronomer will take another picture a night or two later. If the fuzzy image is at the same place as before, it is probably the image of an object far beyond the solar system. But if it moved among the stars, it is most likely an object within the solar system. It is, in fact, very likely to be a comet. After several photographs of the moving object are available, the size and shape of its orbital path can be calculated using a method like the one that Gauss developed for the missing asteroid problem. Then the astronomer can decide whether he has discovered a new comet or has relocated one discovered on a previous swing by the Sun.

Many comets are discovered by professional astronomers who stumble on them while looking for something else. However, just as many are discovered by amateur astronomers who spend their spare moments sweeping the sky with low-power telescopes, looking for out-of-place fuzzy images. Amateur astronomers have discovered at least as many comets as the professionals.

FIGURE 10.16   Parts of a comet.

When comets are first discovered, they are often on their way in toward the Sun. The fuzzy object is a cloud of gas known as the **coma** of the comet. As the comet moves closer to the Sun, the coma will typically increase in brightness and size. Ultimately the comet will grow a **tail** (Figure 10.16).

Usually one of the several comets that pass through the inner part of the solar system each year will become bright enough to be seen with the naked eye. Then we see a beautiful object stretching across the sky. In the very brightest of these comets, a starlike spot can be seen within its coma. This is the comet's **nucleus**.

Orbits of Comets

The orbits of comets are typically ellipses, just like planetary orbits. The Sun lies at one focus, and the comet's motion obeys the law of areas. The difference between cometary orbits and planetary orbits is the degree of ellipticity of the former: comets move in very elongated ellipses (Figure 10.17). Long-period comets move in ellipses which bring them to within roughly one astronomical unit of the Sun at closest approach, whereas they are thousands of astronomical units from the Sun at their greatest distance. Applying Kepler's law of areas, we must conclude that such comets move very rapidly when they are near the Sun, and very slowly when at their greatest distance away. Let us take a hypothetical long-period comet with a period of 100,000 years and a perihelion distance (closest approach) of one astronomical unit as an example. When this comet is at its greatest distance (aphelion) from the Sun, it is 4,300 astronomical units away. Thus it moves 4,300 times faster at perihelion than at aphelion.

Comets fall into one of two groups, the long-period comets, with periods as long as two million years, and short-period comets, with periods between the shortest known (Comet Encke, 3.3 years) and 200 years. That is, if a comet has a period over 200 years, it is defined as a long-period comet. Short-period comets make up 20% of the comets with well-determined periods.

Interestingly, the orbits of long-period comets do not all lie in the flat system of the planets. The orbits of these comets can be inclined up to 90° to the plane of the ecliptic. Furthermore, the comets move in either a direct or a retrograde direction in their orbits. Thus long-period comets form a spherical halo around the pancake-thin solar system.

Short-period comets, on the other hand, behave more like the planets. Their motion typically is direct and in orbits that lie close

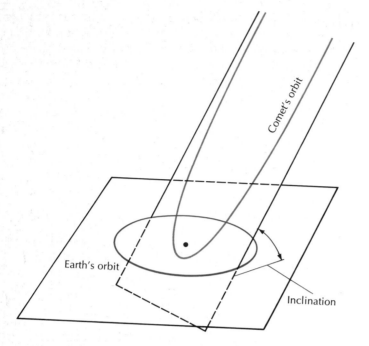

FIGURE 10.17 The elongated orbit of a comet.

Comet's orbit

Earth's orbit

Inclination

to the ecliptic plane. The mean revolution period of short-period comets is seven years. As we will discuss below, there is an important reason for the differences between the orbits of long-period comets and short-period comets.

Comets' Names

A comet is usually named for its discoverer or discoverers. Thus the comet that caused so much public interest around Christmas 1973 was called Comet Kohoutek for its discoverer, Lubos Kohoutek (Figure 10.18).

A group of Japanese amateur astronomers devote much of their time searching the sky for comets. Their efforts have been well-rewarded with such comets as Comet Ikeya-Seki, a spectacular naked-eye comet in 1965.

Comets are also given number designations, because some of the multiple-discoverer names do get out of hand (for example, Comet Honda-Mrkos-Pajdusakova). The comet's number designation is based on the year of discovery and the order of discovery in that year. Thus Comet Kohoutek, the sixth comet found in 1973, is called also Comet 1973f. The first comet discovered in 1980 will be

FIGURE 10.18 Comet
Kohoutek in December
1973. (Joint Observatory
for Comet Research, op-
erated by NASA and the
New Mexico Institute of
Mining and Technology.)

1980a, the second 1980b, and so on. Later the comets will be renamed according to the order in which they pass perihelion. Comet Honda-Mrkos-Pajdusakova was the third comet to pass perihelion in 1954 and is therefore also called Comet 1954 III. We can tell whether the name is based on discovery or perihelion passage, depending on whether a lowercase letter (a, b, c, . . .) or a Roman numeral (I, II, III, . . .) is used in the designation.

The Nature of Comets

The Greek philosophers believed that comets were phenomena of the Earth's atmosphere. Aristotle, for instance, discusses them in his treatise *Meteorologia,* or *Meteorology.* (Incidentally, people who work on meteorites are called meteoricists to distinguish them from meteorologists, who study the atmosphere).

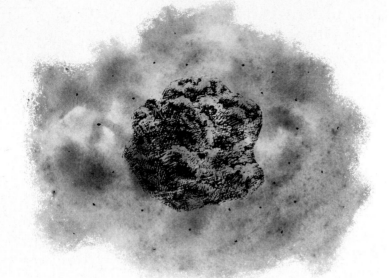

Tycho Brahe tried to measure the parallax of a comet in 1577. He had observers at several sites in Europe measure the position of the comet relative to the stars. If it were an atmospheric phenomenon, it would be at a different place among the stars as seen from each location; that is, it would show a parallax. The comet did not show a parallax, and Tycho concluded that it was much further away than the Moon, for he could measure the Moon's parallax.

We have already talked about Newton, Halley, and Halley's Comet in Chapter 3. They showed, by calculating the comet's orbit and successfully predicting its return in 1758, that it orbits the Sun just like the planets. The work of Tycho, Newton, and Halley thus proved that comets are celestial objects.

During the intervening years, a picture of comets has been built up which culminated in Whipple's so-called "dirty iceberg" model in 1950. Whipple hypothesized that the nucleus of a comet is a chunk of frozen material with solid meteoroid particles imbedded in the ice (Figure 10.19). The composition of the ice is assumed to be similar to the gasses found in the giant planets: methane, ammonia, water, carbon dioxide, and probably other things as well.

As the nucleus moves toward the Sun from the far reaches of its orbit, the Sun's heat causes the ices to **sublime**; that is, they go directly from the frozen state to the gaseous state without first becoming liquid. The gas forms a cloud around the cometary nucleus, which we see as the coma (Figure 10.20).

Spectroscopic study of the comas of many comets reveal simple constituents such as H, OH, O, CN, C_2, C_3, CO^+, NH, NH_2, CH,

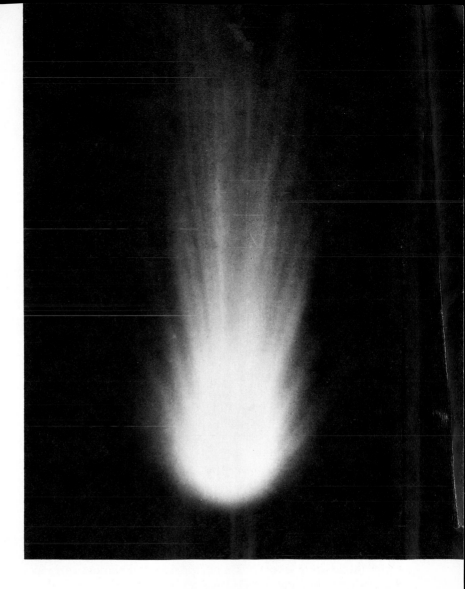

FIGURE 10.20 Head of Brook's Comet in October 1911. (Lick Observatory photograph.)

N_2^+, as well as heavier elements, such as Fe, Ni, and other metals. The former group of constituents are called **radicals**. They are created when the short-wavelength photons from the Sun tear apart the complex parent molecules found in the ices of the nucleus. Studies of Comet Kohoutek revealed some more complex constituents, such as ionized water, H_2O^+, and methyl cyanide, HCN.

The tail of a comet is created by solar photons and the solar wind, which blow material out of the coma. Each solar radiation (photon and wind) causes a different type of tail. Plate II shows a typical two-tailed comet. One tail is uniformly bright, gently curving, and yellowish. The other is straight, shows a turbulent structure, and is bluish.

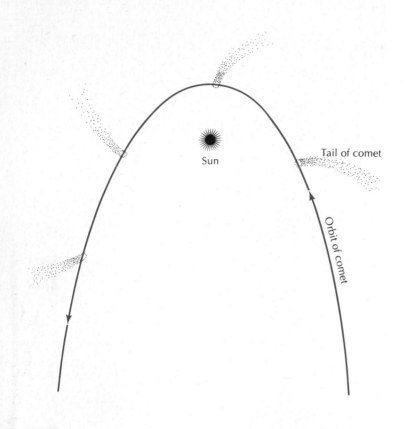

Sun

Tail of comet

Orbit of comet

The yellowish tail consists of dust: tiny meteoroids freed from the nucleus as the ice sublimes. The photons from the Sun strike a dust particle and bounce off, and in the process give the dust particle a tiny push away from the Sun. The yellowish color is the reflected sunlight which reaches us. The intense stream of photons continues to push the dust away from the Sun, forming a tail.

The bluish tail consists of radicals and molecules blown out of the coma by the solar wind. This gas tail is bluish because of strong emission by the molecular ion CO^+.

The tail of a comet always points away from the Sun, since the "breezes" which cause the tail blow from the Sun (Figure 10.21). As a comet leaves the solar system, its tail swings around, and it leaves the system tail-first.

The larger meteoroids that are freed from the nucleus are not affected by the solar photons' shove (Figure 10.22). They continue to move around the Sun in the comet's orbit. If the Earth passes through the orbit, these particles cause a meteor shower (see section on meteors). The Leonid meteor shower which we discussed earlier,

FIGURE 10.22 Comet Arend-Roland on April 25, 1957. This photograph was taken just as the Earth crossed the comet's orbital plane. The thin spike pointing opposite the comet's tail is actually the meteoroid debris along the comet's orbit illuminated by sunlight. (Lick Observatory photograph.)

for instance, is associated with the comet 1866 I, which has a 33-year orbital period.

Since a comet's nucleus slowly sublimes, the comet cannot last forever. Study indicates that cometary nuclei seldom exceed a few tens of kilometers in diameter, and lose several meters of diameter at each pass through the solar system. Most comets do not last more than 100 close approaches to the Sun, according to experts. How, then, do we explain short-period comets? If the solar system is five billion years old, how can we see a comet like Halley's Comet, which has a 75-year period? It should only last 7,500 years. Let's talk about the origin of comets.

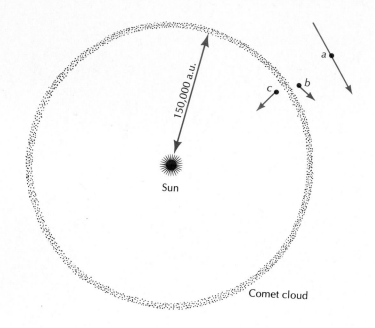

FIGURE 10.23 Oort's theory of comets. The solar system is surrounded by a giant cloud of comets. Every few million years a passing star (*a*) tears comets from the cloud. Some comets (*b*) then leave the system forever; others (*c*) fall in toward the Sun.

150,000 a.u.

Sun

Comet cloud

The Origin of Comets

One solution to the comet problem was proposed by Jan Oort in 1950. Oort suggested that the solar system is surrounded by a spherical shell, containing billions of comets, out at distances as great as 150,000 astronomical units from the Sun (Figure 10.23). This distance is a significant fraction of the average distance between the Sun and other stars. According to Oort, on a time-scale of millions of years, stars pass close to this cloud and give a few comets a tug. Some of these comets will fall in toward the Sun and become long-period comets. Since a star can pass the cloud any which way, we expect the orbits of long-period comets to be oriented at random, as they are.

Occasionally, a long-period comet will pass very close to one of the giant planets, especially Jupiter (Figure 10.24). If the encounter is just right, Jupiter will slow down the comet and trap it in a short-period orbit. Or Jupiter could propel the comet out of the system forever, of course. The theory says that the short-period comets we see today were captured into their orbits by a Jupiter encounter rather recently. These comets would be captured in the plane of the solar system.

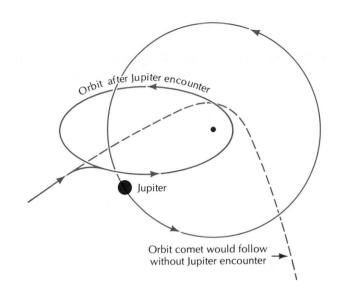

FIGURE 10.24 Capture of a long-period comet into a short-period orbit by encounter with Jupiter.

Orbit after Jupiter encounter

Jupiter

Orbit comet would follow without Jupiter encounter

Very recently, an interesting alternative theory has been proposed. According to this theory, there was a Saturn-sized planet that orbited the sun at 2.8 astronomical units until about 16 million years ago. Then, for some unexplained reason, it exploded. The comets are the debris blown out by the planet. This is at least plausible, since the comets have compositions similar to that of the giant planets, as we mentioned earlier. How did the date of 16 million years for the explosion come about? Many of the long-period comets we see today have orbital periods that cluster around 16 million years. The idea is that the explosion propelled these comets away from the Sun, and they are just now falling back to their origin point. Comets with periods of 15 million years or 17 million years would not be seen now, 16 million years after the explosion. Much more evidence is required to help decide between the Oort theory and the explosion theory. We will discuss other evidence for a missing planet later.

In Chapter 9 we found that there are two groupings of planets: terrestrial planets and Jovian planets. In this chapter, we have discussed asteroids, meteors, and comets, as well as regularities of the solar system. What does all this mean? To close out the chapter, we will try to tie all the discussion together.

THE ORIGIN OF THE SOLAR SYSTEM

How did our solar system begin? What processes formed the planets and set them moving in their very regular orbits? This is a question that has bothered scientists for many centuries. It is also a question that brings forward an unnecessary conflict between science and theology. There is no question that the scientific theories cannot be reconciled with a literal interpretation of the Old Testament story of creation. However, a scientific study of the origin of the universe and of the solar system is not incompatible with belief in God. A possible concept is that God created the raw materials and natural law. Then the system evolved from these raw materials according to the natural law. The scientific study of the origin of the system is a study of the natural law that led to the system.

The solar system began from a dense, rotating cloud of gas (or **nebula**) with a chemical composition like that of the Sun today. The nebula had to be dense, so that it could collapse under its own gravitational influence (Chapter 15). Each little blob of mass in the cloud attracted every other blob according to the universal law of gravitation put forward by Newton. The gravitational force between the masses pulled them together and the cloud shrank (Figure 10.25). As the cloud got smaller, its speed of rotation increased, since angular momentum had to be conserved.

As the central part of the nebula collapsed and compressed, it got hotter. In the end, this central part of the cloud formed the Sun. Most of the mass of the cloud wound up in the Sun. Today the Sun has only a few per cent of the angular momentum of the entire solar system, since it rotates so slowly. But if we believe this nebular hypothesis that the Sun formed from a collapsing gas cloud, it should have a much larger fraction of the angular momentum of the solar system than it does. In short, the Sun should rotate much faster than it does.

The Sun's slow rotation was considered to be an argument against the nebular hypothesis many years ago. However, the discovery of the solar wind provided an effective means by which the Sun's rotation may have been slowed down. As the solar wind flows outward from the Sun, it carries away angular momentum. Acting for several billion years, the wind could have carried away the majority of the Sun's original angular momentum. Thus the discovery of the solar wind was important for the theory of the origin of the solar system.

The direction of spin of the original cloud is responsible for the fact that the Sun, the planets, and the smaller bodies in the system almost all rotate and revolve in the same direction. The rapid

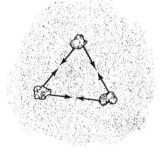

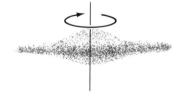

FIGURE 10.25 The solar system forms from a large cloud of gas that collapses. Rotation in the original cloud leads to the rotations and revolutions in the solar system.

spin of the nebula would have caused it to become very oblate, thus accounting for the flatness of the solar system.

Interstellar space is filled with grains: tiny material particles that were probably formed in the outer layers of cool stars and subsequently blown out into space. These grains would have existed in the original solar-system nebula. In theory, the grains would have been cold, and gasses in the cloud would have condensed and frozen on their surface, causing fluffy structures (Figure 10.26). With time, these fluffy grains would collide and stick together, resulting in ever larger clumps. The theory predicts that, as the clumps grow, they settle toward the equatorial plane of the rotating nebula, forming a thin sheet of particles.

Recent studies have shown that such a sheet is unstable, and will break up into large clumps a few kilometers in size. These large

FIGURE 10.26 Planets form because of the collision and sticking together of small grains (a) in the primordial solar system. The growing particles fall toward the plane of the original cloud (b) forming a loose disk of material. The disk breaks up into asteroid-sized bodies (c), which cluster together (d), collide (e), and coalesce (f) into planet-sized bodies (g). The planet-sized bodies have enough gravity to collect gas from the nebula (h). The result is a primordial planet (i). (From *The Origin and Evolution of the Solar System* by A. G. W. Cameron. Copyright © 1975 by Scientific American, Inc. All rights reserved.)

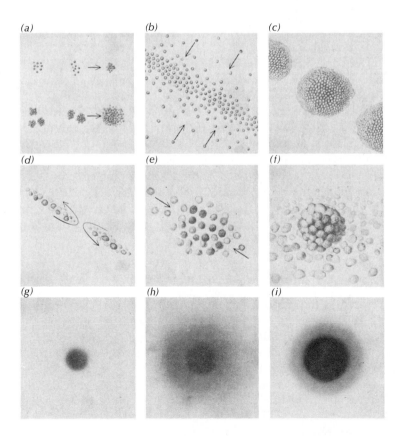

clumps would be near the midplane of the nebula and would be rather numerous. The next step in the process has these clumps collecting into gravitationally bound clusters, which also collide and intermingle. After many collisions, planet-sized solid bodies are built up in the ecliptic plane.

Very near the primitive Sun, it would have been too hot for hydrogen and helium, for instance, to freeze out on the grains. As a result, the planet-sized bodies that coalesced near the primitive Sun would have a relatively small amount of volatile materials. We would expect these planets to consist of the less-volatile light elements, like silicon, aluminum, and magnesium. Farther out from the primitive Sun, there was less heat available to vaporize the volatile materials. The bodies formed there would have had a high percentage of these substances. Thus the location of each planet determined its composition. This is why there are terrestrial and Jovian planets.

Probably when the final chunks that formed the planets began to collide, their energy of motion was converted to heat and the planets began as hot liquid bodies. The present-day internal structures were formed by gravitational differentiation (see above) before the bodies cooled.

Here and there in our Galaxy, we see stars known as T Tauri stars (see Chapter 14), which we believe to be stars in the early stages of their lives. These stars are observed to lose mass very rapidly, in a process we call the T Tauri wind. The T Tauri wind may have blown away the last vestiges of the solar-system nebula after planets were formed.

We have described in very brief, loose terms a theory of the formation of the solar system. There are many problems and questions which cannot be discussed without devoting an entire book to the subject; so we will have to stop with this summary. There is one remaining question that is interesting to mention in closing the chapter.

Bode's Law and the Missing Planet

It was long thought that Bode's law gave a clue about the origin of the solar system, and if so, the nebular hypothesis would need to account for this regular spacing of the planets. There is evidence today that this is not the case: the primitive solar system did not follow Bode's law.

A series of papers written since 1972 by M. J. Ovenden of the University of British Columbia shows that in our solar system, planetary perturbations, acting for billions of years, will naturally alter any original arrangement of distances from the Sun into a Bode's-law arrangement. Once the planets are in that arrangement, the effects of perturbations are minimum, and further changes will be very small. These papers state that the solar system is not in a stable configuration today. This is a surprising result, because the time needed to reach a stable configuration is short compared to the age of the system according to Ovenden. The system therefore should have had time to become stable.

Ovenden concludes that the system would be almost stable if there were a Saturn-sized planet in the asteroid belt. He then asked, what would the solar system look like with such a planet present, and how long would it take that configuration to change into the configuration that exists today? His conclusion is that it would take roughly 16 million years to reach our present arrangement. Does this mean that a large planet was present in the past, and that it disrupted 16 million years ago?

This is an interesting possibility. As we mentioned above, it could account for the comets, for instance. The asteroids might be the remains of the core of the planet. In fact, the largest asteroids could conceivably be the planet's moons. No doubt, Ovenden's theory will stir a lot of interest for the coming years.

We have now completed our survey of the solar system. Before we proceed to the rest of the universe, we will take one final look at the Earth, and talk about solar-terrestrial relations.

REVIEW QUESTIONS

1. What are three "regularities" of the solar system?
2. Compare and contrast terrestrial versus Jovian planets.
3. What is the Titus-Bode law? How did the discovery of asteroids contribute to the "law"?
4. What is a meteor shower?
5. Differentiate between the terms meteoroid, meteorite, and meteor.
6. Describe the visual appearance of a comet. How do cometary orbits differ from planetary orbits? What is the internal structure of a comet?
7. Describe the modern theory of the origin of the solar system.

11

THE SUN AND THE EARTH

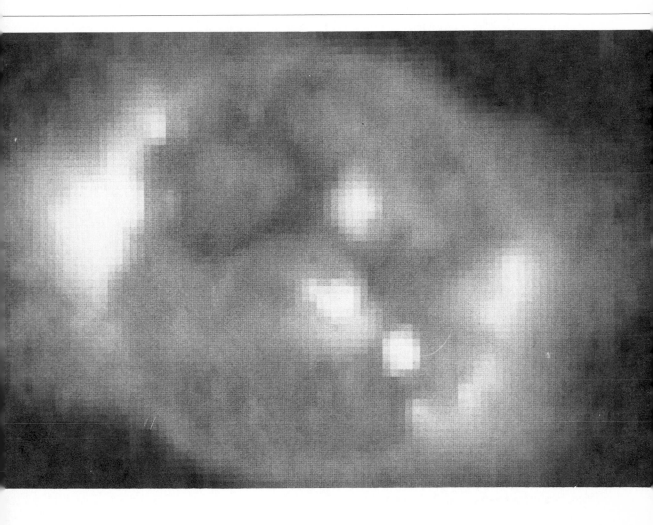

Most of us enjoy loafing on the beach on a balmy summer day, soaking up the warmth of the Sun. It is easy to be lulled asleep by the soothing sounds of the waves gently breaking on the shore. Under such idyllic circumstances, we might even contemplate the meaning of it all. Where does the Sun's energy come from? How does it affect our environment and our lives?

There is a great deal of activity around us as we bask at the beach, and the Sun is the source of that activity. The gentle breeze and the rolling waves take energy from the Sun. The gulls, soaring and diving for fish, are just one small link in the food chain that converts energy from the Sun into life-sustaining energy. Without the Sun, the Earth would be a cold body like Pluto, lifeless, traveling forever through the darkness of space.

The total output of energy from the Sun seems to be very steady. It is this output that is responsible for our planet's hospitable climate. The Sun varies, too. The short-wavelength radiation changes with time in a slow, steady rhythm. In addition, flares cause a short-term, large increase in the short-wavelength solar ultraviolet and X-ray radiation. Fortunately for us on Earth, the variable part of the radiation accounts for only a tiny fraction of the total solar output.

In this chapter, we will talk about the energy and particle output of the Sun. We will discuss how this output may help create our climate, and how variations in the output affect weather and climate. In addition, the solar wind interacts with the Earth's magnetic field, creating a magnetosphere; we will describe the chief characteristics of this feature of the planet.

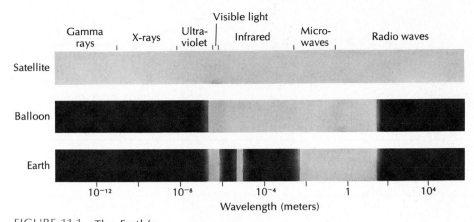

Gamma rays | X-rays | Ultra-violet | Visible light | Infrared | Micro-waves | Radio waves

Satellite

Balloon

Earth

10^{-12} 10^{-8} 10^{-4} 1 10^4

Wavelength (meters)

FIGURE 11.1 The Earth's atmosphere is an effective shield for most wavelengths of radiation (dark shading). As one goes to balloon or satellite altitudes, more and more radiation is accessible. (From *Observatories in Space* by A. I. Berman. Copyright © 1963 by Scientific American, Inc. All rights reserved.)

SOLAR PHOTONS

Photons from the Sun are the main source of heat input to the Earth's atmosphere. When we study the Sun from ground-based observatories, we find a few narrow bands of wavelengths that can get through the atmosphere. The visible part of the spectrum is one such band, and carries the majority of the Sun's energy. This radiation heats the ground and affects the weather near the surface.

Atmospheric scientists are interested in the wavelengths that do not reach the ground (Figure 11.1). These photons are absorbed in the atmosphere, heating the gas and determining its chemistry. The most interesting region of the solar spectrum is the so-called XUV region, which includes X-rays and ultraviolet. These XUV photons have sufficient energy to break apart the molecules and ionize both the atoms and molecules in the region of the atmosphere where they are absorbed.

The XUV part of the spectrum is studied from space vehicles that get above the absorbing layers of the atmosphere. Unlike the visible part of the Sun's spectrum, the XUV does not have absorption lines. Instead, the spectrum is composed entirely of emission lines. The character of the spectrum changes around 1800 Ångstroms. At longer wavelengths it is an absorption spectrum; at shorter wavelengths it is an emission spectrum. The emission lines come from the ultrahot rarified gas in the Sun's chromosphere and corona.

If one looks at an XUV picture (Figure 11.2) of the Sun made by a space vehicle above the region of the Earth's atmosphere where

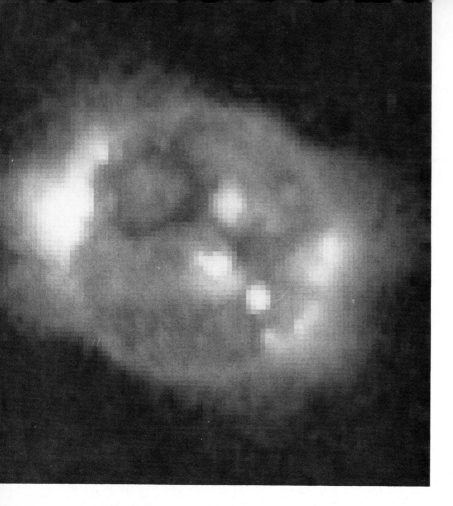

FIGURE 11.2 An image of the Sun at an extreme ultraviolet wavelength. Most of the radiation at this wavelength arises in the Sun's hot corona. The bright areas are active regions. (National Aeronautics and Space Administration.)

these short wavelengths are absorbed, it is obvious that the radiation does not come from the entire disk, as it does in visible light. Instead, these short wavelengths originate predominantly from active regions. This is an important fact: it means that the amount of XUV radiation varies with solar activity. When the sunspot number is large, the number and intensity of active regions is large, and the quantity of XUV radiation emitted is large. In addition, there is frequently one band of longitudes on the solar disk that has larger active regions than other longitudes. Therefore the level of XUV varies as the Sun rotates. The Earth receives more XUV from the Sun when the longitude of greatest activity faces the Earth than when it faces away from the Earth. Figure 11.3 illustrates the variation of two different wavelengths of solar radiation as the Sun rotates. Since the Sun rotates roughly once per month, the period of the variations in the graphs is roughly one month.

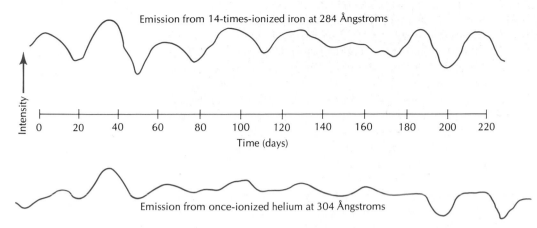

Emission from 14-times-ionized iron at 284 Ångstroms

Intensity →

| | | | | | | | | | | | |
0   20   40   60   80   100 120 140 160 180 200 220

Time (days)

Emission from once-ionized helium at 304 Ångstroms

FIGURE 11.3 Daily measurements of the total intensity of the Sun in two extreme ultraviolet wavelengths. Time is indicated in days from an arbitrary starting day. Notice that the high points in the curve occur roughly every thirty days, approximately the rotation period of the Sun.

One big question that both solar scientists and Earth scientists wish to answer is whether the total solar radiation varies. The so-called solar constant is the total energy that the Sun emits in the visible as well as the ultraviolet, infrared, radio, and XUV. The overwhelming majority of this energy lies in the visible part of the spectrum. Charles Greeley Abbot devoted a long lifetime to the measurement of the solar constant. He claimed to have detected a tiny variation with a 273-month period (note that this is the length of two sunspot cycles). Many scientists doubt Abbot's results, and considerable effort will be expended in coming decades to improve the measurement of the solar constant.

These measurements are important for an understanding of the Earth's climate. The solar constant determines the Earth's average temperature. A small decrease in the heat that the Earth receives from the Sun could plunge the planet into another ice age. In fact, a major question that remains to be answered is whether long-term variations of the Sun are responsible for the major periods of glaciation that have occurred on Earth every 100 million years or so.

We have now talked briefly about solar photons. Next let's look at the Earth's atmosphere, which is strongly influenced by these photons.

THE EARTH'S ATMOSPHERE

Within about 100 kilometers of the Earth's surface, the atmosphere is constantly stirred by winds, and its composition is quite uniform. If we take a sample of air near the surface and remove the water vapor, we will find it to consist of about 78% molecular nitrogen (N_2), 21% molecular oxygen (O_2), 0.9% argon, 0.03% carbon dioxide (CO_2), 0.0005% helium, and other gasses in tiny amounts. The real atmosphere is not dry, of course, but consists of between 0.1 and 1% water vapor.

Above 100 kilometers, the atmosphere is still: there are no winds. The composition of this part of the atmosphere varies with height, because of the physical process called gravitational differentation (Chapter 10). This process pulls the more massive molecules, such as oxygen, closer to the Earth than the light constituents such as helium.

The hydrogen in the upper reaches of the atmosphere comes from water vapor, which is carried to the 100-kilometer level by winds and is then torn apart (dissociated) by high-energy solar photons. Some of the hydrogen resulting from the dissociation of water then moves up into the upper atmosphere. Because hydrogen is the lightest atom, gravity exerts the smallest force on it. As a result, hydrogen is found farther from the Earth than nitrogen and oxygen atoms (Figure 11.4).

The XUV radiation from the Sun interacts with the atoms and molecules in the atmosphere. The photons can dissociate the molecules, and ionize both atoms and molecules. The energy of the photons ultimately becomes energy of motion of the atoms, molecules, and free electrons. As we discussed in Chapter 7, temperature is related to the energy of motion of molecules in a gas. Thus, if photons increase the energy of motion, they increase the temperature; that is, they heat up the atmosphere. Photons of various wavelengths are absorbed at different heights in the atmosphere; so they add their heat at different layers. This heat is conducted to lower layers.

Figure 11.5 shows how temperature changes with height in the Earth's atmosphere. Note that temperature is plotted across the bottom of the graph, and height is plotted up and down. The temperature makes a couple of "wiggles" with increasing height, defining the various layers of the atmosphere.

At a height of about 35 kilometers is a layer of ozone (O_3) molecules. This layer shields the ground from ultraviolet radiation at wavelengths between 2000 Å and 3000 Å. The existence of this layer is important for our survival, since these ultraviolet wavelengths are harmful to living beings. The ozone layer is formed when sunlight dissociates O_2 molecules to give free oxygen. Other O_2 molecules interact with the free O atoms to give O_3. This process occurs in a so-called triple collision (Figure 11.6). That is, an oxygen atom, O, an oxygen molecule, O_2, and any other atom or molecule must collide for the ozone-forming reaction to occur. The third body is not changed chemically: it is a catalyst for the reaction.

Ozone can only be formed low in the atmosphere, where the density is high enough for triple collisions to occur often. The top of the ozone layer is the level where such collisions are rare. The bottom of the layer is where all the solar photons that can dissociate oxygen have been used up. There are no more photons for additional dissociations. The atmosphere is relatively warm in the ozone layer because of the large amount of solar radiation absorbed there.

At about 100 kilometers, solar radiation forms a layer of ionized atoms, called the **ionosphere**. The ionization exists in a layer because, at the top of the layer, the atmosphere is so rarefied there are few atoms to ionize, and at the bottom all the ionizing photons are used up. The free electrons in the ionosphere act as a mirror to reflect radio waves from a station back toward Earth. This reflection process allows radio transmission over long distances.

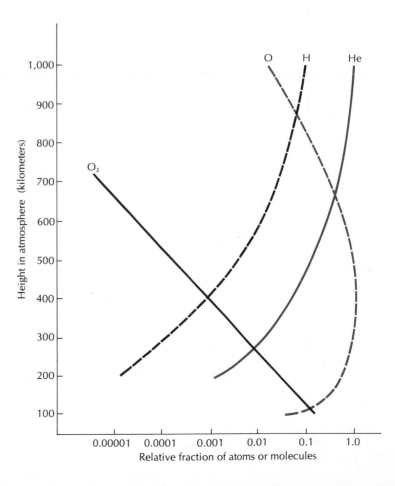

FIGURE 11.4 Composition of the Earth's atmosphere with height. The relative fractions of the light elements hydrogen and helium increase greatly with height, whereas the heavier oxygen decreases in relative abundance.

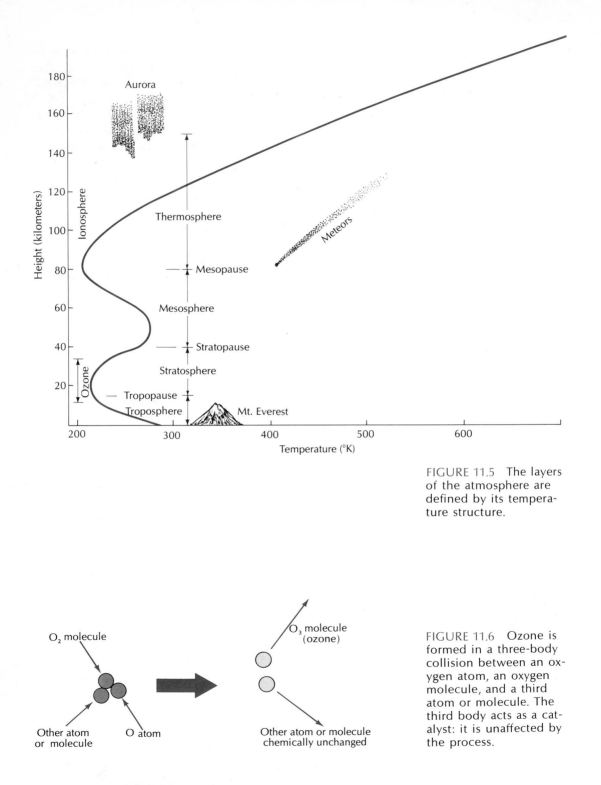

FIGURE 11.5 The layers of the atmosphere are defined by its temperature structure.

FIGURE 11.6 Ozone is formed in a three-body collision between an oxygen atom, an oxygen molecule, and a third atom or molecule. The third body acts as a catalyst: it is unaffected by the process.

The Sky Is Blue

Why is the sky blue? Molecules in the Earth's atmosphere scatter blue light much more efficiently than other colors. Thus photons of blue light, traveling toward the Earth's surface from the Sun, interact with molecules in the atmosphere and wind up traveling in all directions (Figure 11.7). The bright blue sky results.

At sunset, the Sun often appears red. This is because the blue, green, and yellow light is scattered in the long path through the atmosphere, and only the orange and red light reaches us. The Sun does not appear red when it is high in the sky because the light has to travel a shorter path through the atmosphere than at sunset, and the scattering is less.

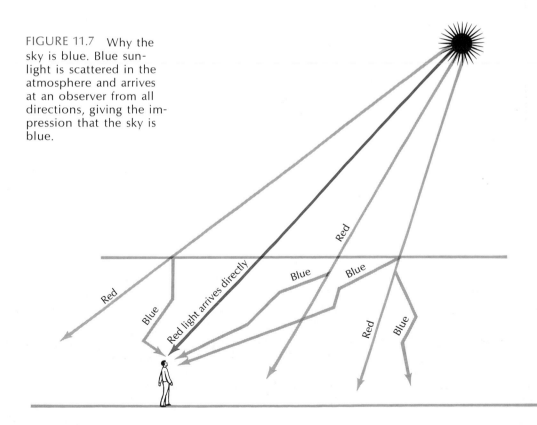

FIGURE 11.7 Why the sky is blue. Blue sunlight is scattered in the atmosphere and arrives at an observer from all directions, giving the impression that the sky is blue.

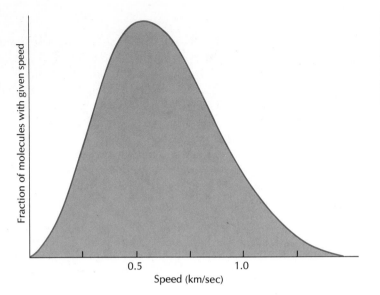

FIGURE 11.8 A plot of the relative speeds of oxygen molecules at room temperature. Note that most molecules have a speed around 0.5 kilometers/second.

Escape of the Atmosphere

Gas molecules do not all move with the same speed. Actually, the temperature of the gas is a measure of the average speed of the molecules. Some molecules move faster than the average, and some move slower (Figure 11.8). A few molecules have high-enough speeds that they can escape from the Earth's gravitational influence.

To understand this, let's think of someone throwing a ball up into the air (Figure 11.9). The person throws the ball upward; it rises a certain distance, stops, then falls back to be caught. The height the ball reaches depends on the speed with which it is thrown. The greater the speed, the higher the ball goes. Even the best pitchers in baseball cannot throw the ball much faster than 160 kilometers/hour (or 0.04 kilometers/second). However, if we could throw the ball upward at 11 kilometers/second, it would never come back down. It would continue to travel and leave the Earth entirely. The 11 kilometers/second is called the Earth's **velocity of escape**.

Molecules near the Earth's surface cannot escape even if they move at 11 kilometers/second, however. Before they travel too far, they collide with another molecule and change speed and direction of motion. However, high in the atmosphere, where hydrogen is the chief constituent, the gas is so rarefied that collisions are rare. There, atoms and molecules with escape velocity that are moving upward will leave the atmosphere.

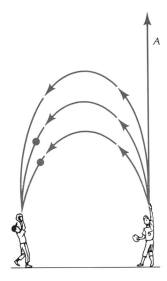

FIGURE 11.9 Escape of a ball from the Earth's surface. The height to which the ball goes before falling back to the Earth depends on its speed when thrown. If the ball could be thrown at 40,000 kilometers/hour (A), it would leave the Earth. Of course, even the best fast-ball pitchers can throw only about 160 kilometers/hour.

The Earth's atmosphere will not escape completely: the planet's gravity is too high, and the average speed of the gas molecules is too small. However, a less-massive planet, like Mercury, that travels in an orbit near the Sun could not retain an atmosphere for long.

With this brief look at the Earth's atmosphere, let's push on and talk about the solar wind in greater depth.

SOLAR WIND

The story of the discovery of the solar wind is interesting, because it is a prime example of how theory and observation have worked closely together to make a fundamental discovery.

Studies of comets showed that photons from the Sun could not be the only agent producing cometary tails (Chapter 10). In the early 1950s, the German astrophysicist Ludwig Biermann stated that his studies of comets led inescapably to the conclusion that the Sun constantly emits particles in all directions. Biermann came to this conclusion by realizing that the motion of ions in cometary tails away from the Sun and away from the comet's head could not be explained by the pressure of solar radiation, which does explain the movement of dust in the tails. Instead, material particles are required that could impart more momentum to the ions in a collision.

Soon thereafter, Eugene Parker of the University of Chicago showed on theoretical grounds that the solar corona could not be static. He found that, because of the corona's high temperature, it must be constantly expanding, sending its material away from the Sun in what he called the solar wind. The solar wind blows away from the Sun in all directions, even enveloping the Earth. Thus, both observations of comets and theory indicated the presence of this stream of matter blowing through the solar system. Direct measurement came quickly.

In 1962, NASA launched the Mariner II spacecraft toward Venus. For over 100 days this first successful interplanetary spacecraft sped toward its target, measuring particles in space as it went. The results telemetered to anxious scientists on earth confirmed the existence of the solar wind. During the 100 days that measurements were made, the average speed of the solar wind was around 500 kilometers/second, and it contained a few particles per cubic centimeter. Thus the indirect inferences from the cometary studies and coronal theory were verified by direct measurements in the solar wind; these are called in-situ measurements.

As often happens in scientific investigations, a new and exciting discovery was made. The speed of the solar wind was found to vary between about 300 kilometers/second and 800 kilometers/second in the three months that Mariner II took observations. Furthermore, disturbances in the Earth's magnetic field (geomagnetic disturbances) occurred soon after changes were measured in the solar wind.

Since the time of Mariner II, numerous in-situ measurements have been made of the speed, density, chemical composition, temperature, and other properties of the solar wind. Today we know a great deal about the outflow from the Sun. We know, too, that the solar wind exerts a profound effect on the Earth's environment as it blows past.

THE MAGNETOSPHERE

The Earth has a magnetic field. This fact was demonstrated by William Gilbert in 1600. A magnetized sphere, alone in space, has a magnetic field that is called a dipole. It looks like the field of a bar magnet (Figure 11.10).

If we place a bar magnet under a sheet of paper and sprinkle iron filings on the paper, they align themselves with the field of the magnet and trace out its shape. Scientists have introduced a concept known as the **line of force**. A line of force is one of many lines drawn in a magnetic field whose directions are the direction of the field at each point. The iron filings spread over the paper covering a bar magnet are always parallel to lines of force.

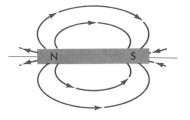

FIGURE 11.10 The magnetic field of a bar magnet.

Satellites orbiting around the Earth have traced out its magnetic field (Figures 11.11 and 11.12). Near the surface of the planet, the field looks like a simple dipole; but further out from the surface, it becomes extremely distorted. The over-all field of the Earth is comet-shaped, with the Earth in the place of the nucleus and with a tail pointing away from the Sun. Why does the field have such a shape?

The Earth's magnetic field is distorted by the solar wind. The solar wind is an example of a **plasma**. The plasma state of matter is when the matter is gaseous, and the atoms of the gas are ionized. Thus the plasma consists of positively charged ions and many negatively charged free electrons. The Sun is composed of plasma, for instance. Plasma cannot flow across lines of force in a magnetic field. It can only flow along lines of force, for the following reason. When a charged particle moves perpendicular to a line of force, the magnetic field exerts a force on the particle and causes it to orbit

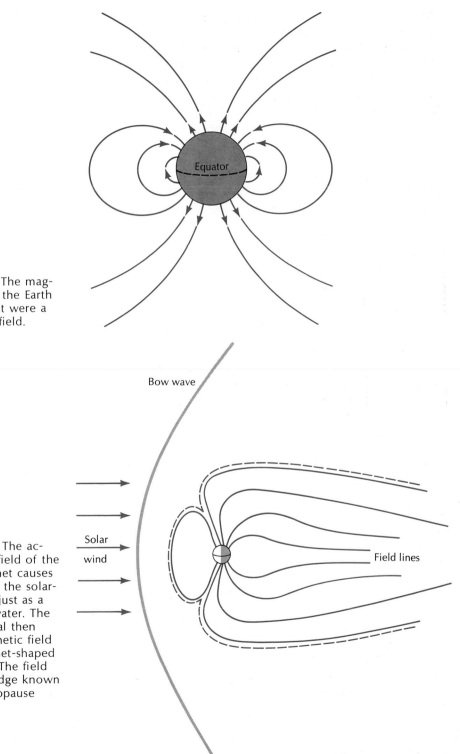

FIGURE 11.11 The magnetic field that the Earth would have if it were a perfect dipole field.

Equator

Bow wave

Solar wind

Field lines

FIGURE 11.12 The actual magnetic field of the Earth. Our planet causes a bow wave in the solar-wind material just as a boat does in water. The moving material then drags the magnetic field out into a comet-shaped configuration. The field has an outer edge known as the magnetopause (dotted line).

around the line of force. When the charged particle moves parallel to the field line, there is no force on the particle from the magnetic field. Charged particles trapped in a magnetic field perform a corkscrew or helical motion around the lines of force (Figure 11.13).

If fast-moving plasma encounters a magnetic field, the particles do not cross field lines. In fact, the particles can actually exert a pressure on the field and change its shape. This is what happens to the Earth's magnetic field. The high-velocity solar-wind plasma encounters the Earth's magnetic field, and "blows" the lines of force back into a comet-like tail. The wind blows around the outer boundary of the magnetic field; this boundary is known as the **magnetopause**. The Earth's magnetic field and the plasma it contains is called the **magnetosphere**.

The most energetic particles in the magnetosphere are the particles in the **Van Allen belts** (Figure 11.14). These belts of trapped particles were discovered by James Van Allen in 1958, using a geiger counter flown on Explorer 1, the first successful American satellite. Scientists are still puzzled by these belts. The particles have much more energy than typical solar-wind particles. We do not yet know where that energy comes from.

Mercury and Jupiter also have magnetospheres (Figure 11.15). Mercury's is less extensive than Earth's, and Jupiter's is much more extensive. Scientists expect to find magnetospheres on Saturn, Uranus, and Neptune when they explore these planets.

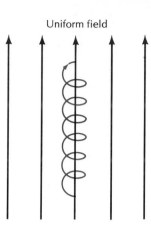

FIGURE 11.13 Motion of an electron in a uniform magnetic field.

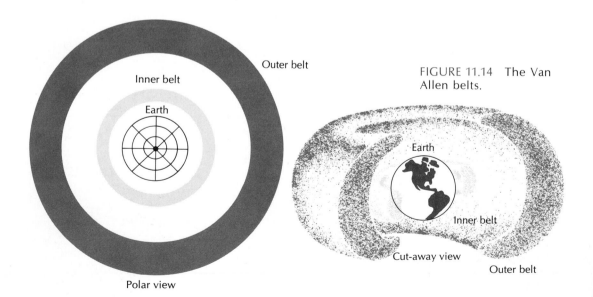

FIGURE 11.14 The Van Allen belts.

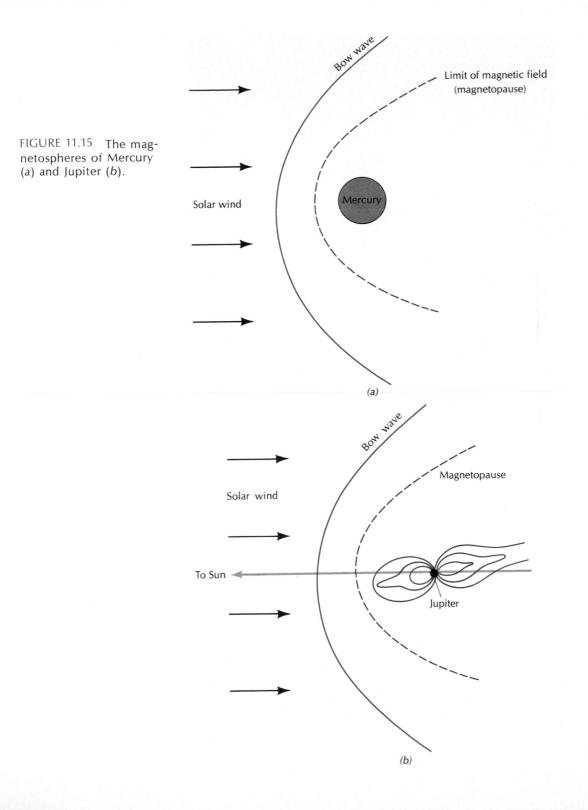

FIGURE 11.15 The magnetospheres of Mercury (a) and Jupiter (b).

FIGURE 11.16 A spectacular aurora observed in August 1950 with a wide-angle camera. (Yerkes Observatory.)

Aurora

The **aurora borealis** or northern lights is a phenomenon of the north polar region of the Earth (Figures 11.16 and 11.17). A glow can be seen in the sky that flickers or slowly pulses. The displays are most often green, but can also be red. The green color is due to the stimulation of atomic oxygen by energetic protons and electrons from the magnetosphere. The red is due to molecular nitrogen and oxygen. The emission that appears as auroras takes place between 70 and 100 kilometers above the Earth's surface.

Auroras are most common during the maximum of solar activity, at which time displays can sometimes be seen from the northern parts of the conterminous U.S. The University of Alaska at Fairbanks is one of the great centers for the study of auroras. At the latitude of that university, aurora displays are common no matter what the solar activity level happens to be.

FIGURE 11.17 An auroral form known as a rayed arc. (Courtesy W. A. Feibelman.)

Geomagnetic Substorms

The magnetosphere pulses with disturbances known as **substorms**. According to current theory, the solar wind flowing by the Earth causes magnetic energy to be stored in the tail of the magnetosphere. Periodically—roughly four times each day—the energy is released in a substorm.

The substorm produces curtain-like auroral forms, and increases in giant electrical currents that flow in the polar ionosphere. These currents produce magnetic fields that cause compass needles to waver for a short time. Energetic particles from the magnetosphere produce X-rays in the Earth's atmosphere. The explosive release of energy in a substorm reminds astronomers of a small solar flare. Scientists are pursuing the similarities in the hope of better understanding both phenomena.

The phenomena of the magnetosphere are varied and complex: far too much so to be easily covered in an astronomy textbook. So, with this brief mention, we will proceed to a description of the terrestrial effects of solar flares.

FIGURE 11.18 A solar flare observed in hydrogen light. (Hale Observatories.)

SOLAR FLARES AND THE EARTH

What happens on the Earth when a flare erupts on the Sun? (See Figure 11.18.) The flare emits large quantities of short-wavelength X-radiation, which reaches the Earth quickly. This radiation increases the degree of ionization in the ionosphere very suddenly, and produces a series of effects collectively called a **sudden ionospheric disturbance** or SID. The most obvious aspect of an SID is that it disrupts transmission of shortwave radio signals, in what is called a shortwave fadeout. The Earth's magnetic field can be affected, too.

The flare produces an increase in the solar wind: both the number of particles and the velocity of the wind increase. When the blast of wind reaches the Earth, roughly 36 hours after the flare, a new series of disturbances occur. The solar-wind pressure against the outside of the magnetosphere increases, causing changes in the level of the magnetic field at the Earth's surface. The result is a magnetic storm, where compass needles waver erratically. The increased solar wind increases the electric field across the magnetosphere, which increases the currents flowing inside the magnetosphere, which in turn increases the brightness of auroras.

The very largest flares produce large quantities of protons with high energies. When some of these protons reach the Earth a few hours after the flare, they are funneled into the polar regions by the magnetic field, where they reach the ionosphere and increase ionization by collisions with atmospheric atoms. The result is a polar-cap absorption (PCA) event, in which radio transmission in the Earth's polar regions is blacked out.

Thus we see that a flare produces spectacular events on the Earth that take place during a period of up to two days. The effects on communications can be very severe.

THE MAUNDER MINIMUM

The sunspot cycle was discovered in 1848 (Chapter 8). Enough data existed in old records to trace the cycle back to about 1750. Before 1750 the record is very spotty. However, a careful study of all old records seems to indicate that the period 1645 to 1715 was unusual, in that sunspots were almost completely absent for the full 70 years (see Figure 8.23). This extended period is called the Maunder minimum, because it was carefully documented by E. W. Maunder in 1894.

Recently John A. Eddy has taken up the case of the missing sunspots, and has compiled a convincing mass of evidence to support it. Eddy's research is a beautiful example of scientific detective work. Basically, his argument boils down to the fact that there were some very careful and curious astronomers observing the sky during the Maunder minimum. A good example is Edmund Halley, who gave Halley's comet its name. One finds in the literature written by these scientists statements like: "I saw a sunspot the other day. It was an exciting event, since I have not seen one for twenty years, and I have looked for spots often." It is great luck that Galileo invented the telescope when he did, for if the invention had been delayed until after 1645, sunspots might not have been discovered until after the Maunder minimum was over in 1715.

The long minimum may not have been unique. Eddy describes much weaker evidence for another minimum lasting roughly from 1460 to 1550, and an even earlier one from 1100 to 1250. Thus the solar cycle may have a very long-period variation, several centuries in length. Eddy points out that we do not have enough evidence yet to say whether the extended minima occur cyclicly.

The Maunder minimum is more than a mere curiosity. The historical record suggests that the winters in Europe were unusually severe during this time. The period from the thirteenth century to the nineteenth century is often called the Little Ice Age, the coldest portion of which corresponds to the Maunder minimum. In

1750, European glaciers reached their largest size since the Quarternary Ice Age. Beginning in 1750, they have shrunk steadily. Tree-ring records in the Great Plains indicate that the Maunder minimum was a period of prolonged drought.

SOLAR-TERRESTRIAL RELATIONS

The relationship between the Maunder minimum and severe winters is just one of the many clues that solar activity influences our weather. There has been a great deal of interest recently in possible connections between solar activity and terrestrial weather. Those who think that the connections are real can point to many relationships that have been shown to be statistically significant. Critics, on the other hand, rightly argue that statistical connections are not enough: physical mechanisms to account for the connections have to be found. We cannot talk about all the suspected relationships here, but let's look at two very interesting ones.

As we discussed in Chapter 8, if we consider the magnetic nature of sunspots as well as the sunspot number, the sunspot cycle is 22 years long. A review of Figure 8.23 shows that there is a tendency for sunspot maxima to alternate, high, low, high, low, and so on. We speak of major maxima and minor maxima. If we plot the sunspot cycle with the sunspot number negative during the alternating minor maxima, we get a graph that looks like Figure 11.19. It turns out that at the times of sunspot minima going from negative to positive numbers, there is a major drought in the high plains just east of the Rocky Mountains. The great dust bowl of the 1930s occurred during one of these periods. Another such period fell in 1976, and another major drought is now in progress.

A second example is a relationship between the size of low-pressure systems observed in the Gulf of Alaska and geomagnetic activity. W. O. Roberts and his co-workers find that low-pressure systems observed in the Gulf of Alaska two to four days after a rapid increase in geomagnetic activity are significantly larger than average. A more detailed study has verified this result.

Both these observations present major puzzles. How can sunspot activity affect rainfall in the high plains? How can geomagnetic activity affect low-pressure zones? For the latter, there may not be a direct cause-and-effect relationship, but both events may occur as a result of some complex chain of events.

Mankind is curious about nature. To satisfy our curiosity, we study natural phenomena. In studying the Sun and its influence on

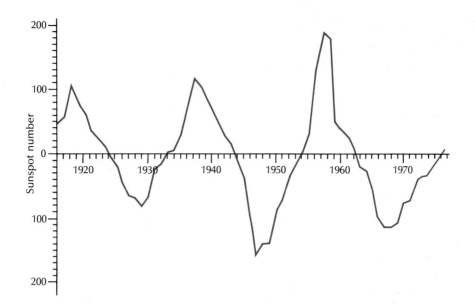

FIGURE 11.19   The double sunspot cycle.

the Earth, the payoff may be more than fascinating science; we may find major clues to help in our attempts to predict the Earth's weather and climate.

We have now reached a major milestone. We have completed our detailed look at the solar system, and are now ready to expand our horizons. Let's look at the rest of the universe.

REVIEW QUESTIONS

1. Describe the short wavelength spectrum of the Sun. In what ways does it differ from the visible spectrum?
2. How does gravitational differentation affect the composition of the Earth's atmosphere? Why is the lower 100 miles of the Earth's atmosphere not gravitationally differentiated?
3. Why is the sky blue?
4. Describe the solar wind. What causes it?
5. What is the magnetosphere? Why is it distorted from a dipole field shape?
6. What is an aurora?

7. How do solar flares affect the Earth?

8. What is the Maunder minimum?

9. Give an example of a suspected terrestrial phenomenon influenced by changing solar activity.

12

INTRODUCING THE STARS

On a clear, moonless night the sky appears to be filled with stars, which are grouped into the familiar constellations. At certain times of the year, we can see the Milky Way stretching across the sky. If we look at it with a pair of binoculars or a small telescope, we will see that the Milky Way is made up of billions of stars just at the limit of naked-eye visibility. All these stars are a part of our Galaxy, an unimaginably huge, flattened object. The naked-eye stars make up only one remote corner of our Galaxy. In the following several chapters we will build up a picture of the Galaxy. We will see that it contains stars, singly, in pairs, and in large groupings, together with a liberal quantity of gas and dust. We will learn about each of the constituents of the Galaxy, and will discuss its size, mass, and shape. Then, we will find that this immense thing is only one of billions of galaxies filling the universe. First, we will study the stars.

Each of the stars is a sun, like our own, pouring its radiation into space. Though we cannot see any other star as closely as we see the Sun, we can discover a lot about them by careful study. We can discover their temperatures, distances, sizes, and sometimes their masses. Among the brightest stars we see in the sky, we will find that some are bright mainly because they are close to us, others because they are extremely bright, though far away. Some of the faint stars we see are tiny, nearby objects; others are extremely bright objects at distances that begin to boggle the mind.

DESCRIPTIONS

To begin this chapter, let's talk about some of the most obvious things we can say about stars. Each star occupies a distinct position in the sky, and each star has a certain brightness. How do we describe these characteristics? We will answer this question in the following pages. First, however, let's put a label on each star.

The Names of Stars

In order to talk about one of the multitude of stars, we must be able to identify it in some unique way. In history, several different naming systems have been devised. The most colorful system, though probably the least practical, gives each star a proper name, which must be remembered.

Most of the proper names of stars come from the Arabic, even though the names of constellations are mainly of Greek origin. The names are often descriptions of the star's place in the constellation. Orion, one of the most conspicuous of the winter constellations, represents a mighty warrior and hunter (Figure 12.1). The bright red star marking the left shoulder of Orion is named Betelgeuse (pronounced beetle-juice), which is a slurred form of *Ibt al Jauzah,* meaning "the armpit of the central one." Rigel is from *Rijl Jauzah al Yusra,* which means "the left leg of the central one." The constellation Libra represents a balance, or scale (Figure 12.2). The two brightest stars in Libra are called Zubenelgenubi and Zubeneschamali, which literally mean the Southern Claw and the Northern Claw. These names are holdovers from Greek times, when Libra was not a separate constellation, but a part of Scorpio, the scorpion. These few examples show the kinds of proper names that have been given to stars. They also show that it is not practical to have to learn thousands of star names. Some other system will have to be used. Astronomers typically use only a handful of proper names: those of the very brightest stars.

In 1603, Johann Bayer published an atlas of the heavens called the *Uranometria*. In the atlas, he named the stars in each constellation with Greek letters (see Appendix E). The brightest star in each constellation he called α, the next β, the next γ, and so on. To identify a star, we can use Bayer's Greek letter together with the genitive form of the constellation name. Thus Betelgeuse is α Orionis (or α of Orion), Rigel is β Orionis, and the two brightest stars in Libra are α Librae and β Librae.

Somewhat after Bayer, John Flamsteed, the English Astronomer Royal, produced an atlas in which he numbered the stars in each constellation, beginning with the brightest. Today, we use Bayer's Greek-letter names as far as they go. In some constellations, like Cygnus, which has many bright stars, we use Flamsteed's numbers after we exhaust Bayer's designations (Figure 12.3).

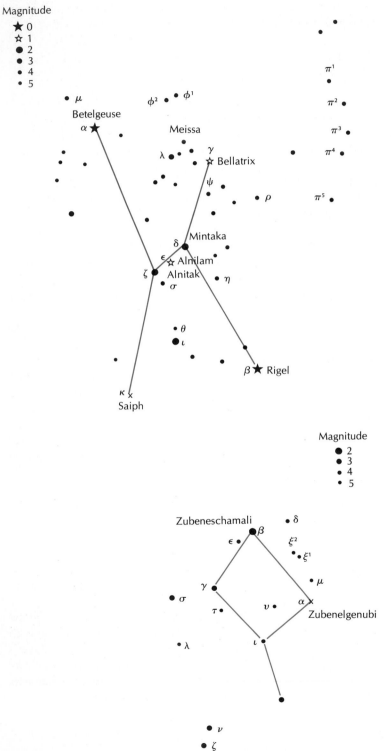

Magnitude

★ 0
☆ 1
● 2
● 3
● 4
· 5

FIGURE 12.1 The constellation Orion, with the names of its brightest stars noted. Bayer's Greek-letter names are also indicated.

Betelgeuse
α

Meissa

λ

γ

Bellatrix

ψ

ρ

μ

φ² φ¹

π¹

π²

π³

π⁴

π⁵

δ Mintaka

ε Alnilam

ζ Alnitak

σ

η

θ

ι

β Rigel

κ

Saiph

Magnitude

● 2
● 3
● 4
· 5

FIGURE 12.2 The constellation Libra. The proper names of two of the brighter stars are shown on the figure. It would not be practical to remember many such tongue-twisting names.

Zubeneschamali

β

δ

ε

ξ²

ξ¹

μ

γ

σ

τ

ν

α

Zubenelgenubi

ι

λ

ν

ζ

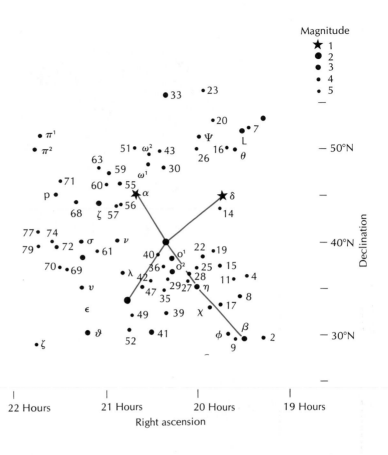

FIGURE 12.3 The constellation Cygnus, using Flamsteed's number designations for all but the brightest stars.

Since the time of these atlases, many catalogs have been compiled which list the positions and other characteristics of literally hundreds of thousands of stars. Fainter stars are often referred to by their number in one of these catalogs.

Positions in the Sky

Positions of stars and other objects on the sky are given by coordinates that are analogous to latitude and longitude on the Earth (Figure 12.4). The celestial equator, midway between the north and south celestial poles, marks the zero point of the celestial coordinate called **declination**, which is analogous to latitude. The declination of a star is its distance north or south of the equator, measured in degrees. The stars of the Big Dipper, for instance, lie between declination about 50° N and 62° N, the N meaning north of the celestial equator.

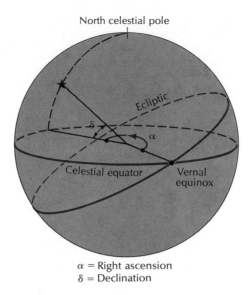

North celestial pole

Ecliptic

δ

α

Celestial equator

Vernal
equinox

α = Right ascension
δ = Declination

FIGURE 12.4 The right ascension and declination of a star are celestial coordinates equivalent to longitude and latitude on Earth.

The coordinate analogous to longitude is called **right ascension**. On Earth, the zero point of longitude is the Greenwich meridian, which passes through Greenwich, England. In the sky, the zero point of right ascension is the great circle that passes through the poles and the vernal equinox, the point where the Sun crosses the equator in its motion along the ecliptic from south to north. Right ascension increases eastward around the sky, and is measured in *hours*. Because of precession of the equinoxes, the vernal equinox moves slowly westward. As a result, the right ascensions and declinations of stars change slowly. In any catalog of star positions, the date on which the positions are valid, called the epoch of the catalog, is always indicated. Figure 12.3 shows Cygnus with right ascension and declination for the year 2000.

Brightness

We can now talk about individual stars by their names and positions in the sky. How do we describe the star's brightness? Historical practice has left us with a seemingly unusual scheme for ascribing a brightness to a star. It all started with Hipparchus.

Hipparchus compiled a catalog of stars in the second century B.C., in which he pigeonholed stars into six brightness classes, or **magnitudes**. Stars of the first magnitude were the few brightest stars in the sky, and stars of the sixth magnitude were those just

Table 12.1 Stellar Brightness Ratios Versus Magnitude Differences

Magnitude Difference	Brightness of Bright Star / Brightness of Faint Star
0.1	1.09
0.2	1.20
0.3	1.32
0.4	1.45
0.5	1.59
0.6	1.74
0.7	1.91
0.8	2.09
0.9	2.29
1.0	2.51
1.5	4.00
2.0	6.31
3.0	15.85
4.0	39.82
5.0	100.0
10.0	10,000.
15.0	1,000,000.
20.0	100,000,000.

visible to the naked eye. Stars were placed in the six magnitude classes very subjectively.

Study of the human organism since Hipparchus' day has revealed a remarkable fact. A series of stimuli that are related to one another in a geometric progression produce a series of responses that are related to one another in an arithmetic progression. This fact is true of our response to light, sound, pain, and so on. What exactly does this mean? Stellar brightnesses serve a good example. Hipparchus' magnitude classes are the arithmetic progression 1, 2, 3, 4, 5, 6: the numbers have a constant difference. The brightnesses of the stars, on the other hand, form a geometric progression; that is, the brightnesses of stars in any two successive magnitude classes are in a constant ratio to one another. First-magnitude stars are 2.512 times brighter than second-magnitude stars; second-magnitude stars are 2.512 times brighter than third-magnitude stars; and so on. Table 12.1 relates brightness ratios to magnitude differences.

The brightest stars have negative magnitudes. Stars 2.512 times brighter than first magnitude are zero-magnitude objects. Even brighter stars may have magnitude –1, and so on. Sirius has magnitude –1.4, the Sun –26.

To measure magnitudes, we go back to the photomultiplier tube discussed in Chapters 4 and 7. We use a telescope to focus the light of a star of known magnitude on a photomultiplier tube and measure the current produced. We then do the same for a star of unknown magnitude. The ratio of the currents is the same as the ratio of the brightnesses, which relate directly to a magnitude difference. Then, since we know the magnitude of one star and the magnitude difference, we can find the magnitude of the other star.

Now, we have talked about star names, positions, and brightnesses. The next step is to find how far away a star might be.

STELLAR DISTANCES

How far away are the stars? This is a question that arises again and again in astronomical problems. As astronomers studied objects that are farther and farther away—stars, star clusters, galaxies, quasars—different methods had to be found to calculate distances. Each new method relied on information gathered using the earlier methods. In the end, an interwoven chain of methods has been developed, with each link depending on those that came before it. Throughout the remainder of the book, we will encounter the links one by one.

Hold a finger at arm's length and look at it first with one eye, then with the other (Figure 12.5). The finger appears to shift back and forth relative to distant objects as you shift from one eye to the other. This is an effect known as **parallax**. The effect can be used to calculate the distance of nearby stars. The distance between our two eyes is replaced by the distance across the Earth's orbit (Figure 12.6). Astronomers look at a nearby star on some date, then look at it six months later, when the Earth has traveled halfway around its orbit. The star appears to shift slightly relative to distant stars. Half the total shift is the star's heliocentric parallax, or more simply, the star's parallax.

The heliocentric parallax of a typical nearby star is a tiny angle. Friedrich Wilhelm Bessel measured the first stellar parallax in the years 1837 to 1840. The star he considered was 61 Cygni (see Figure 12.3) and its parallax was found to be about 0″.3, that is, three-tenths of a second of arc, the size of a penny viewed from 13 kilometers. In subsequent years, more parallaxes were measured, the largest being for the nearest star, Proxima Centauri (0″.76).

The distance to a star can be found from its parallax (see Appendix A). The larger the parallax, the nearer the star; the smaller the parallax, the more distant the star (see Table 12.2). If a star has a

FIGURE 12.5 The concept of parallax demonstrated by viewing one's finger first with one eye, then with the other. The finger appears to shift back and forth relative to distant objects.

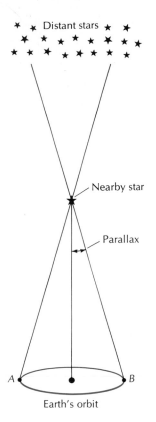

Table 12.2 Distance to a Star Versus Parallax

Parallax of a Star (seconds of arc)	Distance to the Star (parsecs)
1.00	1.00
0.75	1.33
0.60	1.67
0.50	2.00
0.40	2.50
0.30	3.3
0.20	5.0
0.10	10.
0.05	20.
0.02	50.
0.01	100.

FIGURE 12.6 Parallax of a nearby star. When viewed from vantage points on opposite sides of the earth's orbit (A and B), a nearby star appears to change its position relative to distant stars.

parallax of one second of arc, it is 206,265 astronomical units away. We call that distance one **parsec**. The term is derived from *par*allax of one *sec*ond. An older unit of distance is the light year, the distance light travels in one year. We will not use the light year in this book. Astronomers use parsecs for distances to stars in all their scientific work.

The nearest star, Proxima Centauri, has a parallax of 0″.76, as we said above. Its distance in parsecs is the **reciprocal** of this number, that is, 1 ÷ .76 = 1.3 parsecs. Since there are roughly 200,000 astronomical units in a parsec, Proxima Centauri is 260,000 astronomical units away. We found earlier that the solar system is mostly empty. The volume of our Galaxy near the Sun is even more empty: the distance to the nearest star is a quarter million times the distance from the Earth to the Sun, or 3.9×10^{13} kilometers.

We can measure the parallax of stars as far as 20 parsecs away with about 10% accuracy. However, for a star 200 parsecs away, the heliocentric parallax method is no longer satisfactory: such a star would have a parallax of 0″.005, an angle too small to be measured with any certainty. That angle is the size of our penny seen from 780 kilometers away. In 1969 there were 1,049 known stars (or systems of stars) nearer than 20 parsecs from the Sun.

Many of these nearby stars are faint objects, too faint even to be seen without a sizeable telescope. The first star to have its parallax measured, 61 Cygni, is a fifth-magnitude object, just slightly brighter than the faintest naked-eye star. How did Bessel choose 61 Cygni as a likely nearby star? Why not choose Sirius, the brightest star? The answer lies in the star's motion.

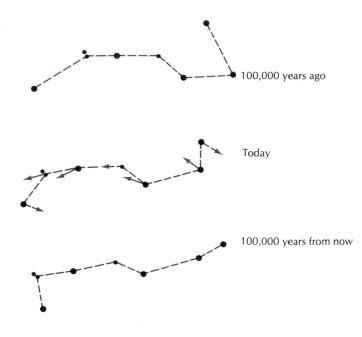

100,000 years ago

Today

100,000 years from now

FIGURE 12.7 The Big Dipper 100,000 years ago, today, and 100,000 years from now. The shape of the constellation changes considerably on this long time-scale because of the motions of the stars.

STELLAR MOTIONS

Proper Motion and Tangential Velocities

The constellations appear the same year after year, century after century. The Greeks referred to the "fixed" stars, because they thought the stars were attached to the celestial sphere. In fact, the stars do move through space, and the constellations change their shapes considerably in a few tens of thousands of years (see Figure 12.7).

In 1718, Edmund Halley measured the positions of the two bright stars Sirius and Arcturus, and compared his poisitions with those listed by Ptolemy. Halley found that, in the more than 1,500 years between the measurements, the stars had moved by substantial amounts. Sirius, for instance, had moved a distance equal to the size of the full moon, and Arcturus had moved almost twice as far. This motion is known as the star's **proper motion**.

Today, many hundreds of proper motions have been measured. The largest is that of a ninth-magnitude star known as Barnard's star. This star moves 10″ each year, or the Moon's apparent diameter every two centuries. More than 300 stars have proper motions in

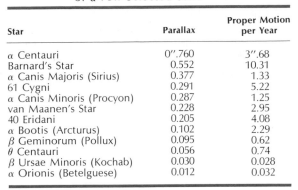

Table 12.3 Proper Motion and Parallax of a Few Selected Stars

Star	Parallax	Proper Motion per Year
α Centauri	0″.760	3″.68
Barnard's Star	0.552	10.31
α Canis Majoris (Sirius)	0.377	1.33
61 Cygni	0.291	5.22
α Canis Minoris (Procyon)	0.287	1.25
van Maanen's Star	0.228	2.95
40 Eridani	0.205	4.08
α Bootis (Arcturus)	0.102	2.29
β Geminorum (Pollux)	0.095	0.62
θ Centauri	0.056	0.74
β Ursae Minoris (Kochab)	0.030	0.028
α Orionis (Betelguese)	0.012	0.032

FIGURE 12.8 The tangential velocity of a star can be calculated if we know the star's proper motion and distance. For several stars with a given proper motion, the more distant one has the greatest tangential velocity.

Tangential velocities

Proper motion is the angular distance that a star moves in one year

Earth

excess of 1″ per year. Table 12.3 lists the proper motions of a few nearby stars.

The proper motion of a star is due to the star's motion perpendicular to the line of sight from our eye to the star. Think about several stars moving with the same speed, but lying at different distances from us. The nearest star will have the largest proper motion, of course. It turns out that 61 Cygni has a proper motion of 5″ per year. This is the largest proper motion of any naked-eye star, and one of the largest of any star. Bessel realized that the large proper motion indicated that the star was nearby; so it was a good candidate for parallax measurement.

The velocity corresponding to the proper motion is called the star's **tangential velocity** (Figure 12.8). It can be calculated if we know the star's distance and proper motion. The tangential velocity of 61 Cygni is 85 kilometers/second. Typical values are smaller than this, around 20 kilometers/second.

Radial Velocity

The motion of a star toward us or away from us—that is, along a radius of the celestial sphere—does not change its position on the sky (Figure 12.9a). To detect a star's radial velocity, we must photograph its spectrum and make use of the Doppler effect (Chapter 7). The typical star, like the Sun, has an absorption-line spectrum. If the star is moving away from us, all its spectrum lines are shifted slightly toward the red end of the spectrum; if it is moving toward us, its spectrum lines are shifted slightly toward the blue (Figure 12.9b). The shifts are small for stars. If a star is approaching us at 20 kilometers/second, a spectrum line at 5000 Ångstroms is shifted roughly 0.3 Ångstrom to the blue of its normal position.

We measure the radial velocity of stars from the Earth, which moves around the Sun with an average orbital speed of 30 kilometers/second. Furthermore, the direction that the Earth moves changes as it swings around the Sun. Thus the radial velocity that we measure for a star will change, depending on the time of year the measurement is made (Figure 12.10). Since astronomers know how

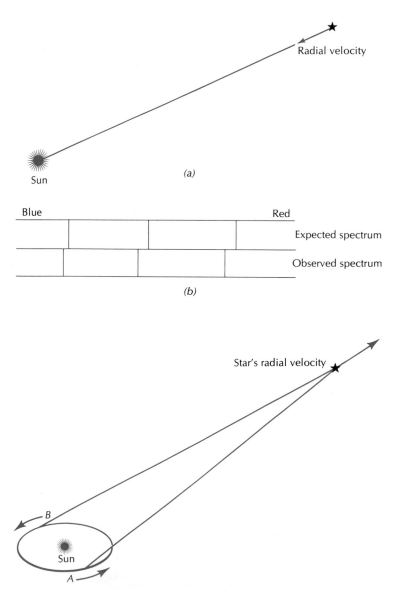

(a)

Sun

Blue

Red

Expected spectrum

Observed spectrum

(b)

FIGURE 12.9 (a) A star's radial velocity is the component of its velocity toward or away from the Sun. (b) If the star is approaching us, its spectrum lines are all shifted slightly toward the blue end of the spectrum.

Star's radial velocity

B

Sun

A

FIGURE 12.10 The radial velocity of a star measured from Earth is a combination of the star's motion and the Earth's motion. At A the Earth is approaching the star, and at B it is moving away from the star.

Table 12.4 Radial Velocities of a Few Stars

Star	Radial Velocity (km/sec)*
α Centauri	−22
Barnard's Star	−110
van Maanen's Star	+238
α Aurigae	+30
o Ceti	+58
γ Leonis	−36

*Here, a + sign means the star is moving toward the Earth and a − sign means the star is moving away from the Earth.

the Earth moves, they can correct measured radial velocities for this motion. Table 12.4 lists a few stellar radial velocities. These velocities are corrected for the Earth's motion. They are the velocities we would measure if we could make the observations from the Sun.

Space Velocity

The star 61 Cygni has a tangential velocity of 85 kilometers/second and a radial velocity of 64 kilometers/second toward us. These velocities are two components of the star's **space velocity**, which is as indicated in Figure 12.11.

We have already pointed out that we must correct measured radial velocities for the motion of the Earth. But if the Sun is like the rest of the stars, it moves too. Thus the solar motion should affect both the radial velocity and the proper motion of all the stars. How can we discover the Sun's motion and correct for it?

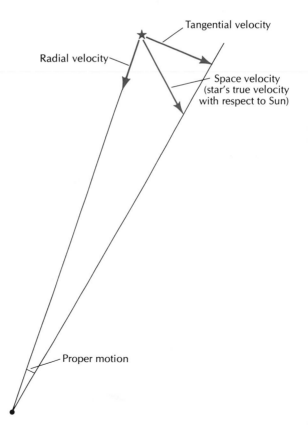

FIGURE 12.11 The tangential velocity and radial velocity of a star are components of its space velocity.

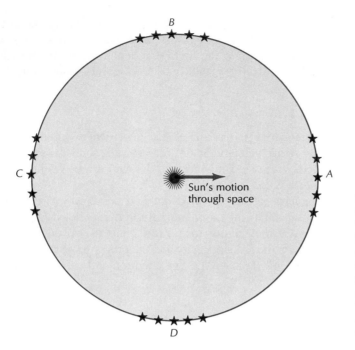

FIGURE 12.12 The solar motion, calculated from a study of average radial velocities of stars over the sky. Stars in direction A appear to approach the Sun on the average, but stars at C tend to move away from the Sun on the average. At B, the average radial velocities of stars is around zero.

Solar Motion

Astronomers assume that the local region of our Galaxy is like a room full of gas molecules. The stars move around in all directions, but if we calculate their average velocity as seen from inside it is zero. It turns out, if we calculate the average radial velocities of stars in various areas of the sky, we find a tendency for stars in one half of the sky to move toward us, and in the other half to move away from us (Figure 12.12). In a band around the sky separating the two halves, the radial velocities average to zero. This effect is due to the Sun's motion.

The Sun is moving toward the middle of the half of the sky where the stars appear to move toward us. The average radial velocity of stars around the point the Sun is approaching is about 20 kilometers/second toward us, whereas stars in the opposite direction appear to move away at that speed. Thus we conclude that the Sun is moving at 20 kilometers/second, toward a point very near the bright star Vega in the constellation Lyra. We call the direction toward which the Sun moves the **apex** of the Sun's motion.

The Sun and the other stars are seen to move, with speeds around 20 kilometers/second being typical. However, a few stars move

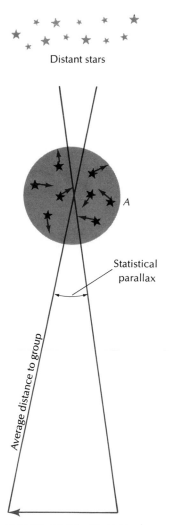

Distant stars

Statistical parallax

Average distance to group

FIGURE 12.13 Statistical parallax of a selected group of stars. The motions of all the stars in the circle A average to zero. Their apparent average motion relative to distant stars is due to the space motion of the Sun, and can be used to find the average distance of the group.

with much greater speeds, as we shall see. The motions of stars help us measure distances greater than those that can be measured by the parallax method.

STELLAR DISTANCES AGAIN

Stellar heliocentric parallax measurements can be carried out only for nearby stars. As we shall see, our Galaxy is very much larger than the volume of space occupied by these nearby stars. Stellar motions offer a way to extend the distance-measuring capability. Since the Sun is moving at 20 kilometers/second, it covers about four astronomical units/year; and in a quarter of a century the Sun moves more than 100 astronomical units. This is 50 times the baseline provided by the diameter of the Earth's orbit, and we should be able to use this baseline to extend our distance scale 50 times farther than that measurable by the heliocentric parallax method.

The basic problem with using the Sun's motion to measure stellar distances is the fact that the other stars also move. It is difficult to disentangle a distant star's real motion from its apparent motion, which is due to the Sun's motion. Thus the solar motion does not lend itself to measuring the distances of single stars. However, the average motions of a large group of stars is zero. If we choose a group of stars that we know are at more or less the same distance, we can be fairly sure that their average apparent motion is due entirely to the motion of our Sun; so we can use this average motion to calculate the average distance to the group of stars (Figure 12.13). As we shall see in the following pages, there are certain criteria that can be used to choose groups of stars that will be at the same distance. This method for calculating distance is called the **statistical parallax** method.

ABSOLUTE MAGNITUDE

There are two factors that determine a star's apparent brightness in the sky: the star's true brightness and its distance. Once its distance is known, a star's true brightness can be calculated. Two stars that can be used as examples are Sirius, which is both the brightest star in the constellation Canis Major and the brightest star in the sky, and Rigel, a slightly fainter star in Orion. Actually, Sirius is only 2.7 parsecs away from us, whereas Rigel is 250 parsecs away. If we take into account the inverse-square law (Chapter 4), we will find that Rigel is almost 2,000 times brighter than Sirius. Thus their true brightnesses are not related to their apparent brightnesses.

Astronomers have cooked up a scheme for describing the true brightnesses of stars. The **absolute magnitude** of a star is the magnitude it would have if it were moved from its true distance to a point 10 parsecs away from us. Since Sirius is nearer than 10 parsecs, its absolute brightness is fainter than its apparent brightness by a factor of $(10/2.7)^2 = 13.7$ or 2.8 magnitudes. Its absolute magnitude is therefore $-1.4 + 2.8 = 1.4$. In the same way, we can find Rigel's absolute magnitude to be -6.8. The star's magnitude as seen in the sky will be called its **apparent magnitude** to distinguish it from absolute magnitude.

Most of the things we have talked about so far do not tell us anything intrinsic about a star. A star's name, distance, position, and motion do not carry any information about its internal workings. Only absolute magnitude is an exception: it tells us the star's real brightness. Now we will focus on the physical state of a star: How big is it? How hot? How dense? What is it made of? What is its mass? A lot of this information can be inferred from the spectrum of the star.

STELLAR SPECTRA

The spectrum of a star is photographed by a spectrograph. This is a device like the spectroscope (Chapter 7). Starlight is collected by a telescope, then passed through the slit of the spectrograph, where it is next dispersed into a spectrum by a prism. The spectrum is finally recorded by a camera. Like the Sun, most stars have an absorption-line spectrum, a continuous spectrum crossed by a number of dark lines (Figure 12.14).

The fact that stars have absorption spectra tells us something right away: they consist of a hot interior where the continuous spectrum is produced, surrounded by a cooler gaseous atmosphere where the absorption lines are produced. Thus, stars are much like the Sun. As we shall see, some stars show evidence of chromospheres and coronas in their spectra just like the Sun.

In general, stellar spectra are like the Sun's spectrum. However, if we look in detail at many different stellar spectra, we will find remarkable differences: some spectra have few lines; some have many; some even show band structures due to molecules. Toward the end of the last century, astronomers at the Harvard Observatory developed a classification scheme for spectra. They devised 16

FIGURE 12.14 Spectrum of the star τ Scorpii between 3250 and 4650 Ångstroms. The negative print of the spectrum shows the continuous spectrum dark and the absorption lines superimposed on it as white. Flanking the star's spectrum is the emission-line spectrum of an iron arc, which is used as a wavelength reference by astronomers. (Hale Observatories.)

3250   **3450**

3450   3650

3650   3850

3850   4050

4050   4250

4250   4450

4650   4450

FIGURE 12.15 The six primary spectral types. The number of stars brighter than eighth magnitude in each class is shown on the right. (Yerkes Observatory.)

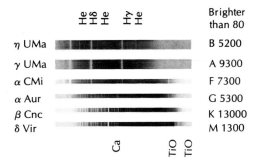

	He Hδ He	Hγ He	Brighter than 80
η UMa			B 5200
γ UMa			A 9300
α CMi			F 7300
α Aur			G 5300
β Cnc			K 13000
δ Vir			M 1300
	Ca	TiO TiO	

classes, which were denoted by the letters A through Q, omitting J. Each class was characterized by a set of descriptions, so that an astronomer could look at a spectrum and decide from its appearance that it belonged, say, in class F.

Between 1918 and 1924, the *Henry Draper Catalogue* of stellar spectra was published by the Harvard Observatory. In one of the great *tours de force* in astronomy, Miss A. J. Cannon of the Observatory classified almost 400,000 stellar spectra to produce that catalog. Some of the spectral classes of the A through Q sequence were combined or dropped, and the sequence was reordered O, B, A, F, G, K, M by the Harvard astronomers (see Figure 12.15).

By 1925, it was recognized that the spectral sequence developed in the *Henry Draper Catalogue* was a temperature sequence. The

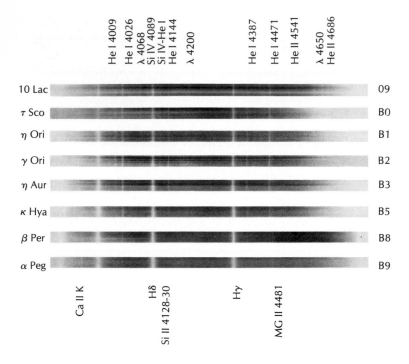

FIGURE 12.16 A closer look at spectral types O9–B9. Note the changes in the lines of HeI, HeII, and MgII. (Yerkes Observatory.)

photospheric temperatures of O-type stars are high (around 30,000° K), whereas the M-type stars are cool (temperatures around 3,000° K). The 6,000° K Sun is a G-type star. As the sequence has been refined, numerical subdivisions have been developed. Thus the Sun is a G2 star. (By the way, to remember the order of spectral types, remember the sentence: "Oh, Be A Fine Girl, Kiss Me.") Figures 12.16 through 12.19 show a representative sample of each type of stellar spectrum. Table 12.5 lists the chief characteristics of each spectrum type along with its temperature range.

How do we know that the differences in the appearances of the spectra are temperature differences? Maybe the Class A stars with strong lines of hydrogen just have more hydrogen than the Sun, with its weak hydrogen lines. In fact, all stars have more or less the same amount of hydrogen.

To produce an absorption line in the visible part of the spectrum, a hydrogen atom must have its electron in energy level 2 (Chapter 7). A photon of the proper wavelength (6562 Å, 4861 Å, 4340 Å, 4102 Å, etc.) will be absorbed by the electron in level 2, which will them jump to level 3, 4, 5, 6, etc. At spectral class A, a large fraction of the hydrogen atoms have their electrons in level 2 and can produce these absorption lines. These electrons are excited to level

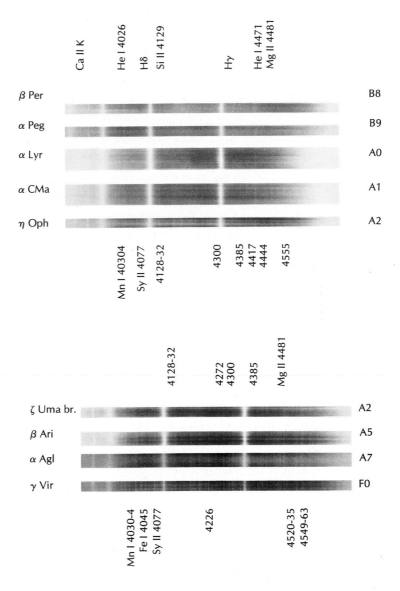

FIGURE 12.17 A closer look at spectral types B8–A2. Note the strength of the hydrogen lines, and the weak CaII K line. (Yerkes Observatory.)

FIGURE 12.18 A closer look at spectral types A2–F0. (Yerkes Observatory.)

2 by collisions. In spectral classes B and O, increasingly large fractions of the hydrogen is ionized, leaving few atoms with electrons in any level. In spectral classes F, G, K, and M, collisions have less energy; so more and more of the hydrogen atoms have their electrons in energy level 1, which produces absorption lines in the ultraviolet. The strengths of the visible absorption lines of hydrogen depend on the number of atoms with electrons in level 2.

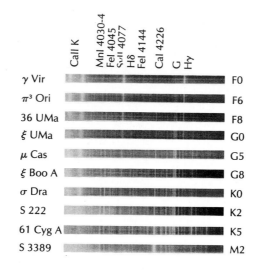

FIGURE 12.19 A closer look at spectral types F0–M2. The hydrogen line Hγ is very weak. Note the great intensity of the CaII K line, and the increasingly strong CaI line at 4226 Ångstroms. (Yerkes Observatory.)

That number, as we have seen, depends in turn on the temperature. Thus the line strengths depend on temperature. Similar arguments apply to other atoms and ions.

For instance, among the strongest lines in the solar spectrum are two lines (called the H and K lines) produced by once-ionized calcium (Ca^+) at 3934 Å and 3968 Å. These lines are even stronger in K-type spectra, then begin to fade in M-type spectra. In hotter F-type stars, much of the calcium is twice-ionized; that is, it has lost two electrons, and cannot produce the H and K lines. In the G- and K-type stars, most of the calcium is once-ionized, and the H and K lines are strong. In the cooler stars, the calcium is becoming increasingly neutral, and fewer ions are present to produce the H and K lines. Spectrum lines from neutral calcium increase in strength from class G0, where they are weak, through class M.

There are some differences in chemical composition between stars. However, these differences are usually differences in trace elements, such as carbon. The temperature effects are usually larger than the effects of abundance differences.

We have now seen that the appearance of a spectrum of a star depends on its temperature. How do we find out the proper temperature to assign to a star? One way is to study the shape of the continuous spectrum on which the absorption lines are superimposed. If we measure the brightness of a star as a function of wavelength with a spectrophotometer (Chapter 7), we can compare the result with the shapes of black-body curves. We assign to the

Table 12.5 Characteristics of Stellar Spectral Types

Spectral Type	Main Distinguishing Characteristic
O	Lines of highly ionized atoms are present, and lines of He^+ are visible in the spectrum.
B	Neutral helium lines are strong. Lines of the ions Mg^+ and Si^+ are present. Hydrogen lines are conspicuous.
A	Hydrogen lines are very strong. Lines of once-ionized and neutral metals (e.g., Fe) are present.
F	Hydrogen lines are weaker than in Class A, but are still clearly visible. The ion Ca^+ has conspicuous lines.
G	The spectrum looks like the Sun's spectrum. The lines due to Ca^+ are very strong.
K	Many strong lines due to neutral atoms. Some molecular bands are present.
M	Many very strong lines due to neutral atoms are present in the spectra of stars of class M. Molecules such as TiO have strong bands in the spectra.

star the temperature of the black body whose curve of brightness vs. wavelength best fits the star's continuous spectrum. In addition, modern astrophysics has developed a quantitative relationship between temperature and the strengths of spectrum lines which allows us to assign temperature to the spectrum types.

We have now looked at stellar spectra and have shown that they tell us something about the star's temperature. Now we know two intrinsic characteristics of stars: absolute magnitude and spectral class.

Colors

Stars have different colors because of their different temperatures: hot class O and B stars are blue, class A and F stars are white, class G stars are yellow, class K stars are orange, and class M stars are red. Astronomers can be more quantitative that that, however. Suppose we take two photographs of a region of the sky. One photograph is made with a photographic plate that is most sensitive to yellow light, the other with a plate that is most sensitive to blue light. A cool, red star will appear brighter on the yellow-sensitive plate than on the blue-sensitive plate, but the opposite is true for a hot, blue star. Two magnitude scales have been devised: a photo-visual magnitude (m_{pv}) and a photographic magnitude (m_{pg}) for the yellow- and blue-sensitive plates. The difference between the two magnitudes is the **color index** of a star:

$$CI = m_{pg} - m_{pv}.$$

A blue star has a negative color index, and a red star a positive one.

The process of obtaining a spectrogram, particularly of a faint star, that will allow an astronomer to decide a star's spectral type accurately can be very tedious. The exposure of the photographic plate can take several hours, and most spectrographs photograph one spectrum at a time. Photographic colors, on the other hand, can be estimated very rapidly. A single pair of photographic plates can yield colors of thousands of stars. Of course, the colors are not a precise measure of spectral type, but are often good enough for statistical studies of stars. Procedures that are slower but more accurate than the photographic method, and more rapid than the spectrographic method, for estimating spectral types are the modern photoelectric methods.

Modern measurements of stars' colors are carried out using a combination of a photomultiplier tube and a filter. The *UBV* system, for instance, uses filters to isolate the ultraviolet (*U*), blue (*B*), and yellow (*V* for visual) wavelength regions. Figure 12.20 shows the sensitivities of the system in the three wavelength regions. A very hot, blue star of spectral class O will be brightest when measured with the *U* filter and faintest with the *V*, since the star emits much more blue light than yellow light. Two color indices are defined in the *UBV* system: $B - V$ and $U - B$. Table 12.6 lists the colors and spectral types of several selected stars. Like the color index mentioned above, the $B - V$ and $U - B$ colors are a rough indication of a star's temperature and therefore of its spectral type. As we proceed through our story, we will see just how useful colors can be.

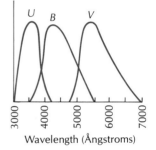

FIGURE 12.20   The sensitivities of the three filters plus photomultiplier tube in the three-color *UBV* system of photometry.

Table 12.6   Colors and Spectral Types of Selected Stars

Name	Spectral Type	V Magnitude	B – V Color
34 Piscium	B8	5.47	−0.06
6 Ceti	F6	4.88	+0.49
θ Andromedae	A2	4.61	+0.06
ζ Tucanae	G2	4.22	+0.58
o Cassiopeiae	B2	4.57	−0.06
η Cassiopeiae	G0	3.45	+0.58
107 Piscium	K1	5.23	+0.83
β Arietis	A5	2.65	+0.13
ν Ceti	K3	5.82	+0.96
BD+0°2989	M0	8.49	+1.49

THE HERTZSPRUNG-RUSSELL DIAGRAM

In 1911, E. Hertzsprung made a plot of the apparent magnitudes of the stars in the Pleiades cluster versus their color indices. The Pleiades is a beautiful group of stars in the constellation Taurus, seen in the fall and winter sky. It is one of the nearest of the open star clusters (Chapter 13). All the stars in the cluster are very close to the same distance from us. This means that the difference between the absolute magnitude and the apparent magnitude of each star is the same for all stars in the cluster. Hertzsprung's plot would have looked the same if he had used absolute magnitudes instead of apparent magnitudes: only the numbers on the vertical scale would have changed. Hertzsprung found that the points representing the stars did not scatter around the diagram. Instead, they all fell along a diagonal line. H. N. Russell extended this work in 1914, when he made a similar plot, this time of absolute magnitude versus spectral type, for every star for which both these quantities were known (Figure 12.21). The diagram is now called the **Hertzsprung–Russell diagram** (or H–R diagram) in honor of the pioneering work of these two men; however, spectrum-luminosity diagram is more correct.

Note that in Figure 12.22 most of the stars, including the Sun, fall on a diagonal line across the diagram. This diagonal is called the **main sequence** because the main bulk of the stars fall on it.

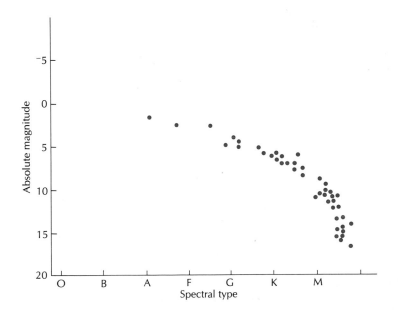

FIGURE 12.21 Hertzsprung–Russell diagram of nearby stars, within about 5 parsecs of the Sun.

There are also some stars, primarily of spectral class A and cooler, above the main sequence. These stars are called giants or supergiants, depending on their brightness. Let's think about stars of spectral class K0. They all have roughly the same temperature, and because of the radiation laws (Chapter 7) must radiate the same amount of energy per square centimeter of surface. Therefore, for the giant or supergiant K0 star to be brighter than the corresponding main-sequence star, it must have more square centimeters to radiate; that is, it must be larger. Similarly, stars such as white dwarfs, which lie below the main sequence in the H–R diagram, are smaller than main-sequence stars.

It turns out that if we compare spectra of a main-sequence star, a giant star, and a supergiant star of the same spectral type, we will see subtle differences which allow astronomers to distinguish between the stars. Thus the spectrum contains size as well as temperature information.

By the way, when we talked about statistical parallaxes earlier, we stated that the Sun's motion is used to measure parallaxes for groups of stars that are chosen by some criteria that tell us the stars are all at the same distance. The H–R diagram tells us how to make

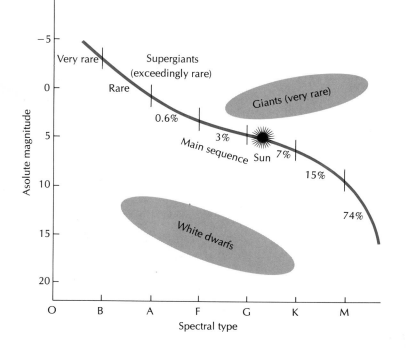

FIGURE 12.22 Hertzsprung–Russell diagram for stars in the Sun's region of the Galaxy. If the diagram were plotted with points for the positions of stars, the points would fall along the main sequence, in the giant region, and in the white-dwarf region. The relative numbers of stars are indicated. White dwarfs are undoubtedly very common. However, they are so small and faint that few are known. M-type main-sequence stars make up most of the stars in our region of the Galaxy.

such a choice. If, for instance, we chose main-sequence stars with specrtral class B and apparent magnitude 8, we would be choosing similar stars with the same absolute and apparent brightness; so they would be at the same distance. Probably, rather than choosing stars by spectral class, astronomers would choose them by color index. A useful variant of the H–R diagram is the color-magnitude diagram, where absolute magnitude is plotted against, for instance, the $B - V$ color. This type of diagram is especially useful for star clusters (see Chapter 14), where many faint stars must be studied. It is easier to measure the colors of these stars than to decide on their spectral classes.

STELLAR SIZES

Over the years, astronomers have used interferometric techniques to measure the sizes of stars (Figure 12.23). In simplest terms, one can look at the method as follows. If we take two telescopes a certain distance apart and combine the light they receive in the

FIGURE 12.23 Comparative sizes of stars, divided into groups with a known object (Sun, Earth's orbit, etc.) for reference.

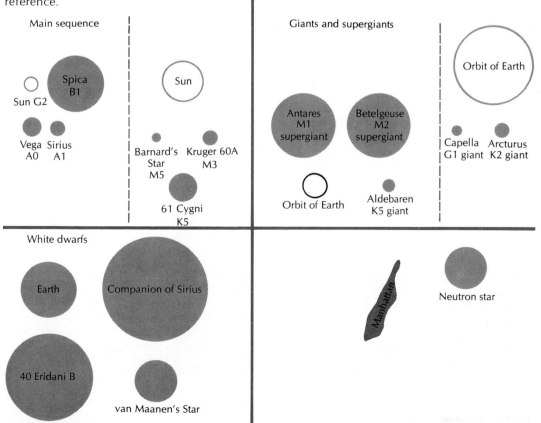

appropriate way, the combination will have the same resolving power as one telescope whose objective is as large as the distance between the telescopes. The method is not affected by astronomical seeing, since the telescopes look along two distinct paths through the atmosphere. One of the most promising devices is an intensity interferometer built by two Australian astronomers. The angular sizes of most stars are much smaller than the resolving ability of even the largest telescope, and therefore can be seen only as a point of light. When we know the angular size and the distance of a star, we can obtain its true linear size.

The first successful image of the disk of a star was obtained in 1974. With a technique called speckle interferometry, astronomers used the 4-meter Mayall reflector at the Kitt Peak National Observatory to photograph the immense star Betelgeuse, which has an angular size of 0.06 seconds, making it 580 times larger than the Sun. It takes a great deal of computer processing to produce the final picture (Figure 12.24). The result is good enough to show the presence of huge bright and dark surface markings on the star. Astronomers believe these markings may be a very few giant cells, like the innumerable granules on the Sun.

We have now talked about the general properties of a star: distance, motion, size, and temperature. In the next section we will look at binary stars and variable stars. Binary stars are important

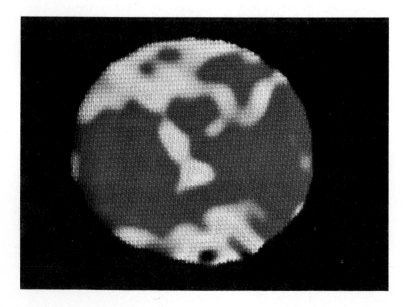

FIGURE 12.24 The supergiant star Betelgeuse. This picture is a reconstructed image (see text). (Courtesy Kitt Peak National Observatory. Copyright © 1975 by the Association of Universities for Research in Astronomy, Inc.)

because they tell us something about stellar masses, and variable stars are interesting beacons that can be used to estimate distances in the universe.

BINARY STARS AND STELLAR MASSES

In the solar system, the mass of a body is calculated by studying its gravitational pull on another body. For instance, we calculated the Earth's mass from the Moon's orbital motion around the Earth. The same principle could be used to calculate stellar masses if we could find a situation where one star's gravity influences the motion of another star. Fortunately, such situations have been found.

Visual Binaries

In the eighteenth century, astronomers discovered the existence of many double stars that appear as one star to the naked eye, but split into a pair of stars when seen through a telescope. A few pairs, such as Alcor and Mizar in Ursa Major (ζ Ursa Majoris), can be seen separately by the unaided eye (Figure 12.25). At first, the pairs were thought to be chance arrangements of two stars seen in the same

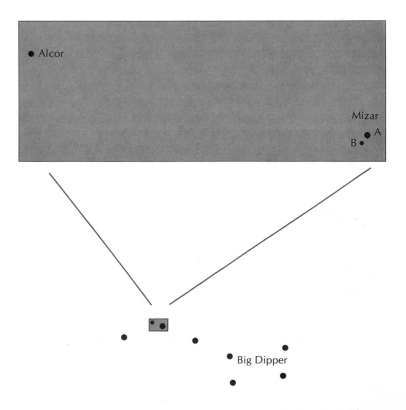

FIGURE 12.25 The star system Alcor and Mizar in the Big Dipper. Alcor and Mizar are a binary system, and Mizar itself is a binary system. Furthermore, all three stars, Alcor, Mizar A, and Mizar B, are spectroscopic binaries; so actually six stars are present.

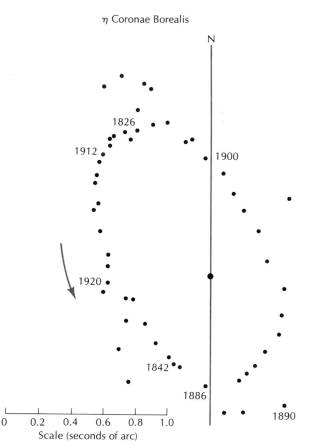

η Coronae Borealis

FIGURE 12.26 The apparent motion of one star about the other in the system η Coronae Borealis. The orbital period of the system is 41.6 years. The angular size of the orbit is indicated by the scale, and the year in which some of the observations were made is indicated. The apparent shape of the orbit is not its true shape, since we probably see the system at an angle.

direction but at different distances. However, William Herschel established that some of these pairs actually show orbital motion about one another. Since his time, orbital motion has been established for many thousands of double stars (Figures 12.26 and 12.27). We call these physical **binaries**, or simply binaries, to distinguish them from the optical doubles, which are just chance pairs of stars.

   If we know the distance to a binary star system, we can calculate the linear dimensions of the orbit of one star about the other, from measured angular dimensions. The orbital period is easily measured from the observed motion of the stars. Once we know the size of the orbit and the orbital period, we can calculate the stars' masses. As with the Earth-Moon system (Chapter 6), we find the sum of the

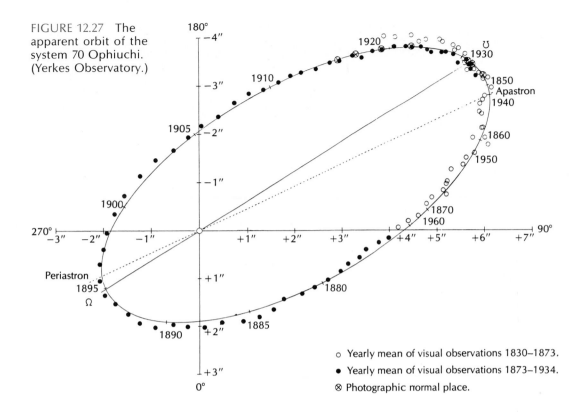

FIGURE 12.27 The apparent orbit of the system 70 Ophiuchi. (Yerkes Observatory.)

o Yearly mean of visual observations 1830–1873.

• Yearly mean of visual observations 1873–1934.

⊗ Photographic normal place.

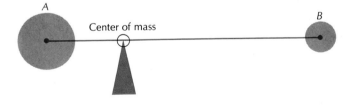

FIGURE 12.28 Center of mass of a binary star system. If we could connect the two stars by a rod, the system would balance at its center of mass.

masses of the two stars. If we study the motion of both stars relative to other stars in the sky, we will find in fact that each star moves around the center of mass of the system (Figure 12.28). At any moment, the center of mass of the system lies at the balance point of the system. Thus if we study each star's motion around the center of mass, their relative distances from that point tell us their relative masses, and allow us to find the individual masses.

FIGURE 12.29 The spectrum of one of the two stars of the Mizar pair. Note that in the upper spectrum all lines are double, whereas in the lower spectrum all lines are single. (Yerkes Observatory.)

Spectroscopic Binaries

Spectroscopic binary stars are star systems that are not split into an apparent double star even in the largest telescope. Instead, the orbital motion of the pair is discovered spectroscopically. The spectrum of the star Mizar is shown at two different dates in Figure 12.29. Note that at one time the lines are single, but at the other the lines are double. The explanation is shown in Figure 12.30. When the lines are double, one star is approaching us, producing blue-shifted lines and the other is receding, producing red-shifted lines. When the stars move perpendicular to the line of sight, their spectrum lines appear to be a single line. The time it takes to go from the single-line state to maximum separation of the lines is one-fourth of the orbital period.

If we cannot separate the two stars in a telescope, how do we find out the size of the orbit? To do so, we find the velocities of the stars in their orbits by measuring the maximum Doppler shifts of the spectrum lines, and we also find the total orbital period of the stars. The speed of each star times the orbital period tells us the distance that star moves in one circuit of its orbit, or the circumference of the orbit. Since the circumference is 2π times the radius, the radius follows at once. (Of course, this assumes the orbits are circles, which they may not be.)

From the orbital period and the radii of the orbits, we can find both the sum of the masses and the individual masses of the two

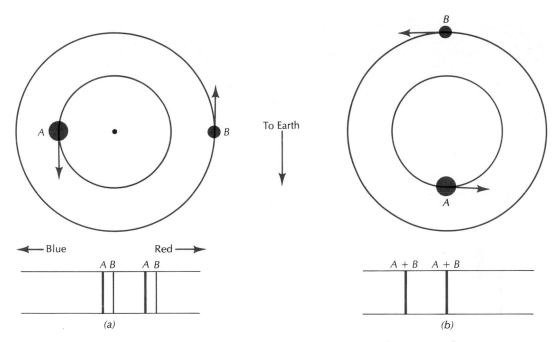

To Earth

Blue ← → Red

A B A B

(a)

A + B A + B

(b)

FIGURE 12.30 Explanation of why a spectroscopic binary star has sometimes a double-line spectrum (a), sometimes a single-line spectrum (b).

stars, just as for planets and visual binary stars. In many spectroscopic binaries, only one star's spectrum can be seen, in which case we can find only the sum of the masses. There is a small hitch in these calculations, we must hasten to point out. We are not necessarily looking at the orbits of spectroscopic binaries edge on. The orbit may be tipped to our line of sight; if so, we do not measure the total velocity, but only the part of it that is directed toward us or away from us. The resultant masses are uncertain because the tilt of the orbit is uncertain. However, there is one situation where we can be sure we know the inclination.

Eclipsing Binaries

If the orbit of a spectroscopic binary is almost exactly edge-on to us, one star will pass directly in front of the other, producing an eclipse. Such a situation is called an **eclipsing binary** system. The first-discovered example of such a system is β Persei (Algol). Figure 12.31 shows how the magnitude of Algol varies as a function of time. Compare this graph, called a **light curve**, with the hypothetical one (Figure 12.32) for binaries with total eclipses or partial eclipses and uniformly illuminated disks.

Algol's partial eclipses repeat every 2.867 days, with the larger minimum lasting about 10 hours. The two stars are quite different in brightness, as we can tell from the relative magnitude changes of

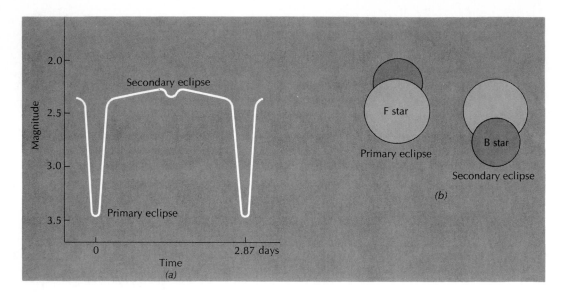

FIGURE 12.31   (a) The magnitude as a function of time (or light curve) of the eclipsing system, Algol. The system is composed of a very bright B star and a slightly larger, but much dimmer, F star. (b) The relative positions of the stars during eclipse.

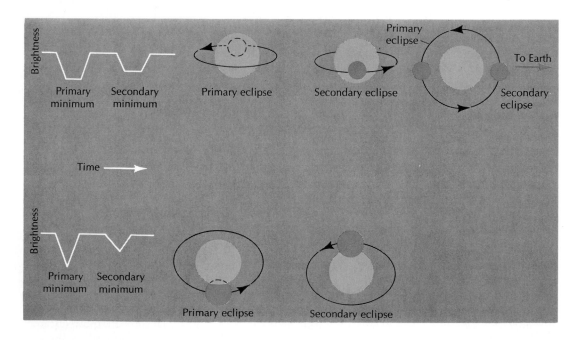

FIGURE 12.32   Hypothetical light curves for two different types of eclipsing binary systems. The stars are assumed to have uniformly bright disks with no limb-darkening.

the eclipses. The fact that the light curve of the eclipses has rounded corners rather than sharp ones as in the hypothetical curve tells us the stars exhibit limb-darkening like the Sun. The magnitude of the star system is not constant even when eclipses are not in progress, because the intense radiation from the B star produces a hot spot on the surface of the F star. Eclipsing binary systems allow us to calculate stars' masses without facing the problem of unknown orbital tilt that we face for spectroscopic systems that do not eclipse.

Binary stars are intrinsically interesting objects. Some systems consist of stars that are very close together, even closer than in the Algol system. In such systems, streams of hot gas can be pulled from one of the stars by the other, producing a spectrum that is both fascinating and difficult to interpret. An example is the system β Lyrae, which has puzzled and interested astronomers for years (Figure 12.33). The influence of the stars in that system on one another is so great that they are elongated into ellipsoidal bodies.

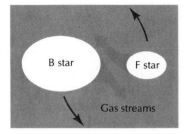

FIGURE 12.33 A picture of the β Lyrae binary system proposed by O. Struve.

MASS-LUMINOSITY RELATION

The masses that have been calculated from the study of many binary star systems over the years are plotted in a diagram of absolute magnitude versus mass in Figure 12.34. Notice that the

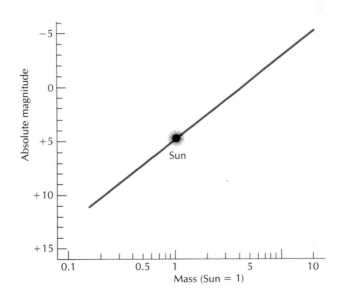

FIGURE 12.34 The mass-luminosity relation derived from binary star systems.

stars all fall on a diagonal line in the diagram. This diagram, together with the spectrum-luminosity diagram, gives astronomers vital clues in their search for understanding of the life cycles of stars, and we will encounter it again in our discussions.

COMMON STARS

Earlier, we learned how to find temperatures, diameters, and masses of various types of stars. Once we know a star's diameter and mass, we can calculate its density. Table 12.7 is a summary of information for typical common stars. In the last column, we list densities in grams per cubic centimeter.

The M0 supergiant, which is modeled after Betelgeuse, has an incredibly low density. Even so, the density in the table is the average value for the entire star: the material in the star's core is much more dense than the average; the outer layers are much less dense. In fact, the outer layers of a red-supergiant star have been compared to a red-hot vacuum.

At the other extreme is the white dwarf. One cubic centimeter of its average material would weigh almost 100 kilograms if it were on the surface of the Earth. The central density of the star is even

Table 12.7 Average Properties of Common Stars

Star	Mass (Sun = 1)	Radius (Sun = 1)	Temperature (°K)	Density (gm/cm³)
Main Sequence				
O5	30	20	35,000.	0.01
B0	15	9	20,000.	0.02
A0	3	2.2	10,000.	0.2
F0	1.5	1.5	7,500.	0.5
G0	1	1	6,000.	1.4
K0	0.75	0.8	4,000.	1.6
M0	0.5	0.6	3,600.	3.0
Giants				
G0	3.0	10	5,200.	0.003
K0	3.5	24	4,100.	0.0003
M0	3.8	75	3,400.	0.00001
Supergiants				
G0	10	100	4,900.	0.00001
K0	12	200	3,800.	0.000002
M0	15	500	3,000.	0.0000002
White Dwarfs				
A0	0.6	0.02	10,000.	100,000

higher than the value in the table. Incidentally, the typical white-dwarf star is only twice the size of the Earth.

With this brief look at the extremes of common stars, let's proceed to discuss variable stars.

CEPHEID VARIABLE STARS: STELLAR MILEPOSTS

The Cepheid variable stars are named after the type star δ Cephei, which was first recognised as a variable star in the eighteenth century. It varies with a period of 5 days, 8 hours, and 53 minutes, and with an amplitude of roughly 1 magnitude. The schematic light curve of a typical Cepheid variable is shown in Figure 12.35. Notice that the star's brightness rises to maximum rapidly, then more slowly subsides to minimum. The periods of Cepheid variables range between one and 50 days, with the most common period being about a week.

What causes the variability of the Cepheids? The answer can be found in their spectra. First of all, the spectral types of the stars change slightly with changing brightness. At maximum light, the spectrum of a typical Cepheid is similar to that of an F-type star, but at minimum it is similar to that of a G-type star; thus, the star's temperature is greatest when it is brightest.

The star's radial velocity also varies with the same period as the light variation. This change in velocity, it is now known, indicates that the star actually pulsates: it grows and shrinks rhythmically. Astronomers at one time thought Cepheids might be eclipsing variable stars, too. However, the American astronomer Harlow Shapley showed in 1914 that this is not possible. By combining the radial-velocity measurements and the orbital period, he could infer the radius of the orbit in which the eclipsing star would have to

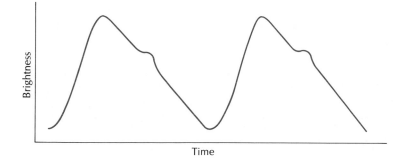

FIGURE 12.35 The schematic light curve of a typical Cepheid variable with a period of roughly one week.

move. The radius came out to be only $\frac{1}{10}$ the size of the supergiant Cepheid. Thus, the eclipsing star would have to orbit inside the Cepheid. Furthermore, the radial velocity showed the greatest motion away from us when the star was faintest, but in an eclipsing system, the greatest motion toward us cannot occur during eclipse, since the eclipsing star is then moving perpendicular to the line of sight toward us.

The observed facts about Cepheids therefore are inconsistent with orbital motion, but are in complete agreement with the pulsation hypothesis. If we measure the average radial velocity of the star as it expands, and multiply that velocity by the time during which expansion takes place, we can find the change in radius that the star experiences during a pulsation. It turns out to be a bit less than 10%.

Cepheids are interesting as stars. However, they are vitally important to astronomy for another reason: they are a tool for measuring distance. In 1912 Henrietta Leavitt was studying the Magellanic Clouds at Harvard College Observatory's South African observation station. The two Magellanic Clouds are actually nearby galaxies, some 60,000 parsecs away, although Miss Leavitt did not know this. She did know that the Clouds were quite distant and that all the stars in either Cloud could be considered to be at the same distance. The difference between the apparent magnitudes and the absolute magnitudes of the stars is therefore constant. If we see several stars or other celestial objects at the same distance, the difference $(m - M)$ between the apparent magnitude (m) and the absolute magnitude (M) is the same for all the objects and depends only on the distance. Astronomers call this difference the **distance modulus**. Table 12.8 shows the relationship between distance modulus and distance. Miss Leavitt found a large number of Cepheid variables in the Small Magellanic Cloud, and discovered a relationship between the pulsation periods and brightnesses of the stars, known today as the period-luminosity relation. The longer the period of the Cepheid, the brighter it is. If you plot the median magnitude (the magnitude halfway between maximum and minimum brightness) against the period, on a logarithmic scale, the period-luminosity diagram for the Small Magellanic Cloud is almost a straight line (Figure 12.36).

How do we use Cepheids to measure distances? If we can calculate the distance modulus for stars in the Small Magellanic Cloud, we can calibrate the period-luminosity relation. That is, we can make it a plot of absolute magnitude versus period. Then, if we see any other Cepheid in the Universe, we can infer its absolute magnitude from its period by using the period-luminosity relation. We

Table 12.8   Distance Versus
              Distance Modulus

Distance Modulus	Distance in Parsecs
–2	4.
–1	6.3
0	10
1	15.8
2	25.1
3	39.8
4	63.1
5	100.
10	1,000.
15	10,000.
20	100,000.
25	1,000,000.
30	10,000,000.

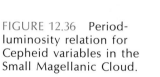

FIGURE 12.36 Period-luminosity relation for Cepheid variables in the Small Magellanic Cloud.

can also find the Cepheid's apparent magnitude from a direct brightness measurement, and therefore can calculate its distance modulus. The distance then follows from Table 12.8.

Cepheids have been discovered in a type of star cluster known as an open star cluster. We measure the apparent magnitudes and colors of the stars in one of these clusters and construct a color-magnitude diagram. The distance modulus of the cluster can then be calculated, because we know the absolute magnitudes of the main-sequence stars from our studies of the stars that are near enough to have measurable parallaxes (heliocentric or statistical). Once we know the distance modulus of the cluster, we can calculate the absolute magnitudes of all its stars, including the Cepheids. If we do this for many clusters that contain Cepheids, we can con-

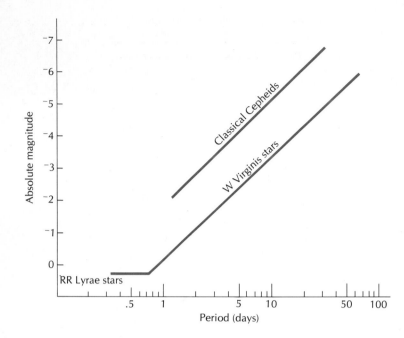

FIGURE 12.37 General period-luminosity relation for classical Cepheids, W Virginis stars, and RR Lyrae stars.

struct a very precise period-luminosity relation. The period-luminosity relation becomes an excellent tool for estimating large distances. The brightest Cepheids can be recognized as such at distances like 5 million parsecs.

The period-luminosity law as a distance indicator has one small problem. There are two types of Cepheids, the classical Cepheids that are like δ Cephei and the W Virginis stars. Stars of the latter type are almost two magnitudes fainter than classical Cepheids for any given period (Figure 12.37). Related to the W Virginis stars are the RR Lyrae stars (also shown in Figure 12.37). These stars all have about the same absolute magnitudes, roughly about +0.5, and pulsation periods of less than a day. We will have further discussion about the relationship between the groups of Cepheids in the next chapter.

REVIEW QUESTIONS

1. Describe the magnitude system by which astronomers describe brightness of stars.

2. Describe heliocentric parallax. Which is more distant: a star with a parallax of 0.″75 or a star with a parallax of 0.″15? What is the ultimate limitation of heliocentric parallax distance determination?

3. Describe: (a) proper motion; (b) radial velocity; (c) space velocity.

4. Describe the Sun's motion through space.

5. Define: (a) absolute magnitude; (b) color index.

6. What is the spectral class of the Sun? In what spectral class are hydrogen lines strongest? Why do the spectra of main-sequence stars differ?

7. Draw the Hertzsprung-Russell diagram.

8. How can one determine (a) stellar sizes; (b) stellar masses?

9. What is the period-luminosity law for cepheid variables? What is its significance to astronomy?

13

DISCOVERING OUR GALAXY

We live in one of many galaxies that populate the universe. We will call it "the Galaxy" or "the Milky Way galaxy." Just what is a galaxy? In this chapter we will attempt to answer that question. It will not be an easy task to give you a feeling for the scale of the Galaxy. It is so immense that our everyday experience cannot encompass any description of the size.

We have already talked about the main inhabitants of the Galaxy: stars. But we need to understand the characteristics of larger groupings of stars known as star clusters. After a treatment of star clusters, we will talk about the dust and gas between the stars. Finally, we will put it all together to see what the Galaxy is like. In essence, we are putting together a puzzle. We begin by turning over each piece and examining it carefully. Then we put them together.

STAR CLUSTERS

Stars come singly, in binary star systems, in multiple-star systems, and in clusters of various sizes and descriptions. Two types of stable star clusters have been recognized: open or galactic star clusters, and globular star clusters.

Open Clusters

Open clusters are irregularly shaped, loose groups of stars containing anywhere from a dozen to several hundred members . One of the best-known open clusters is the Pleiades, in the constellation Taurus (Figure 13.1). This dipper-shaped cluster contains about eight naked-eye stars, but is seen to be rich in stars through binoculars. In fact, several open clusters are delightful objects when

FIGURE 13.1 The Pleiades in Taurus. The open cluster is imbedded in a cloud of dust and gas, which shows up in this time-exposure. (Hale Observatories.)

FIGURE 13.2 Open cluster in the constellation Cancer, photographed by the 5-meter Hale reflector. (Hale Observatories.)

viewed through binoculars. The Praesepe in Cancer is a beautiful example (Figure 13.2). Roughly 900 open clusters are known, in total.

The distances to clusters are found from their color-magnitude diagrams. If we construct the color-magnitude diagram of a galactic cluster it looks like a backward 7 (Figure 13.3). The sloping leg of

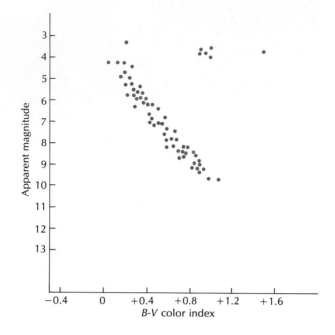

FIGURE 13.3 Color-magnitude version of the H–R diagram for the Hyades open cluster. Note that it looks like a backward 7, with only a few stars on the horizontal part.

the backward 7 is the main sequence, and the horizontal part consists of the giant stars. If we plot the backward 7s for many clusters on one graph, they will occur at different positions (Figures 13.4 and 13.5). The vertical displacements between them depend on the relative distances of the clusters: the clusters with fainter main sequences are further away. Further, we know the absolute magnitudes of the main-sequence stars for the Hyades open cluster, whose distance can be found very accurately by other methods. Thus, we can find the distance moduli and therefore the distances to the other clusters.

Two results of great interest came out of this study. One was the discovery of interstellar dust, which dims starlight; we will talk about this later. The other is that the horizontal part of backward 7 falls at different absolute magnitudes for different clusters. As we will see, this fact helps us estimate the ages of the clusters.

We have now looked at galactic clusters and have seen how to calculate their distances. From the distances we find that their diameters are between 2 and 10 parsecs. The star densities in open clusters work out to be almost a hundred times the density near the Sun. Let's next look at the second type of star cluster, the globular clusters.

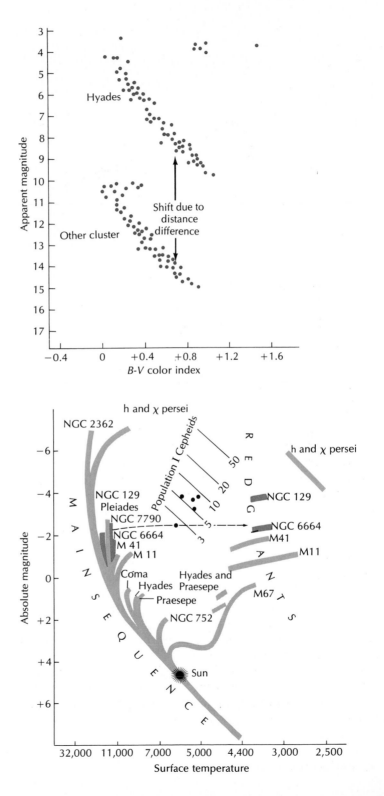

FIGURE 13.4 Idealized color-magnitude version of the H–R diagrams of two open clusters plotted on the same chart. The vertical displacement between the main sequences of the clusters results from their different distances.

FIGURE 13.5 Composite H–R diagram for several open clusters. The horizontal scale of this chart is temperature; however, the chart would look similar if spectral type or color were used. The main-sequence parts of all the clusters' H–R diagrams have been fitted together, and the absolute magnitudes of the stars have been indicated. (From *Pulsating Stars and Cosmic Distances* by R. P. Kraft. Copyright © 1959 by Scientific American, Inc. All rights reserved.)

FIGURE 13.6 The glob-
ular cluster Messier 92
photographed with the
3-meter (120-inch) reflector at Lick Observatory.
(Lick Observatory photo-
graph.)

Globular Clusters

Globular clusters are spherical aggregates of stars that contain
anywhere between a few tens of thousands of stars and a million
stars (Figures 13.6 and 13.7). The stars are concentrated toward the
central core of the cluster, as can be seen from Figure 13.6, which is
a typical rich cluster. Over a hundred of these clusters are known.

In the centers of these clusters, densities may reach as high as
1,000 stars per cubic parsec. If we lived on a planet orbiting a star
near the center of a large globular cluster, the sky would be spectacular. Most of the 100,000 or more stars in the cluster would be
visible to the naked eye, and there would be hundreds of stars
brighter than Sirius. In fact, the sky would be so filled with stars
that it would never get completely dark on the planet. Even at
midnight, it would be like twilight.

Distances to the globular clusters have been calculated by using
the RR Lyrae variable stars observed in them. Since the typical RR
Lyrae star (see Chapter 12) has an absolute magnitude known to be
around +0.5, its distance modulus can be found from its apparent

FIGURE 13.7 A globular cluster with a relatively loose appearance. This photograph was made with the same telescope as Figure 13.6. (Lick Observatory photograph.)

magnitude, and its distance follows quickly. Distances are quite large, typical values being between 2,000 and 100,000 parsecs. Of course, sizes follow immediately from the distances. The angular sizes of the clusters are small and the skinny-triangle method works well. Typical radii range between 20 and 150 parsecs.

Associations

A third type of stellar aggregate is the **association**. Associations are low star-density objects that would not stand out from the background of stars, were it not for one fact. Most of the stars in associations are hot, blue O and B stars. If we were to make a map of the stars in the sky, coloring the stars according to their spectral types, the associations could be picked out easily as unusual concentrations of blue stars (see Figure 13.8). Basically, that is how they are found.

We will have more to say about the ages of star clusters and associations later. We do get a clue about ages from their dynamic

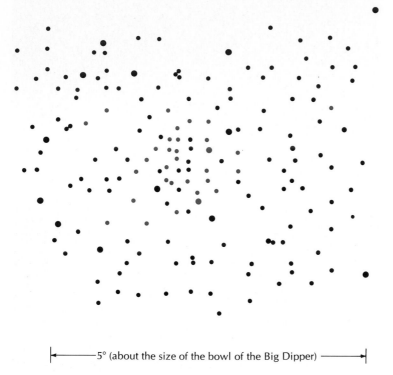

5° (about the size of the bowl of the Big Dipper)

FIGURE 13.8 An association viewed on the sky. The stars that belong to the association are indicated in color. If the stars were not labeled in this way, the association would look like any other region of the sky. Unlike open clusters, associations are not extra-rich in stars.

stability. Astronomers study the gravitational interactions between the stars in these groupings, and find that globular clusters are highly stable objects that will look more or less the same for many billion years. Galactic clusters are less stable. The stars in one of them will probably trickle away in a few hundred million years. Associations are unstable groups of very young stars which will not be recognizable as an association after a few million years.

We have now described star clusters and associations. Let's next talk about matter between the stars.

INTERSTELLAR MATTER

Dust

Robert J. Trumpler calculated the distances to galactic clusters by estimating individual stars' absolute magnitudes from their spectral types. He then used the observed angular diameters of the clusters

FIGURE 13.9 Because of the presence of obscuring dust in the Galaxy, galactic clusters seem farther away than they actually are. The size of the distance error increases with distance.

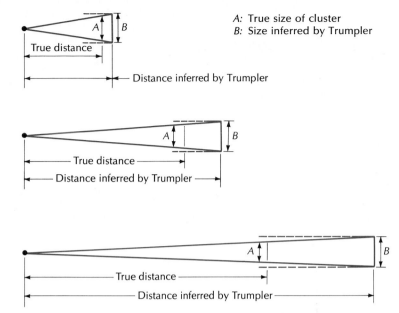

A: True size of cluster
B: Size inferred by Trumpler

and their distances to find their linear diameters. He came up with an immediate puzzle in the results. The linear sizes of galactic clusters depended on their distance from the Earth. The farther the cluster was away, the larger it was. Why should a cluster's size depend on its distance from us?

Trumpler saw the answer to the puzzle right away. The clusters are all the same size, on the average. His method of calculating distances produced incorrect distances; it made the clusters seem too far away. For nearby clusters, the distance error was small, and their sizes came out about right. As the distance increased, the relative size of the error increased, and the diameter of the cluster inferred from the erroneous distance increased. Trumpler blamed the whole problem on dust between us and the clusters. The dust absorbs the light from the cluster stars, makes them appear fainter than they actually are, and so makes us think they are farther away. If the amount of dust in each volume of space is the same, then there is more dust between us and distant clusters than between us and nearby clusters. The total amount of dimming, and thus the error in distance, depends on distance (Figure 13.9).

From the sizes of the errors, Trumpler concluded that the light of a star was cut in half for every 1,000 parsecs of its distance; that is, a star at 1,000 parsecs would appear to have only half its true

FIGURE 13.10 Close-up of the dust cloud around the Pleiades star Merope (see also Figure 13.1). In a color picture, the cloud would appear blue, since the dust is scattering blue light of a blue star. (Hale Observatories.)

apparent brightness, a star at 2,000 parsecs would appear to have one quarter of its true apparent brightness, and so on.

Trumpler's hypothesis that dust is present between the stars has been borne out by other observations. Like the scattering that causes the blue sky, the interstellar dust obscures the blue light from stars more than it does the red light. We can study the spectrum of a star and discover that it shows absorption lines characteristic of, say, a B-type star; yet the star appears red. Like the reddening of the sun at sunset, this reddening occurs because the dust selectively dims the blue light from the stars. This fact points outs that we must exercise care when we use color indices rather than spectral types to study groups of stars.

The reddening of starlight does allow astronomers to estimate the amount of dimming present, because the amount of reddening and the dimming increase together. The color excess of a star is the difference between its observed color index and the color index it would have with no reddening. The latter can be inferred from its spectrum type. There is a simple relationship between color excess and the total change of the star's magnitude due to absorption by the dust.

The blue light scattered from the star sometimes illuminates the dust, if it is in a dense-enough cloud. The Pleiades cluster is imbedded in a dust cloud, and if we take a long-exposure photograph, we can see the dust in the vicinity of the individual stars (Figure 13.10). We can tell this is simply scattered starlight from its color and from the absence of an emission-line spectrum from the

FIGURE 13.11 The southern Milky Way, photographed with a single-lens reflex camera attached to a larger telescope for guiding purposes only. The exposure took 45 minutes. The mottled appearance of the Milky Way is due to the presence of dust. (Courtesy J. C. LoPresto, Edinboro State College, Edinboro Pennsylvania.)

cloud, as we will see below. These clouds are called **reflection nebulae**. The word nebula (plural, nebulae) comes from the Latin and means cloud.

The dust is not spread uniformly through the Galaxy. It occurs in relatively dense clouds here and there. Photographs of the Milky Way show a nonuniform star density. Yet astronomers believe that the distribution of stars should be smooth. The apparent nonuniformity is due to clouds of dust and gas that blot out the light of distant stars. If dust were absent, Figure 13.11 would not show the mottling apparent on the Milky Way, which stretches diagonally

FIGURE 13.12 The Horsehead Nebula in Orion. (Hale Observatories.)

across the picture. The dark patches of absorbing material are called **dark nebulae**. Another beautiful example is the Horsehead Nebula (Figure 13.12).

Gas

The evidence for gas in the space between stars is more direct. Photographs of the sky show many bright patches of nebulosity. One of the most spectacular is the Orion Nebula (Figure 13.13). This nebula can be seen even with binoculars, surrounding the stars of the sword hanging from Orion's belt. Through binoculars, just

FIGURE 13.13 The Orion Nebula. (Lick Observatory photograph.)

the hint of a glow can be seen. Through a small telescope, the Orion Nebula is a lovely sight. The cloud appears as a soft, greenish glow surrounding the distinct points of light of the stars. This is one example of a bright diffuse nebula. We will return to these shortly.

Dark Clouds. There is also gas associated with dust in dark clouds. This gas does not glow brightly: its presence can be inferred from the existence of absorption lines in the spectra of distant stars (Figures 13.14 and 13.15). The gas clouds are masses of cool gas through which the starlight passes, and the atoms in the gas absorb their characteristic spectrum lines. It is clear from the spectra that

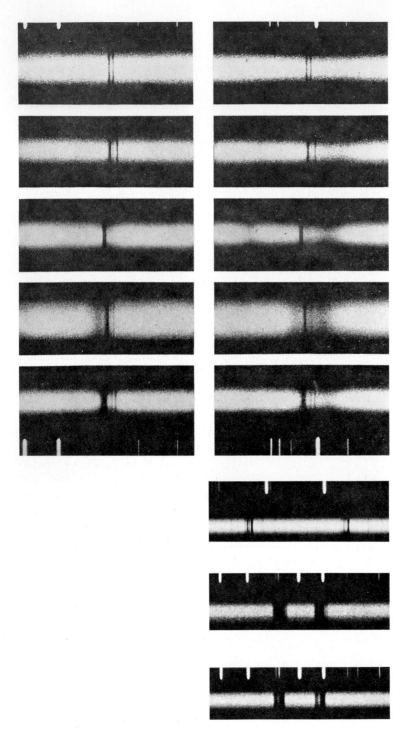

FIGURE 13.14 Interstellar CaII lines in the spectra of five different stars. The CaII K line is at the left and the CaII H line at the right. Several sharp lines can be seen in each case, indicating the presence of several clouds with different velocities. (Yerkes Observatory.)

FIGURE 13.15 Interstellar lines of sodium. (Hale Observatories.)

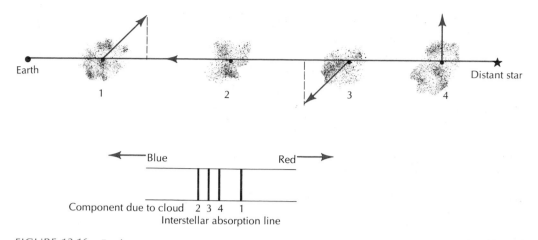

FIGURE 13.16 Explanation of multiple interstellar lines in the spectrum of a distant star. Each cloud produces one absorption line. Since each cloud has a different velocity from the other clouds, each line is Doppler-shifted by a different amount away from the zero-velocity position of the spectrum line.

there is more than one absorbing cloud between us and some of the stars (Figure 13.16). Each cloud has its own peculiar radial velocity, producing a Doppler-shifted absorption line. The motions allow us to infer the presence of more than one cloud. If the clouds did not move, all the absorption would fall at one wavelength, and we could not separate the effects from different clouds. The elements sodium, calcium, potassium, iron, and titanium, and the molecules CN and CH, have been identified in the visible-light spectra of many stars.

The study of natural radio radiation emitted from these clouds has provided some fascinating information during recent decades. The first cosmic radio noise was detected in 1931, at a wavelength of about 2 meters. This radiation had a continuous spectrum, and arose from free electrons in the bright clouds of diffuse nebulae.

In 1944 the first radio-wavelength atomic emission line was predicted by H. C. van de Hulst at Leiden Observatory. The emission arises from what is called a hyperfine transition in the ground state of neutral hydrogen. Let's look at a hydrogen atom in its ground state (Figure 13.17). We will see an electron orbiting around a proton in the smallest orbit possible. If we look more closely, we will see that both the electron and the proton spin. For our purposes, we can view this spin as a rotation of the charged particles on an axis. Because of the spin, each particle generates a magnetic field, and the fields interact with each other. When the electron is in its ground state and is spinning in the same direction as the proton, it has a tiny amount more energy than when it is rotating in the direction opposite to that of the proton. The electron in the same-direction rotation state can spontaneously jump to the

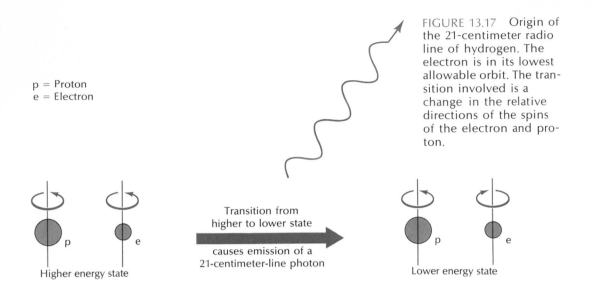

p = Proton
e = Electron

Transition from higher to lower state

causes emission of a 21-centimeter-line photon

Higher energy state

Lower energy state

opposite-direction rotation state by reversing its direction of spin, and in the process emits a photon with a wavelength of 21 centimeters. The probability of the emission is very small. A hydrogen atom capable of emitting 21-centimeter radiation may sit around for several million years before it emits.

Despite this low probability, the 21-centimeter radiation was detected in 1951 by two Harvard physicists. The detection of the hydrogen line shows that there must be a truly immense number of hydrogen atoms capable of emitting the rare line.

Over the years, many other radio lines have been discovered (Table 13.1). The radical OH (a radical is an incomplete molecule; see discussion of comets, Chapter 10) has been detected, as well as the molecules formaldehyde (HCHO), methyl alcohol (CH_3OH), acetaldehyde (CH_3CHO), water, and carbon monoxide, to mention a few. Since water can be viewed as the compound H-OH, the presence of H and OH together virtually guarantees the presence of some water. The presence of the complex organic molecules, such as formaldehyde, was a surprise. It seems clear now that these molecules form naturally, given the proper ingredients and the right set of conditions. Does this have any bearing on life in the universe?

We have now talked about the dark clouds in interstellar space. We have seen that they contain dust, hydrogen and other elements, and a vast array of molecules. Why are these clouds dark, whereas the bright diffuse nebulae are not? What differentiates the two types of nebulae?

Table 13.1 Some Interstellar Molecules Discovered by Radio Techniques (not a complete list)

Molecule	Formula
Hydroxyl radical	OH
Silicon monoxide	SiO
Water	H_2O
Ammonia	NH_3
Methylidine	CH
Hydrogen cyanide	HCN
Carbonyl Sulfide	OCS
Formaldehyde	H_2CO
Isocyanic acid	HNCO
Cyanoacetylene	HC_3N
Formic acid	HCOOH
Methyl cyanide	CH_3CN
Acetaldehyde	CH_3CHO

Bright Clouds. Why are bright clouds bright? The answer is relatively simple, on the surface. The bright clouds contain hot O and B stars that emit copious amounts of ultraviolet light. The ultraviolet causes the clouds to fluoresce by processes similar to those that occur in minerals that fluoresce under an ultraviolet lamp. Examples of bright clouds can be seen in Figures 13.18, 13.19, and 13.20.

The hot O and B stars in gaseous nebulae emit a large fraction of their radiation at wavelengths shorter than 912 Ångstroms. A photon with a wavelength shorter than 912 Ångstroms will ionize a hydrogen atom if it is absorbed by a bound electron in any energy level. Later, the electron may be captured by another proton to form a new hydrogen atom, in a process known as recombination. The electron will jump downward through the energy levels of the atom, and will emit photons of various wavelengths, depending on the levels at which it stops. If it stops at level 2, for instance, one of the photons it emits will be a Balmer-series photon in the visible part of the spectrum.

Since many, many hydrogen atoms participate in the ionization and recombination process, virtually all possible photons are emitted: the nebula emits a bright-line or emission-line spectrum. Since one ultraviolet photon is absorbed in the ionization process, and many photons are emitted in the recombination process, the net effect is to convert a short-wavelength ultraviolet photon into several longer-wavelength photons, some of which are in the visible part of the spectrum. A series of similar, but more complex processes leads to lines of atoms and of ions of nitrogen, oxygen, and neon in the spectra of these nebulae.

The number of ionization processes and recombination processes occurring in a nebula are such that most of the hydrogen is ionized at any one instant. The temperature in a bright gaseous nebula is around 10,000° K, and the densities are exceedingly low, around 100 atoms per cubic centimeter. These ionized nebulae are sometimes called HII regions, and the cool, dark clouds are called HI regions.

A photograph of these hot nebulae in color (Plate VI) shows the presence of green and red hues. The red is due to the well-known Hα line of hydrogen, and the green, so conspicuous in the Orion nebula when it is seen through a small telescope, is due to emissions by ionized oxygen.

HII regions also emit radio radiation. Since the hydrogen is mostly ionized, they do not emit 21-centimeter radiation. Instead, they emit what is called thermal radiation: radiation that is due to the thermal motions of the electrons and has a continuous spectrum

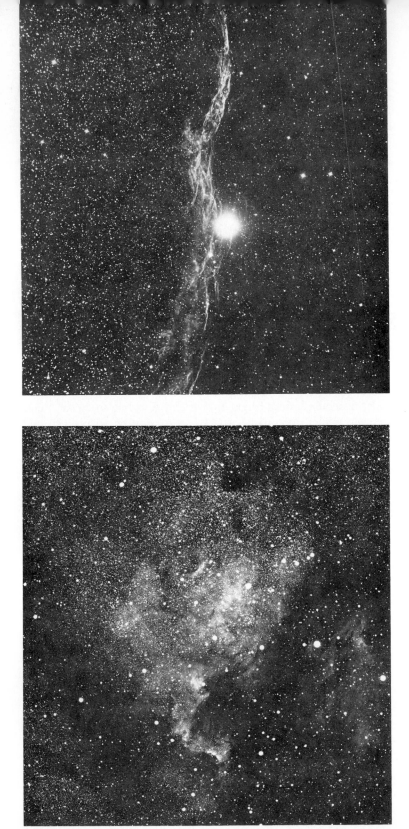

FIGURE 13.18 Part of the Cygnus Loop. The entire loop (see Figure 14.26) is a circle made up of lacy segments like this one. (Hale Observatories.)

FIGURE 13.19 North America Nebula in Cygnus. (Lick Observatory Photograph.)

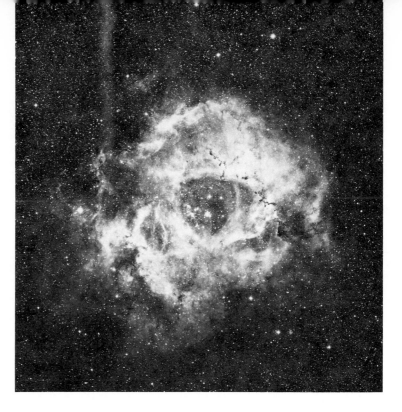

FIGURE 13.20 The Rosette Nebula in the constellation Monoceros. (Hale Observatories.)

with a shape that is characteristic of the temperature. The emission is due to a process called *Bremsstrahlung* (from the German for deceleration radiation). When an electron whizzes past an ion (for example, a proton), the attraction between their charges causes the electron to change its direction of motion. In the process, it can emit a photon of any wavelength. The loss of the energy that the electron gave the photon causes it to slow down.

HII regions also emit radio-wavelength spectrum lines produced by atomic transitions in the small amounts of neutral hydrogen that they contain. These lines originate from transitions between very high levels. For instance, transitions between levels 109 and 108 and between levels 156 and 155 have been observed.

A great deal of new information about the interstellar gas has come from ultraviolet studies by satellites, such as the third Orbiting Astronomical Observatory, called Copernicus. Both HII regions and the dark clouds have been observed. The cool, dark clouds were found to emit radiation produced by the H_2 molecule. The observations indicate that temperatures in the clouds are only about $80°$ K. A similar value can be inferred from the 21-centimeter observations. Interestingly, the evidence we now have seems to say that most of the material in interstellar space is in the form of molecules, mostly H_2. We cannot fully understand the conditions between the stars until we discover the chemistry of this interstellar matter.

Observations by Copernicus also allow a study of the abundances of heavier atoms in the interstellar clouds. These abundances, which turn out to be different from the abundances in the atmosphere of the Sun, may tell us something about how dust grains condense in these clouds. Astronomers are working on this problem today.

If you go back to astronomy texts written a century ago, you will find references to white nebulae. These objects appeared to be clouds just like the Orion nebula, but did not show the colors of the typical gaseous nebulae. One or two of these white nebulae showed a spiral structure when viewed through the giant telescopes of the day, like Lord Rosse's 72-inch reflector, built around 1850. As we shall see, these white nebulae are distant galaxies.

We have now looked at the pieces of our jigsaw puzzle, and it is time to begin fitting them together. In the process, we will learn what our Galaxy is all about.

THE GALAXY

William Herschel, near the end of the eighteenth century, undertook to discover the shape of our Galaxy and to find the Sun's place in it. He reasoned that we live in a highly flattened aggregation of stars (Figure 13.21). The Milky Way, seen stretching across the sky, is a result of that shape. When we look across the plane of the Galaxy, we see many more stars than when we look perpendicular to the plane. Herschel assumed that all stars have the same absolute brightness. From counts of the numbers of stars on a fixed-size area of the sky in many different directions, he concluded that we are

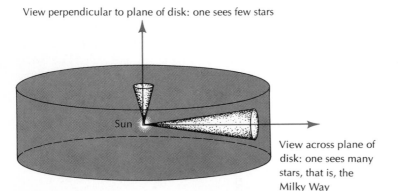

View perpendicular to plane of disk: one sees few stars

Sun

View across plane of disk: one sees many stars, that is, the Milky Way

FIGURE 13.21 William Herschel's concept of our Galaxy. The Sun is at the center of a flat disk-shaped assemblage of stars.

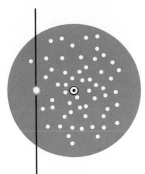

Top view

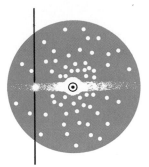

Edge-on view

▣ Schematic locations
of globular clusters

FIGURE 13.22 The distribution of globular clusters form a spherical cloud around the center of the Galaxy.

positioned near the center of the Galaxy, and that its thickness is one-fifth its diameter.

By the beginning of this century, astronomers had arrived at a size for the Galaxy. They used the average absolute magnitudes of stars whose distances had been measured up to that time, and concluded that the diameter was about 700 parsecs.

In the period between 1917 and 1920, Harlow Shapley published a whole new picture of the Galaxy. He believed that the system was about 100,000 parsecs in diameter and 10,000 parsecs thick, and was lens-shaped rather than grindstone-shaped. He placed the Sun 15,000 parsecs from the center of the system. In Shapley's model of the Galaxy, the center lay in the direction of the constellation Sagittarius as seen from Earth. Shapley calculated the Sun's place in the Galaxy and the size of the Galaxy from studies of globular clusters. Of the more than 100 globular clusters known, almost one-third are in or near the constellation Sagittarius. He assumed that the clusters form a spherical halo around the Galaxy, with its center at the center of the Galaxy.

Figure 13.22 is a schematic view of the Galaxy, showing the positions that globular clusters might occupy in a top view and an edge-on view. Note that, as seen from the Earth, most of the clusters should appear in one half of the sky, and should be most concentrated in the direction of the center of the system. Since the globular clusters contain RR Lyrae variable stars, Shapley could use the period-luminosity law to find their distances and thus establish the size of the Galaxy.

When Trumpler established the presence of dimming dust from a study of open clusters, it became clear that Shapley's distances to the globular clusters were also too large, and the Galaxy was in fact smaller than originally thought. Modern estimates place us about 10,000 parsecs from the center of the Galaxy, quite near the edge.

Whereas the globular clusters form a spherical halo around the Galaxy, the open clusters lie near the plane of the system, as does most of the nebulae and dust. We can not see very far across the dusty midplane of the Galaxy. As a result, most of the 900 or so known open clusters are within a few thousand parsecs of the sun (Figure 13.23). If we compare the small volume these 900 clusters

FIGURE 13.23 Distribution of open clusters in the Galaxy. Since the clusters lie in the dusty galactic plane, we see only the nearest 900 or so.

900 or so known open clusters lie
in this sphere centered on the Sun

▨▨▨▨▨ Volume occupied by open clusters

occupy with the huge volume of the entire Galaxy, we must conclude that we can see only a tiny fraction of the total number of clusters. There may be as many as 15,000 galactic clusters in the entire Galaxy.

Shapley's picture of the Galaxy was not universally accepted among astronomers at first. However, as time progressed, it gained credibility. One bit of research that helped things along was a study of the white nebulae. In the early part of this century many photographs of these white nebulae were made at Lick Observatory in California, and they often appeared to exhibit spiral structure. Most astronomers thought they were spiral-shaped gaseous clouds in our own Galaxy, of a type that avoided the Milky Way. However, in 1923 Edwin P. Hubble—working with the 2.5-meter (100-inch) Hooker reflector that began operating at Mount Wilson Observatory in 1917—resolved the outer parts of the great nebula in Andromeda into stars. Thus, it is not a nebula at all. What really turned the tide, however, was Hubble's discovery that among the stars resolved were a few Cepheid variables (Figure 13.24). Using the period-luminosity law, he concluded that the nebula was

FIGURE 13.25 The Great Andromeda galaxy. (Hale Observatories.)

275,000 parsecs away and 10,000 parsecs in diameter, somewhat smaller than our Galaxy, but nevertheless similar in size and content. (See Figure 13.25.)

By this time astronomers accepted the hypothesis that the universe is filled with galaxies, of which our Galaxy is only one example. (See Figures 13.26 and 13.27.) They then turned their attention to studying the detailed structure of our Galaxy. In the following pages, we will discuss this structure.

FIGURE 13.26 The galaxy NGC 3031. If we were to view our Galaxy from a great distance, it would appear similar to this one. (Hale Observatories.)

FIGURE 13.27 The galaxy NGC 4594. Seen edge-on, our Galaxy would look like this galaxy. (Hale Observatories.)

Rotation and Mass of the Galaxy

Earlier, we talked about the motions of the stars in the solar vicinity, and pointed out that they move with speeds around 20 kilometer/second. There are, in addition, a group of stars called high-velocity stars which move with speeds of 60 kilometers/second, or greater. Unlike the low-velocity stars, which move in almost any direction, the high-velocity stars all tend to move toward one hemisphere of the sky. The Swedish astronomer Bertil Lindblad came forward in 1926 with a reasonable hypothesis to explain the high-velocity stars. He asserted that the high-velocity stars belong to a nearly spherical halo of stars surrounding our flattened Galaxy, not unlike the halo formed by the globular clusters (which also show high velocities). Lindblad then stated that the high-velocity stars are actually low-velocity objects, and that the highly flattened part of the Galaxy rotates with a high speed through the halo. The apparent motion of the so-called high-velocity stars toward one hemisphere is due to the fact that the Sun actually moves rapidly toward the opposite hemisphere. Lindblad also pointed out that additional evidence for this hypothesis is provided by the relative degrees of flattening of the two star systems. The highly flattened main body of the Galaxy rotates most rapidly, the extreme flattening being due to the fast rotation. The more spherical halo of high-velocity stars rotates slowly, and therefore is not as highly flattened.

Jan H. Oort quickly proved Lindblad's hypothesis, and showed an unusual feature of the rotation of the Galaxy. The Galaxy does not rotate like a phonograph record, with all stars completing one rotation in the same time. It rotates like our solar system. Stars near the center of the Galaxy complete one rotation much faster than stars near the edge, just as Mercury revolves around the Sun in 88 days, Earth in one year, and Pluto in 250 years.

This differential rotation of the Galaxy shows up in the average motions of distant stars as a function of position around the Milky Way. In Figure 13.28 we show two views of the local region of our Galaxy. In part *A* we show the local region rotating around the center of the Galaxy. In part *B*, we magnify the local region of space. The arrows show the relative motions of several stars if we assume we are looking at the circle as it moves along with the Sun. Note that stars around positions 4 and 8 move toward us on the average, stars at positions 2 and 6 move away from us on the average, and for stars at 1, 3, 5, and 7 the radial velocities average to zero. It is this observation that tells us that the Galaxy rotates

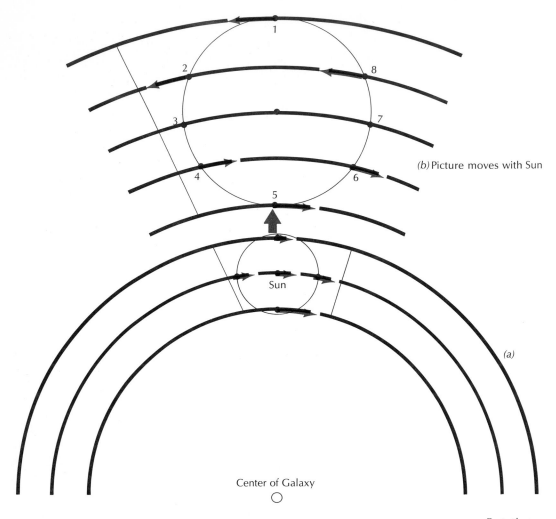

(b) Picture moves with Sun

(a)

Center of Galaxy

FIGURE 13.28   Rotation of the Galaxy (see text).

differentially. If the Galaxy were to rotate like a phonograph record, all the radial velocities would average to zero.

At the Sun's distance from the galactic center, the rotation rate is
roughly 200 kilometers/second. Even so, it takes the Sun
200,000,000 years to rotate once around the center of the Galaxy.
The Galaxy at the Sun's distance from its center has rotated once
since the Triassic Era, when dinosaurs ruled our planet.

Roughly speaking, we can assume that the Sun revolves around
the Galaxy under the influence of the gravity of all the stars, gas,
and dust that are nearer to the center of the Galaxy than we are.

Thus, just as we estimated the Sun's mass from the Earth's orbital parameters, we can estimate the mass of the Galaxy from the Sun's motion around its center. Its distance from the center is 10,000 parsecs, or roughly 2×10^9 astronomical units, and its period is 2×10^8 years. These numbers give a mass of 2×10^{11} Suns. The Galaxy contains as much mass inside the Sun's distance as two hundred billion Suns. A more careful study, accounting for the fact that the mass of the Galaxy is not concentrated at its center, yields a mass of 1.6×10^{11} Suns, not far from the rough estimate.

The work of Oort and Lindblad showed that our Galaxy rotates around a point which agrees with the center as determined by Shapley. The agreement of these results established the Sun's off-center position in the Galaxy once and for all.

We now know that our Galaxy is a flattened disk of stars, gas, and dust, surrounded by a halo of globular clusters and stars. It is about 25,000 parsecs across, and the Sun is 10,000 parsecs from the center. Astronomers next tried to determine if our Galaxy has a spiral structure.

The Spiral Structure of Our Galaxy

We know that a great many other galaxies show spiral structure. The big question that astronomers asked early in this century was: does our Galaxy show spiral structure, too? The answer to their question is yes, although finding that answer has been difficult. We have been hampered by being inside the Galaxy. Distant portions of the system are obscured by the dust and gas in the plane of the Galaxy. Cecilia Payne-Gaposchkin aptly compared the task of mapping the spiral structure of the Galaxy with mapping New York City using only observations made from the corner of 125th Street and Park Avenue.

The spiral arms in other galaxies give us some clues about what we should look for in our Galaxy. Spiral arms are traced by the hot O and B type stars, and by HII regions. The O and B stars make good spiral traces, because they are bright enough to be seen at large distances.

A group of astronomers led by W. W. Morgan of the Yerkes Observatory began the attempt to trace spiral arms using O and B stars in the late 1940s. They succeeded in locating three sections of arms. The so-called Orion arm is in the direction of the constellations Cygnus, Cepheus, Cassiopeia, Perseus, Orion, and Monoceros. It extends 150° along the Milky Way, roughly halfway around the sky. The Sun lies on the inner edge of the Orion arm. The Perseus arm is roughly parallel to the Orion arm, but farther

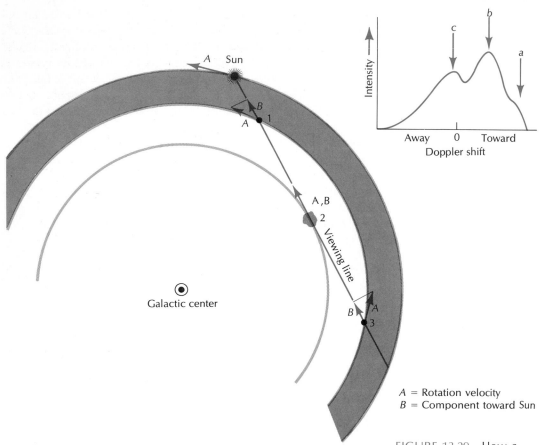

FIGURE 13.29 How a 21-centimeter-line profile tells us about our Galaxy (see text).

A = Rotation velocity
B = Component toward Sun

from the center of the Galaxy, and the Sagittarius arm is nearer the center of the Galaxy. (These names, by the way, are more or less arbitrary.)

This picture has been modified somewhat, and greatly extended, by studies of the distribution of neutral hydrogen using its 21-centimeter radio radiation. Dust and gas between us and a cloud of neutral hydrogen do not affect the radio signal we receive; so we can observe most of the neutral hydrogen in the system. Only the hydrogen on the far side of the nucleus of the Galaxy is difficult to observe. The procedure is as follows.

The astronomer points his radio telescope in a particular direction along the Milky Way and measures the intensity of the 21-centimeter hydrogen emission received. The receiver is swept in frequency around the 21-centimeter frequency (roughly 1,400

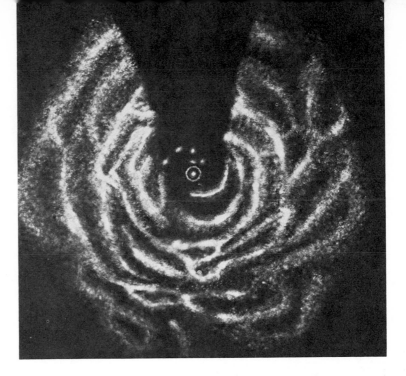

FIGURE 13.30 The spiral structure of our Galaxy as inferred from 21-centimeter studies. (Courtesy G. Westerhout, University of Maryland.)

Megahertz, that is, 1,400 million cycles per second), and a profile of intensity versus frequency is compiled. All the emission is not exactly at the 21-centimeter wavelength, because of Doppler shifts caused by the differential rotation of the Galaxy. Thus, the measured intensity plot can also be considered as intensity versus Doppler shift.

Figure 13.29 shows a schematic picture of an intensity versus Doppler-shift profile, and the line of sight through the Galaxy from which the emission arises. The maximum Doppler shift occurs for material at point 2, where the line of sight passes as close to the center of the Galaxy as possible. That material contributes at b on the emission intensity plot. Material at points 1 and 3 in the Galaxy have the same Doppler shift; so their emissions are mixed together. The peak a in the intensity plot comes from the spiral arm just inside the Sun's distance from the Galactic center and peak c arises from local hydrogen. Studying many such plots from many directions helps us partially remove the ambiguity in the emission from points 1 and 3. We use information on the rotation and size of the Galaxy to relate distances to Doppler shifts. In the end, we build up a picture of the spiral structure of the Galaxy, as seen in Figure 13.30. Incidentally, in this picture, it seems that the Orion arm is not a distinct arm parallel to the Perseus and Sagittarius arms, but is a spur linking the two arms.

We now know what our Galaxy looks like, and how it rotates. It is a spiral galaxy, and, as we shall see when we look at other spirals, has a structure which we would call intermediate. The arms are moderately well-developed and are moderately loosely wound. If we could see it from a distance, it would resemble the Andromeda galaxy. The rest of the story presents some important details that we will need to know in our discussion of the life cycles of stars.

STELLAR POPULATIONS

During the blackouts of the lights of Los Angeles occasioned by World War II, Walter Baade was able to use the 2.5-meter reflector at Mount Wilson to study galaxies under ideal sky conditions. The observatory, in the San Gabriel Mountains overlooking the city, is hampered by lights otherwise. Baade resolved the stars in the nuclear regions of the Andromeda galaxy and in its two elliptical companions. Surprisingly, the stars in these regions differed from stars in the solar neighborhood, but looked much like the stars in the globular clusters.

Baade hypothesized the existence of two stellar populations, which he called Population I and Population II. Population I stars include the galactic star clusters and all the stars in the spiral arms of the Galaxy; Population II includes the globular cluster stars and the high-velocity stars in the galactic halo. There are Cepheid variables in both Populations, by the way. The classical Cepheids belong to Population I, whereas the W Virginis stars (see Chapter 13), which seem superficially similar, are Population II stars.

The existence of these two populations of Cepheids is important, because it led to an early error in measuring the distances to galaxies. Astronomers first estimated the distance to nearby galaxies by comparing Cepheids in the galaxies with the fainter W Virginis stars in globular clusters. When the work of Baade pointed out the error, the result was to double what we thought to be the distance to these nearby galaxies. The Andromeda galaxy, for instance, is now believed to be about 650,000 parsecs away and to be about the same size as our Galaxy.

Like all attempts to cast things into black and white, the concept of two populations of stars has proven to be an oversimplification. There is a spectrum of populations. Basically, the differences in the populations result from their relative ages and their metal contents. The stars in the halo of the Galaxy—high-velocity stars and globular clusters—are very old, and contain very small amounts of metals. The stars in the disk of the Galaxy—the Sun and nearby stars, for instance—are intermediate in age and contain more heavy elements (perhaps created in reactions in older stars). The stars within the spiral arms are very young, and are rich in heavy elements. Astronomers believe the open clusters belong to this group. Thus, astronomers sometimes speak of a Halo, a Disk, and an Arm population of stars. In Chapter 14, we will return to stellar populations and see how their differences depend on their ages and metal contents. The differences tell us something about the life history of our Galaxy.

THE CENTER OF THE GALAXY

The center or nucleus of our Galaxy is a wild place. We cannot see it directly in visible wavelengths because of the tremendous amount of intervening dust. However, the distribution of infrared radiation from the nucleus at wavelengths around 2.2 microns (1 micron is 10^{-6} meters or 10,000 Ångstroms) looks just like the distribution of visible light from the nucleus of the Andromeda galaxy. Of course, we see the nucleus of that galaxy free of problems of dust, because we view it at an oblique angle. Astronomers believe that the above comparison shows that the 2.2-micron radiation from our Galaxy tells us about the distribution of stars here. The conclusion: the center of our Galaxy contains a million stars per cubic parsec. Compare that density to the density of one star per cubic parsec near the Sun.

Longer-wavelength infrared observations (for example, around 20 microns) show several intensely bright sources near the center of the Galaxy. It is thought that they may be several extremely bright stars surrounded by thick clouds of dust. The dust grains are heated by the starlight and emit infrared radiation.

The center of the Galaxy is also a copious emitter of radio radiation of all kinds. It emits synchrotron radiation, thermal radiation, 21-centimeter radiation, and molecular radiation. The main synchrotron source is only about a dozen parsecs in diameter. Incredibly, there are several tiny sources of thermal radiation inside the synchrotron source.

Finally, the center of the Galaxy is a strong source of X-rays. The X-radiation is taken to indicate the presence of an extremely hot mass of gas at the center of the Galaxy.

In putting this picture of the nucleus of the Galaxy together, astronomers have concluded that an extremely violent event occurred there roughly a million years ago. Perhaps the event was the explosion of a monstrous star which formed from the very dense material there. Today, we believe that a star that is more than about 50 times more massive than the Sun cannot be stable. If a starlike object with thousands of solar masses formed, it would explode as soon as its core reached high-enough temperatures for nuclear reactions to begin. As we shall see, there is evidence for similar processes in the nuclei of other galaxies. With those few brief words, we must leave the fascinating topic of the nucleus of the Galaxy.

We have now put together a picture of our Galaxy. It is an immense, flat, rotating disk of dust, gas, and stars, surrounded by a halo of stars and star clusters. The Galaxy is so huge that light takes 80,000 years to cross its 25,000-parsec diameter, traveling at 300,000 kilometers/second. We, on our puny planet orbiting an ordinary star, live at the outskirts of the Galaxy. Poor mankind. We began with the belief that we were the center of the universe. At each step of our growing knowledge of the universe, we become less central. Frankly, whenever I think of our insignificant place in the universe, and contemplate the fact that we live on one of a hundred billion stars in the Galaxy, I become more firm in my belief that we cannot be the only intelligent creatures in the Galaxy—but more about this later. Let's now turn our attention to the life histories of stars.

REVIEW QUESTIONS

1. Describe: (a) open clusters; (b) globular clusters; (c) associations.
2. How was the presence of interstellar dust, not in dark clouds, determined?
3. How can one show the presence of a cool interstellar gas? Hot interstellar gas? Molecules in space?
4. Describe the shape of our Galaxy. How was our position in the Galaxy determined? How does the Galaxy rotate? How do we determine its spiral structure?
5. Compare and contrast Population I and Population II stars.
6. Describe the center of the Galaxy.

LIFE HISTORIES OF STARS

The life history of a star resembles that of a living organism. The star is born out of the interstellar matter, passes through infancy, youth, middle age, and old age, and finally dies. To be able to trace the life history through all these stages is a difficult task: one that has not yet been fully completed. First of all, we must understand the processes which occur deep inside stars. However, the radiation that reaches us from any star has been emitted by atoms in its outermost layers: what has come to be called the star's atmosphere. Light does not reach us directly from the star's interior; the gas is too opaque for that. We have already described our one means of probing the interior of stars when we talked about nuclear reactions in the Sun (Chapter 8).

Nuclear reactions suspected to occur in the Sun and other stars emit neutrinos, which pass through the body of the Sun to empty space. Some eventually reach the Earth. The problem is that attempts to measure these neutrinos have so far come up with fewer neutrinos than are predicted by theory. Thus there is reason to doubt the correctness of our picture of the interior structure of stars. Our whole discussion must proceed with a black cloud hovering over our heads. We will talk about the best modern ideas; but we must remember that somewhere there lurks a flaw. As these words are being written, astrophysicists are deeply involved in searching for that flaw. Perhaps by the time you read this, it will have been found. If the flaw is with the experiment that has been attempting to detect neutrinos, the whole picture may not change. If the flaw is with the theory of stellar structure, much research will be required to build a new theory.

How are stars formed? How do they change as time progresses? How do they die? These are questions that we will address in this chapter. Our earlier discussion of the characteristics of stars—brightnesses, masses, and radii—outlined important information that we will need as we study the life histories of stars. Our consideration of the structure of the Milky Way galaxy and the stars, star clusters, gas, and dust it contains will also turn out to be vitally important; so, as we progress through the topic here, we will have to keep these earlier facts in mind. To start our study, we will take a close look at the internal structure of a star.

THE INTERNAL STRUCTURE OF A STAR

If we cannot see the internal structure of a star directly, how do we know about it? The answer, in simple terms, is that we construct a model of the structure based on what we know about the physical processes that must be occurring in that star. The model is expressed in precise mathematical terms. The detailed computations of the model are carried out on a large, modern computer. The model is a creation of the human intellect, constrained to obey known physical law and to look superficially like a star. Here is one of the many areas where our scientific genius shines. Let's look at how the model is made.

There are several important assumptions in the theoretical model of a star. We assume, for instance, that physical laws which have been discovered and studied on Earth also hold under the extreme conditions in a star's interior. We believe that gasses in the star follow the same laws as gasses on earth; and that nuclear reactions in the star behave like nuclear reactions studied in earthbound physics laboratories and in hydrogen-bomb explosions.

Perhaps the most important assumption made about stars is that they are in a steady state, that is, that any changes which take place in the star as time passes are so very slow that we can assume the star is not changing as we calculate the model. For most of the stars that we can see, this assumption is clearly true. We have been making detailed studies of stars throughout this century and have found that, with one or two rare exceptions, the stars have not measurably changed in size, temperature, or other characteristics. This is true of the Sun, on the average, too. We say on the average, for we know the level of activity changes with the sunspot cycle. But the Sun's size and total energy output appear to be rock steady.

Hydrostatic and Thermal Equilibrium

A star is assumed to be in hydrostatic equilibrium (Figure 14.1). If we look at a bubble of gas at some distance from the center of a star, we find two forces acting upon it. Gravity tends to pull the bubble toward the center of the star.* The second force acting on the bubble is pressure. Pressure decreases outward in the star. The pressure at the point on the bubble nearest the center of the star is greater than the pressure at the point on the bubble nearest the surface. Thus the outward push on the bubble exceeds the inward push, and there is a net upward force on the bubble. The same thing holds true for bubbles in the ocean. When a diver exhales, the bubbles of air rise to the surface because the water pressure pushing them upward exceeds the pressure pushing them downward. For the gas bubble in the star, the downward pull of gravity just balances the net upward push due to pressure, and the total force on the bubble is zero. We assume that the bubble is at rest. With zero net force acting on it, it will remain at rest. The words in this paragraph can be expressed as mathematical equations.

The condition of thermal equilibrium says that, as energy flows through the star from its source in the star's core to the surface, there is no increase or decrease in the amount of energy contained in any part of the star (Figure 14.2). This condition tells us that the temperature at any point in the star stays constant as time passes. A buildup or decrease in energy at a point would lead to an increase or decrease of temperature at that point. However, the fact that energy flows outward through the star indicates a temperature decrease outward from the core of the star to the surface. Energy always

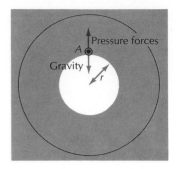

FIGURE 14.1 Hydrostatic equilibrium. Pressure forces pushing upward on the bubble A are just balanced by the force of gravity on it. Thus the bubble experiences no net force.

---

*Only the part of the mass that is closer to the center of the star than the bubble exerts a pull on the bubble toward the center of the star. The total pull of all the mass farther from the center of the star than the bubble sums to zero.

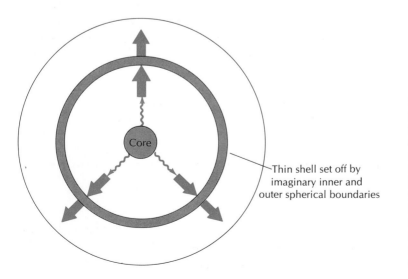

Thin shell set off by imaginary inner and outer spherical boundaries

FIGURE 14.2 Energy flow through a star in radiative equilibrium. Look at a thin shell of material set off by imaginary inner and outer spherical boundaries. The energy flowing in the bottom of the thin shell exactly balances the energy flowing out the top. There is no net change in the amount of energy inside the shell.

flows from a hot region to a cold region, and the rate of flow depends on the temperature difference.

The condition of thermal equilibrium leads to another equation, which relates the outward flow of energy and the temperature structure of the star to the properties of the gas. The form of the equation depends on whether radiation processes or convective gas motions carry the energy.

Energy Transport

There are two main types of processes which carry energy outward through stars: radiative processes and convective processes (Figure 14.3). Both radiative and convective processes occur in most stars. Which process carries energy depends on the opacity of the gas. If the gas is relatively transparent, photons will flow toward the surface of the star, being absorbed and reemitted here and there. This physical process is known as **radiative energy transport**. In a region of a star where the gas is very opaque, photons can no longer flow outward. At the bottom of such a layer, energy would build up, increasing the temperature of the gas, and causing a progressive change in the star's structure with the passage of time, if some process of energy transport other than radiative transport did not occur. Typically, however, **convective energy transport** occurs in opaque layers of stars. In convective transport, an opaque bubble of gas at the bottom of the region absorbs photons, then begins to rise. As soon as the bubble moves a small distance upward, it is warmer and less dense than its surroundings, and as a result is buoyant. The bubble will continue to rise as long as it remains warmer and, as a result, less dense than its surroundings.

FIGURE 14.3 Energy transport. (a) Photons flow outward through the star, being absorbed and reemitted by the material. (b) Mass motion carries energy through the star. Bubbles of gas absorb photons and rise through opaque layers of material. At the top the photons are radiated into more transparent material.

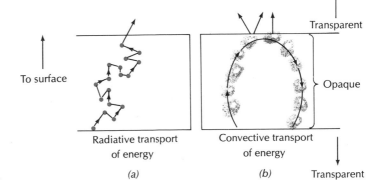

To surface

Radiative transport of energy

(a)

Transparent

Opaque

Transparent

Convective transport of energy

(b)

Finally, the bubble reaches a level where the stellar material is again transparent. It will then radiate away its energy until it cools to the temperature of the surrounding gas, and loses its identity. The net result of the process is that the bubble has carried energy up through the opaque layer of gas.

Energy Generation

We have already discussed the problem of energy generation by nuclear reactions in Chapter 8. Astronomers believe that two different reaction chains can occur in main-sequence stars: the proton-proton chain and the carbon-nitrogen cycle. The result of the two cycles is the same: the transformation of hydrogen to helium. The rate at which one cycle or the other generates energy depends on the temperature of the interior and the density of the particles that participate in the reactions. The rate can be expressed as an equation.

As we have gone through the discussion here, we have mentioned several equations that express the equilibrium state of the star and the nuclear-reaction rates. These equations, together with a knowledge of the mode of energy transport, the behavior of the gas at different temperatures and pressures, the opacity of the gas, and certain boundary conditions, such as the surface pressure and temperature (from observations, say), specify the interior structure of the star. All one needs is a big computer to solve the equations and calculate pressure, density, temperature, and other quantities as a function of position in the star. A table of these quantities is called the model of the star.

Main-Sequence Stars

Models calculated for stars on the main sequence of the H-R diagram reveal two possible structures, depending on the mass of the star (Figure 14.4). Massive main-sequence stars have relatively opaque cores, in which radiative processes are not able to carry away the vast amounts of energy generated. The nuclear energy, created by the carbon-nitrogen cycle, is therefore carried outward through the core by convection. The outer part of the massive star's interior, on the other hand, consists of relatively transparent gas, and radiative processes do carry the energy outward to the surface. Stars with masses more than twice that of the Sun have this type of structure.

Less-massive main-sequence stars have structures like the Sun's. Energy is generated in their cores by the proton-proton chain. The

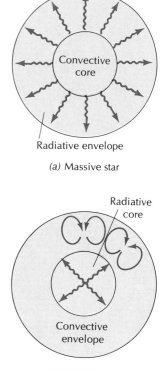

FIGURE 14.4 Two types of main-sequence stars. (a) Very massive stars have opaque cores, where convection must carry energy outward. Their outer envelopes are in radiative equilibrium. (b) Less-massive stars like the Sun have cores in radiative equilibrium and envelopes in which convection carries energy.

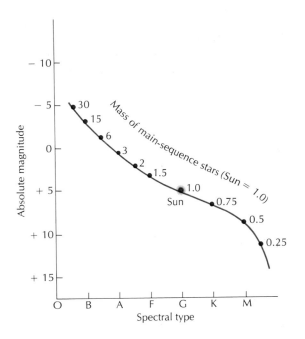

FIGURE 14.5   Positions of stars on the main sequence as a function of their mass.

energy is then carried outward by radiative processes until it reaches the opaque gasses of the convective zone, where gas motion then takes over and carries the energy to the star's surface.

The position a star occupies on the main sequence depends on its mass (Figure 14.5). The more massive a star is, the brighter it is. The range of brightnesses of main-sequence stars is quite large: the brightest stars are as much as 100,000 times brighter than the Sun; the faintest stars may be 25,000 times less bright than the Sun. The range in masses is considerably smaller, however: from roughly 30 times the Sun's mass to about one-tenth the Sun's mass. The relationship between brightness and mass of main-sequence stars gives rise to the mass-luminosity diagram.

The main sequence shown in Figure 14.5 is the so-called Zero-Age Main Sequence (ZAMS). That is, it is the main sequence formed by stars that are just beginning to use their hydrogen as a nuclear fuel. As we shall see, stars do change, incredibly slowly in most cases, as they use up their supply of hydrogen.

An interesting sidelight is the difference between the outer layers of the upper main-sequence stars and those of the lower main-sequence stars. If, as we conjectured in Chapter 8, the chromosphere and the corona of the Sun are created by mechanical energy from the convection zone that is deposited in the outer atmosphere, then we might expect all lower main-sequence stars to have convection zones, chromospheres, and coronas, and all upper main-sequence stars to lack outer convection zones and to be devoid of these structures. Astronomers are using satellites to search for evidence of chromospheres and coronas in stars. More study is required before any definite conclusions can be drawn.

STELLAR EVOLUTION

We are going to talk about the way a star changes from youth to old age, then return and talk about the birth of stars.

What happens to the Sun as it begins to deplete its nuclear fuel? The answer to this question has been found by calculating a series of solar models in which an increasing amount of hydrogen in the core has been replaced by helium. Each model has a different rate of energy generation and different physical characteristics at its center. Each model lies an increasing distance above and to the right of the Sun's ZAMS position. Each model can be assigned an age by asking how long it would take to go from one model to the next, if we assume that the calculated rate of hydrogen burning of the first model produces the additional helium in the core of the second model.*

If we compare the series of solar models with the present-day Sun, we will find one with predicted characteristics (temperature and brightness) that look most like the observed characteristics of the Sun. That model turns out to be one that has been burning its hydrogen fuel for about five billion years. This age agrees with our estimate of the age of the solar system, which makes us reasonably comfortable with the result.

The present model for the Sun is roughly 1.6 times brighter, 1.04 times larger, and 600°K hotter at the surface than the zero-age model. The additional models let us predict that the Sun will continue to evolve slowly for another five billion years, then will begin a rapid evolution away from the main sequence. The rapid evolution begins when a sizable fraction of the hydrogen in its core is exhausted. Thus, the Sun's main-sequence lifetime is about ten billion years.

At the upper end of the main sequence we find hot O-type stars with 30 times the Sun's mass and 100,000 times its luminosity. To

---

*The nuclear-fusion reactions that convert hydrogen to helium are often called "hydrogen burning." This is a loose usage of terminology. In fact, the physical process is not "burning" in the usual sense of the word.

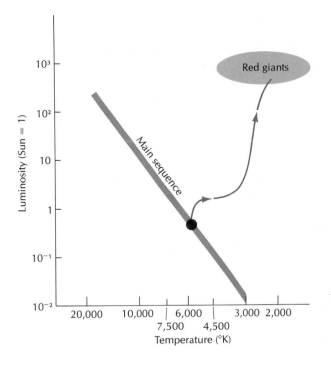

FIGURE 14.6 Evolution of a solar-like star to the red-giant region of the H–R diagram.

keep shining 100,000 brighter than the Sun, that star must use its fuel 100,000 times faster than the Sun does. But it has only 30 times more fuel than the Sun. Thus the O-type star should remain on the main sequence for a period that is 100,000 ÷ 30, or roughly 3,000, times shorter than the period for the Sun. Hence an O-type star will remain on the main sequence for only about 3 million years.

What happens when a star begins to evolve rapidly away from the main sequence? Stars of all masses converge toward the red-giant region of the H–R diagram (Figure 14.6). When the Sun reaches the red-giant phase, it will be almost a hundred times larger than its present size and will engulf the planets Mercury and Venus, and might engulf Earth, too (Figure 14.7). Let's look briefly at what happens inside the star.

When a main-sequence star has exhausted the hydrogen at its very center, nuclear reactions cease there, temporarily. However, hydrogen burning does continue in a shell surrounding the center. In that shell, hydrogen still remains to be consumed. Inside the shell is a core of almost pure helium. As time progresses, more hydrogen is converted to helium in the shell, and as a result the helium core grows in mass. Major readjustments now begin to

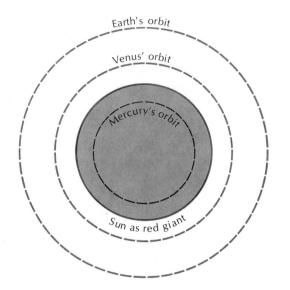

Earth's orbit

Venus' orbit

Mercury's orbit

Sun as red giant

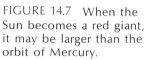

FIGURE 14.7 When the Sun becomes a red giant, it may be larger than the orbit of Mercury.

occur in the star's internal structure. Without nuclear-energy generation in the core, it cannot continue to support itself against the tremendous pressure of the overlying layers. The core begins to contract. As it does so, it gets hotter. Some of the heat is radiated away, adding to the total energy output of the star. In fact, the contracting core causes the energy output of the star to rise rapidly.

Interestingly enough, the outer parts of the star are affected in an unexpected way by the increase in the star's energy flow. These outer layers swell up and cool. As the core of the star becomes increasingly smaller, hotter, and denser, the remainder of the star becomes larger, cooler, and more tenuous. The star becomes a red giant.

Some very strange things happen in the core of the red-giant star. As the density of the matter increases, it becomes **degenerate**. In an ordinary gas there is a simple relationship between pressure, temperature, and density, known as the ideal-gas law. For instance, if you add heat to an ordinary gas, the pressure will increase. If the gas is not confined, it will expand. The nature of an ordinary gas can be explained by picturing the atoms and electrons as billiard balls that move about freely, colliding together at random. In a degenerate gas, such as is found in the core of a red-giant star, the electrons are degenerate, but the atoms and ions form an ordinary gas. The electrons cannot move about at will. The motion of one electron affects the motions of all other electrons. If one adds heat to the degenerate gas, the pressure actually decreases, and the gas contracts—just the opposite from the behavior of an ordinary gas.

In the core of a red-giant star, the degenerate material contracts and heats up until it reaches an astounding temperature of 100,000,000° K. At this point, helium burning begins. That is, nuclear reactions occur which convert helium to heavier elements, such as carbon. When the nuclear reactions start, they add heat to the degenerate gas. The added heat causes the pressure to decrease and the core to compress. The compression in turn heats the gas more, the temperature rises, the nuclear reactions go faster and add more heat to the gas. The entire process is unstable, and it goes faster and faster. A run-away nuclear reaction known as the **helium flash** occurs. Finally, however, the temperature becomes so very high that the gas passes from the degenerate state to the ordinary gas state. Then, as nuclear reactions add even more heat to the gas, pressure increases, causing the gas to expand and cool. The runaway then stops, and the reactions slow down to a steady rate. The red giant has now reached its maximum size.

Observational Evidence: Ages of Star Clusters

Do we have any evidence to support the theory we have discussed so far? The answer is yes. Earlier we talked about the color-magnitude diagrams for galactic star clusters and how one can fit their lower main sequences together to find relative distances. This process showed that the upper main sequences of these clusters were substantially different (Figure 14.8). The galactic clusters known as the double cluster in Perseus (or h and χ Persei) have a main sequence which contains hot blue stars, the hottest being around spectral class B0. The brightest stars in the Pleiades cluster, on the other hand, have spectral types around B8, and the brightest stars in the Hyades are of spectral class A5. In the cluster M67, even the A-type stars are missing from the main sequence.

The theory predicts the shapes of these observed color-magnitude diagrams, and shows that the differences are due to the clusters' ages. The double cluster is new-born by astronomical standards, and its hot, massive O stars are the only objects that have had time to evolve. The gap between the top of the main sequence and the red giants is due to the rapidity of the evolution across that region of the diagram. Stars change so quickly there that the odds of seeing a star in just that phase of its evolution are very small. The successively older clusters have shorter and shorter upper main sequences, until clusters like the ancient cluster M67 have even lost their A-type main-sequence stars.

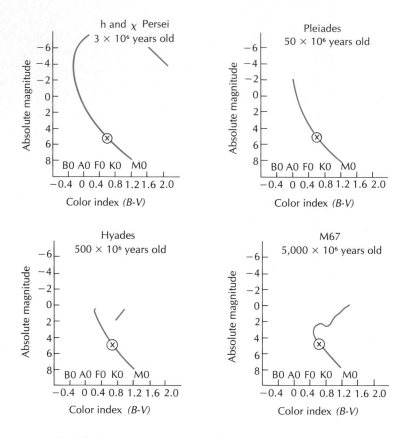

FIGURE 14.8 Color-magnitude version of the H–R diagrams of four different open clusters. The ages of the clusters derived from the theory of stellar evolution are indicated.

⊗ Marks the position the Sun would occupy on each chart

Figure 14.9 shows a schematic color-magnitude diagram for a typical globular cluster. Note that this diagram is similar in part to the diagram for the old galactic cluster M67. Globular clusters are believed to be extremely ancient, perhaps as old as 10^{10} years. The horizontal branch at absolute magnitude 0 represents stars that have passed the red-giant stage and are evolving to their deaths.

The gap in the horizontal branch lies at another point where stars evolve rapidly. Interestingly, the RR Lyrae stars fall in this gap. Undoubtedly, the instability that results in their pulsation is related to that rapid evolution.

A globular cluster, with its hundreds of thousands of stars, has relatively few horizontal branch stars. A galactic cluster like M67, with a total of under 100 stars, has far too few stars for the horizon-

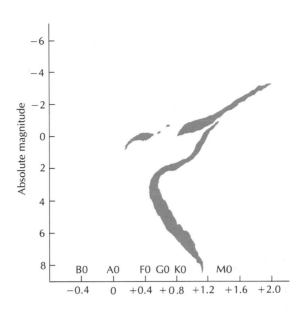

FIGURE 14.9 Color-magnitude version of the H–R diagram for a typical globular cluster.

tal branch to show up. Surely, though, some stars have evolved through that part of the color-magnitude diagram in this cluster. What happens to stars after they leave the horizontal branch?

LATE STAGES OF STELLAR EVOLUTION

Figure 14.10 is a temperature-versus-luminosity version of our earlier color-magnitude diagrams, with the position that globular-cluster stars occupy shaded in. The track that the temperature and luminosity of a star probably follows in this diagram after it completes the red-giant stage of its life is indicated. The reason we do not see stars that are evolving downward from the horizontal branch in globular clusters is once again because the evolution downward in the diagram is so rapid.

Finally, a star reaches the white-dwarf stage, and its evolution slows down. The star then continues a slow evolution along a line sloping downward and to the right. Its radius remains constant as it slowly cools. Its total energy output comes from the remaining heat in its interior: its nuclear fuel has been exhausted. In the end, the star probably becomes a black dwarf: the cool cinder left after all its internal heat has been radiated away.

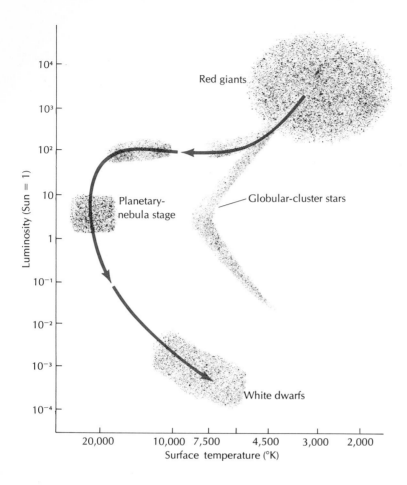

FIGURE 14.10 Stellar evolution after the red-giant phase; the positions of globular-cluster stars in the diagram are shaded.

White Dwarfs

White dwarfs are faint stars with absolute magnitudes typically less than 10, making them at least 100 times fainter than the Sun. Most of these stars show the characteristics of spectral class A, which means that they have temperatures around 10,000° K and appear to be white. The typical white color and low luminosity account for their names.

The discovery of the first white dwarf is an interesting story. The motion of the bright, nearby star Sirius had been found to be slightly irregular as early as 1844 (Figure 14.11). Its proper motion of roughly 1 second of arc per year was not along a perfectly straight line across the sky, but rather along a slightly wavy path.

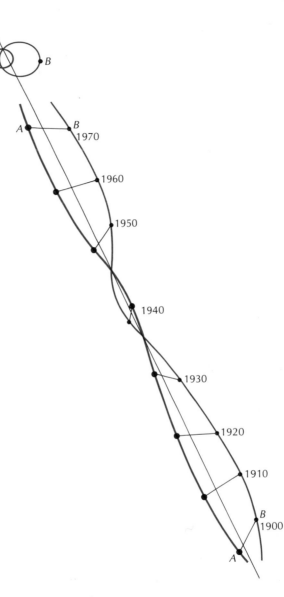

Astronomers concluded that Sirius was a double star, and that its "wavy" proper motion is a combination of an orbital motion around another star and a straight-line proper motion. The other star would have to be quite massive; so why couldn't it be seen?

In 1862, a new refracting telescope with a 46-centimeter (18-inch) objective lens was installed by Alvin Clark, its builder, at the

Dearborn Observatory in Illinois. To test the telescope, Clark turned it toward Sirius and saw the star's faint companion, ten magnitudes fainter than Sirius itself (Figure 14.12).

It was not long before the strange characteristics of Sirius B—as the companion came to be called—were discovered. The best picture of Sirius B available today places its effective temperature at 32,000° K, its radius at 5,400 kilometers, slightly smaller than the Earth, and its mean density at roughly 3×10^6 grams per cubic centimeter. The numbers found in 1862 were similar. Astronomers of the late nineteenth century, when this all came to light, were justifiably skeptical. However, new discoveries of white dwarfs soon poured in, and by today almost 2,000 have been found. There is no longer any doubt about their character.

The combination of tiny size and large mass in white dwarfs means that their surface gravities are truly stupendous. The gravity at the surface of Sirius B is half a million times greater than that at the surface of the Earth. An 80-kilogram (180-pound) man would weigh 40,000 *tons* on Sirius B—if he could exist there.

The spectra of white dwarfs show unmistakable evidence of the high densities in the star's atmosphere. By the same token, the spectra also show the characteristics of the usual spectral types, too. Thus, the nomenclature D0, DB, DA, DF, DG, DK, DM has been

FIGURE 14.12 Three exposures of Sirius, taken with the Lick 3-meter reflecting telescope. The companion is easily visible in the two shortest exposures. (Lick Observatory photograph.)

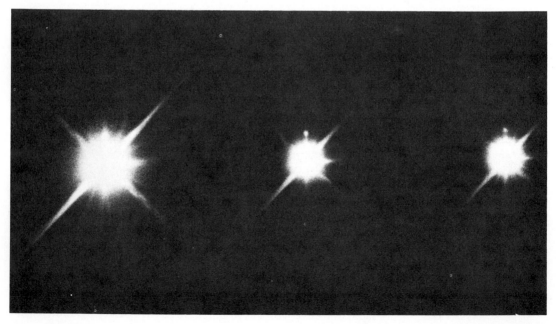

developed to name the spectral types of white dwarfs. (Other names are used, too, but we need not discuss them here). In a list published in 1968, 72% of the white dwarfs tabulated were class DA; that is, they had the characteristics of spectral class A.

The question of the internal structure of white dwarfs was tackled in the 1930s by S. Chandrasekhar. He showed that the stars consist of degenerate matter that can no longer support nuclear reactions without becoming unstable. Chandrasekhar also showed that there is an upper limit to the mass of a white dwarf. A white dwarf with a mass greater than 1.5 solar masses would be unstable, and could not continue to exist. As we shall see, this is an important fact.

The nature of the evolution of stars from the horizontal branch to white dwarfs is not very well understood. It is very difficult to carry out the calculation of the evolution at these late stages. However, we have one observational clue about this phase of evolution in fascinating objects known as planetary nebulae.

Planetary Nebulae

Planetary nebulae are roughly spherical clouds of gas surrounding very hot central stars. The ultraviolet light from the central stars ionizes the gas and causes it to shine by processes similar to those that occur in other bright gaseous nebulae. The name of these objects comes from their resemblance to the disks of planets (see Figures 14.13 through 14.16). Basically, they appear as rings or combinations of rings and disks. The ring nebula in Lyra, Figure 14.13, is a lovely sight through a small telescope. A telescope with a

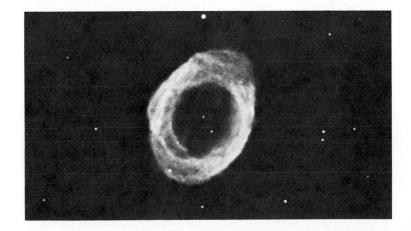

FIGURE 14.13 Ring-shaped planetary nebula in the constellation Lyra. This planetary appears as a faint ring even in a small telescope. (Lick Observatory photograph.)

FIGURE 14.14 Spectacular planetary nebula, NGC 7293, in Aquarius. (Hale Observatories.)

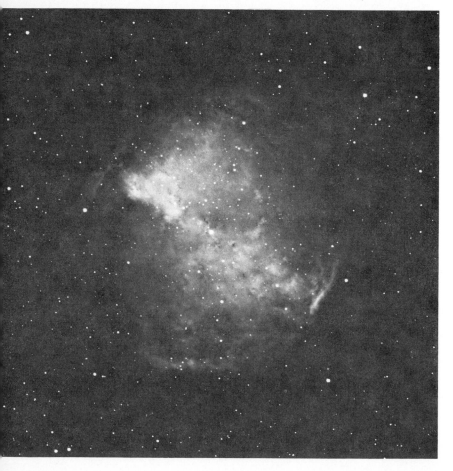

FIGURE 14.15 Planetary nebula in the constellation Vulpecula, known as the Dumbbell Nebula. (Lick Observatory photograph.)

FIGURE 14.16 The Owl Nebula, another planetary. (Hale Observatories.)

9-centimeter (3.5-inch) aperture and a moderate-power eyepiece (say, 50 times magnification) will show the nebula clearly to be a faint, pale ring.

The physical processes which occur in the planetary nebulae to make them emit radiation have been studied extensively over the years. However, we will not concern ourselves with that fascinating problem here. Rather, we will concentrate on the central stars of the nebulae.

The planetary nebulae belong to the Disk population of stars, along with the Sun. This fact has been inferred from their distribution in the Galaxy. Compiling all available information on distances, spectra, and brightnesses, we place the central stars of the planetary nebulae in the H–R diagram as indicated in Figure 14.17.

We now believe that a star passes through the planetary-nebula stage just before it becomes a white dwarf. In a rather gentle event, the star blows off an outer layer of material that contains enough hydrogen to indicate that the material did not undergo extensive nuclear reactions in the past. This material forms the planetary nebula, and its hydrogen content is clearly indicated by studies of the spectra of these nebulae. The nebula itself contains 0.1 to 0.2 solar masses and expands at a rate around 30 kilometers/second. Astronomers have several ideas on what the process might be that

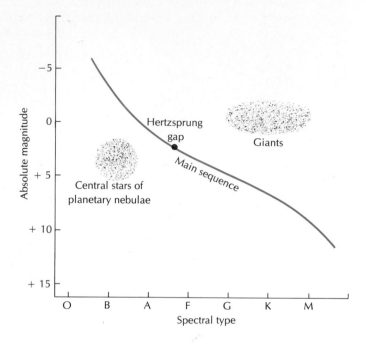

FIGURE 14.17 Hertz-sprung–Russell diagram showing the positions of the central stars of planetary nebulae (see also Figure 14.10).

ejects the outer layers from the star to produce the nebula, but have not yet firmly decided on any one answer to this question.

There are many questions to be answered about the planetary-nebula stage of stellar evolution. Some astronomers believe that all stars that have one to four solar masses go through this stage. Is this true? Do the more massive stars in that range experience more than one ejection? Is this stage a way for the more massive stars to get below the Chandrasekhar mass limit for white-dwarf stars? Astronomers are seeking the answers to these questions.

White-dwarf stars seem to be the major players in the current theoretical explanation of a type of explosive variable star known as a **nova**. The word *nova* (plural, *novae*) means "new" in Latin. These explosive stars increase in brightness from faint stars, visible only with a telescope, often to naked-eye visibility (Figure 14.18). Early astronomers thought they were new stars in the sky.

Novae. In August 1975, a new star was seen in the constellation Cygnus (Figure 14.19). When it reached its maximum brillance, it was a second-magnitude object. Thus Nova Cygni 1975 appeared on the scene. During September the nova faded, reaching seventh magnitude by the end of the month. Astronomers searched older

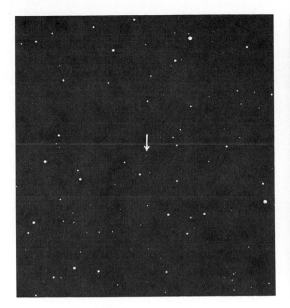

FIGURE 14.18 Nova Herculis 1934. The two exposures show the object before and after outburst. (Yerkes Observatory.)

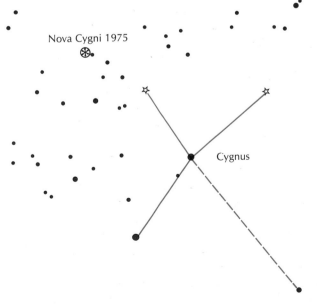

FIGURE 14.19 Position of Nova Cygni 1975 in the sky. Compare with Figure 12.3.

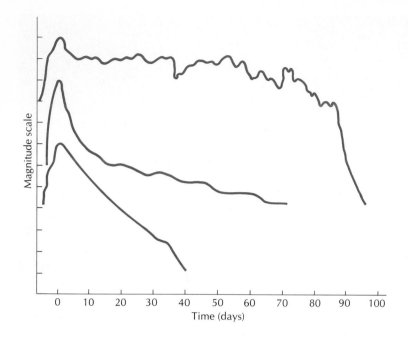

FIGURE 14.20 Schematic light curves of typical novae.

photographic records to see if they could find a star at the location of the nova, but could find nothing. They concluded that the star was fainter than magnitude 21 before the outburst. To reach its peak magnitude, the nova had to brighten by more than 19 magnitudes or by more than a factor of 4×10^7. The typical nova brightens in a few days to 10^4 to 10^5 times its pre-outburst luminosity. Thus Nova Cygni 1975 seems to have undergone an unusually large outburst.

After reaching maximum brightness, the star fades, rapidly at first, then more slowly (Figure 14.20). Eventually many novae settle down to look like the pre-explosion star, seemingly none the worse for wear. As this statement implies, for many novae the star can be identified on pre-outburst photographs, where it appears to be a hot, blue object.

The explosion spews material out of the star at speeds up to 2,000 kilometers/second. After the cataclysmic event, the material can sometimes be seen as an expanding cloud around the nova, which permits us to estimate the distance to the nova (Figure 14.21). If we measure the size of the cloud on some date, and assume that it grew to that size at a constant rate starting on the day of the outburst, we can calculate the angular rate of expansion of the cloud. If we also take a spectrogram of the nova, we can

FIGURE 14.21 Distance to a nova can be inferred by studying the expanding shell of material around it. (a) The linear rate of growth of the shell can be calculated from the Doppler shift of its spectrum lines, and the angular rate of growth can be measured by comparing photographs made at two different times. (b) The result is a triangle, with one side and three angle's known. The third side, distance, can be found easily.

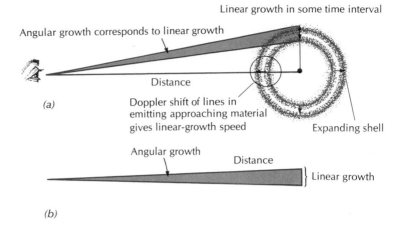

Linear growth in some time interval

Angular growth corresponds to linear growth

(a)

Distance

Doppler shift of lines in emitting approaching material gives linear-growth speed

Expanding shell

Angular growth

Distance

Linear growth

(b)

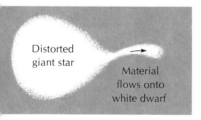

Distorted giant star

Material flows onto white dwarf

FIGURE 14.22 Model of a nova (see text).

calculate the linear rate of expansion from its spectrum lines. We can then ask, how far away is the star if the measured linear rate of expansion produces the observed angular rate of expansion? From the distances estimated in this way, we find that novae reach absolute magnitudes between −6 and −9, and that the pre-outburst stars have absolute magnitudes around +5.

The exact cause of the nova phenomenon was a mystery until it was discovered that all novae are members of close spectroscopic binary systems (Figure 14.22). We believe that the hot, blue dwarf star is a very old object, about to become a white dwarf and rapidly approaching the extremely high densities found in white dwarfs. The material in the star is believed to be made of heavier elements; the hydrogen was all consumed in nuclear reactions earlier in the star's lifetime. The second star in the system is an ordinary giant star. However, it is so large that its surface gas, at the point between the two stars, may be more strongly influenced by the gravity of the dwarf star than by that of the giant star itself. Material will then flow from the giant star onto the dwarf star, where it is compressed to degeneracy and heated by that star's strong gravitational force. Eventually enough compressed degenerate material will be present at very high temperatures that nuclear reactions can start. The reactions will then proceed explosively in the superdense material. If this picture is correct, the process may begin again after the cataclysm subsides and new material begins to flow, leading to later outbursts.

Interestingly, there is a class of stars known as dwarf novae that show smaller outbursts on time-scales of a few weeks or months.

The larger the outburst in dwarf novae, the longer the time between outbursts. These stars are also members of close binary systems. Perhaps ordinary novae are just the extreme of a sequence of similar explosive stars that all work by the same process. If we wait long enough, say, a few thousand years, Nova Cygni 1975 may explode again. Much more study is required to see if this speculation is correct.

Although the violence of the nova explosion is incredibly huge by any terrestrial standards, it pales into insignificance next to the unimaginably intense power of the supernova, discussed in the next section.

We have talked about one scenario for the late stages of stellar evolution here, and have discussed how stars with relatively small masses may become white dwarfs. However, there is at least one other way that stars can die. If we think of the gentle planetary nebula event as a dying whimper, then the supernova explosion is a dying bang.

SUPERNOVAE, PULSARS, AND NEUTRON STARS

Stars with masses less than about four solar masses lose mass by the planetary-nebula process and become white dwarfs. What happens to the upper main-sequence stars, which have 10 or 20 solar masses at the end of their lives? Astrophysicists have studied nuclear-fusion reactions, such as the fusion of two ^{16}O nuclei, which will take place when a star reaches the incredible temperature of $1.5 \times 10^{9\circ}$ K (that is, 1.5 billion degrees). It is believed that this extreme temperature may occur in the core of a very massive star in its late stages of evolution. The oxygen-burning reaction is very sensitive to temperature, and if, as the reaction proceeds, the temperature climbs much above the 1.5×10^9 K level, the reaction will suddenly run away, perhaps leading to a supernova explosion.

Once in a while, a novalike outburst will occur in a distant galaxy, producing a stellar object that rivals the entire galaxy in brilliance for a few weeks (Figure 14.23). Such an outburst is a **supernova**. These rare events, which may occur once every century or so in any galaxy, can emit as much energy in a week as our Sun releases in tens of millions of years. (See Figure 14.23.)

Scientists have searched historical records and have found evidence for seven supernova events in our Galaxy between A.D. 185 and 1604. If the average time between outbursts is at all meaning-

FIGURE 14.23 Supernova in the galaxy NGC 5236. The lefthand picture was taken in June 1959, long before the outburst occurred. The stars in the picture are all foreground stars in our own Galaxy. In the right-hand frame, taken in May 1972, the supernova is present. It is even brighter than the much-closer foreground stars. (Hale Observatories.)

ful, then they have occurred roughly every two centuries. However, it has been more than 370 years since the last one was seen. Perhaps one will flare up any time now.

A supernova occurred in A.D. 1054, as recorded by Chinese and Japanese skywatchers. Their records indicate that the object was so bright that it was visible for more than 23 days in broad daylight. Today the site of the event is marked by the **Crab Nebula** (Figure 14.24 and Plate V). Even though the records were not absolutely precise about the location, they were good enough to leave no doubt of the association between the nebula and the supernova. For instance, we can actually see the nebula grow by comparing photographs made throughout this century. Its present size and rate of growth indicate it must have started expanding nine centuries ago, in complete agreement with the time of the explosion recorded in the Orient. Incidentally, we can estimate the distance to the Crab by comparing its angular expansion rate and its linear expansion rate inferred from Doppler shifts in its spectrum, just as we do for

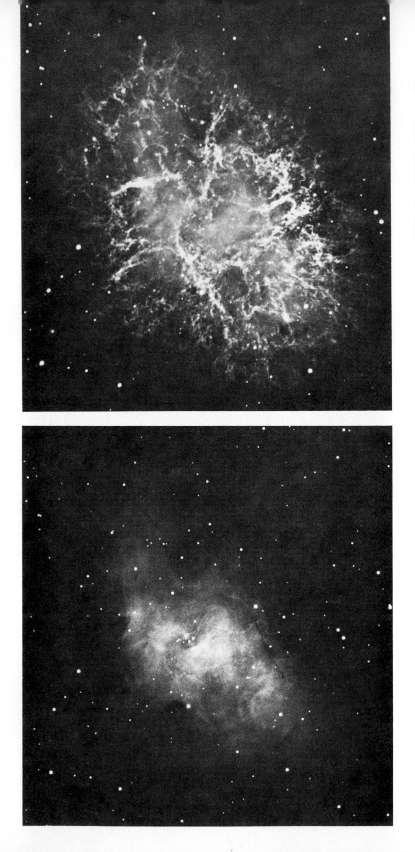

FIGURE 14.24 Two views of the Crab Nebula. (a) The filamentary structure emits mostly the red line of hydrogen, Hα. (b) A filter which blocks this light shows the amorphous central region of the nebula, which emits a continuous spectrum. (Lick Observatory photograph.)

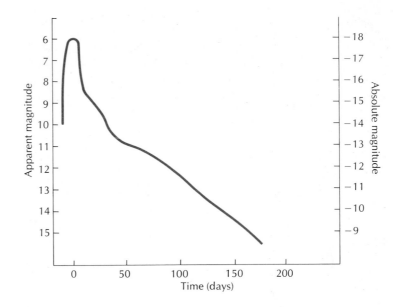

FIGURE 14.25 Light curve of a supernova that was observed in the Andromeda galaxy.

expanding shells around regular novae. It turns out to be about 1,700 parsecs. Of course, there are no magnitude estimates available for the supernova when it was at maximum brightness, but astronomers are sure it had to be brighter than about magnitude -3 or -4 to be seen in broad daylight, and, for some other reasons, favor $m = -5$ as its most likely maximum brightness. The distance modulus corresponding to its 1,700-parsec distance is $m - M = 11.2$, making the absolute magnitude of the supernova at maximum around -16. Additional evidence (see Chapter 13) points to the presence of enough interstellar dust between us and the Crab Nebula to dim its light by almost 2 magnitudes. Thus, the Crab supernova may have reached an absolute magnitude of -18 at its brightest, 1.5×10^9 times brighter than the sun. (See Figure 14.25.)

Two supernovae occurred during the critical period of astronomical history when Kepler and Tycho were studying the heavens. One, observed in 1572, is known as Tycho's supernova, and the other, observed in 1604, is known as Kepler's supernova. These names are more than honorary. The two noted astronomers made extensive series of observations of the stars. Remnants of both these supernovae have been found and studied (Figure 14.26). Like the Crab Nebula, they are intense sources of radio radiation.

The Crab Nebula consists of two parts (Figure 14.24), an almost structureless white cloud, surrounded by a filamentary structure

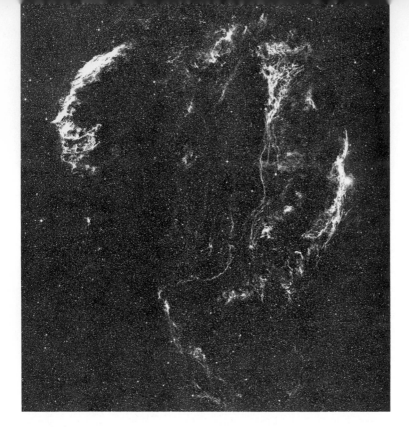

FIGURE 14.26 The Cygnus Loop, an old supernova remnant. (Hale Observatories.)

that produces an emission-line spectrum, dominated by the red Hα line of hydrogen. As we saw in Chapter 13, this radiation comes about when ultraviolet light ionizes the hydrogen. The electron is then recaptured at a high level in the hydrogen atom, and produces emission lines as it jumps from level to level toward the ground level.

The radio radiation comes from the structureless, white central part of the nebula. The radiation emitted there is called **synchrotron radiation**, because it was first observed in electron accelerators in physics laboratories known as synchrotrons (Figure 14.27). In these accelerators, electrons are given velocities very close to the velocity of light. The synchrotron has a strong magnetic field, which causes the electrons to move in a circular orbit around the accelerator. These high-velocity electrons emit a continuous spectrum of radiation that includes wavelengths from the X-ray region to the radio region. The shape of the spectrum as a function of wavelength is very unusual, a fact which allowed astronomers to discover that the radiation from the Crab Nebula is synchrotron radiation.

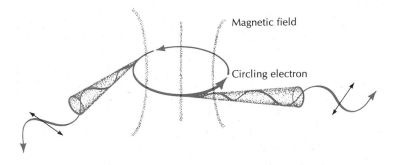

FIGURE 14.27 An electron circling with very great speed in a magnetic field emits synchrotron radiation. The radiation is emitted in the plane of the electron's orbit, and is polarized in that plane.

Magnetic field

Circling electron

Since the Crab emits synchrotron radiation, we know two things. It contains high-velocity electrons, and it contains magnetic fields. The magnetic fields are probably fields that existed within the star before the explosion, but they were somehow made stronger in the cataclysm; astronomers feel that the known material and fields could not fit into a star, because the fields are too strong.

The high-velocity electrons present a problem. If left alone, they would radiate away their energy in much less time than the nebula has existed (900 years). Thus we believe that some process is at work, constantly accelerating electrons in the nebula. The acceleration process, whose exact nature is unknown, is caused by the unusual star left after the supernova explosion. The Kepler and Tycho supernova remnants also emit synchrotron radiation, by the way.

Supernova explosions may or may not be caused by explosive oxygen burning. It is clear, though, that the cataclysmic event probably results from explosive nuclear-fusion reactions in a large fraction of a star's mass. Much interest has recently been focused on explosive reactions of this kind. Astronomers feel that quantities of heavy atoms (e.g., neon, magnesium, silicon, sulphur, and others) are created in these reactions, and are then spewed out into interstellar space with the supernova remnant. Stars of a later generation, formed out of this enriched material, will begin their lives with more heavy elements than the previous generation of stars. Thus young stars, like the Sun, consist of material in which heavy elements are much more abundant than in the old Halo population stars. The solar material has been through at least one earlier generation of stars, and perhaps more.

Supernovae do more than enrich the heavy-element content of the Galaxy. They also leave behind a cinder that, as we will see, is one of the weirdest objects in the universe.

Pulsars

When we look up at the sky from the Earth's surface, we see that the stars seem to twinkle. As has been discussed earlier, this scintillation is due to the fact that the light of the stars is passing through Earth's turbulent atmosphere (Chapter 4). Interestingly enough, radio signals from distant celestial objects also scintillate. The rate of twinkling is relatively slow, with roughly one "twinkle" each second. This phenomenon, known as interplanetary scintillation, is due to the fact that the radio signals pass through the moving interplanetary medium.

In 1967, a study of the scintillations of radio sources was begun at Cambridge, England, using a new receiver that could detect very rapid changes in the radio intensity of a source. One of the sources observed by the group of scientists did not seem to twinkle randomly. Instead it appeared to vary with a regular period of around one second. More detailed study revealed that the object emitted pulses of radio radiation at an incredibly regular interval. The object was called a **pulsar**. Today more than two hundred pulsars are known.

In the time since 1967, many new facts have been discovered about these strange objects. Pulsars have been discovered in the Crab Nebula and in a nebula in the constellation Vela, which, like the Crab, is a supernova remnant. The Crab Nebula pulsar "beeps" once every 0.033 seconds (that is, 30 times per second). Most of the known pulsars have periods between 0.25 and 2 seconds.

The Crab Nebula pulsar has been found to emit radio, optical, X-ray, and gamma-ray wavelengths. Figure 14.28 shows an interesting pair of photographs of the Crab pulsar. A television system has been devised that can look at the pulsar 30 times a second in visible light, and can photograph it when the pulsar is "on" or when it is "off." These pictures were made with that system.

When the period of the Crab Nebula pulsar was measured by very precise timing techniques, it was found to be 0.033099324 seconds on June 29, 1969. Furthermore, it was found to be slowing down at a rate of only about 1.3×10^{-5} seconds per year. One of the few physical processes that can maintain such a precise rhythm is the rotation of a massive object. But what could be rotating? Even a tiny white dwarf cannot rotate 30 times per second without flying apart.

After much argument, astronomers have decided that pulsars are **neutron stars**, truly one of the most unbelievable objects in nature. A typical neutron star is roughly 10 kilometers in diameter, and has a mass probably between 0.1 and 1.0 solar mass. The average

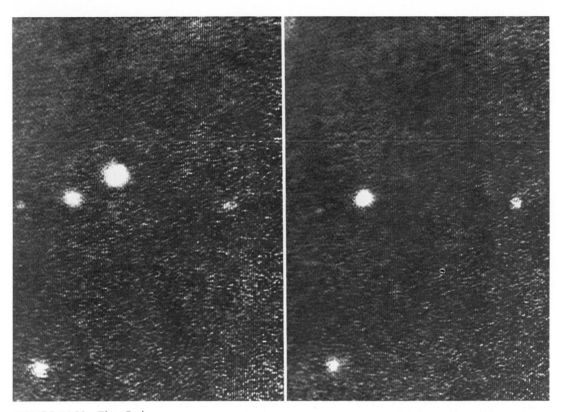

FIGURE 14.28 The Crab Nebula pulsar photographed with a special television system. The lefthand picture was exposed only when the pulsar was "on", and the righthand picture was exposed only when the pulsar was "off." The brightness change between "on" and "off" is large. (Lick Observatory photograph.)

density of a neutron star that has 0.1 solar mass and is 10 kilometers in diameter is 5×10^{14} grams per cubic centimeter. That, by the way, is more than 10^8 times denser than a white dwarf. To put the number in perspective, at neutron-star densities, a grain of sand 0.1 millimeter in diameter would weigh 2,000 tons. At these high densities, matter cannot exist in its usual form. All the atoms, and atomic nuclei too, are disrupted. The star is composed entirely of neutrons, except for a thin outer shell of solid, dense normal matter.

How do we know what a neutron star is like? In the 1930s several physicists, including J. Robert Oppenheimer, calculated the structure of a neutron star, using techniques similar to those outlined at the beginning of the chapter. They found that such a star could exist and be stable with a mass less than about that of the Sun. More detailed studies since that time support their conclusion.

The two most rapidly beeping pulsars are located in supernova remnants. This, and other evidence, indicates that a neutron star is the remains of a star after it has experienced a supernova explosion.

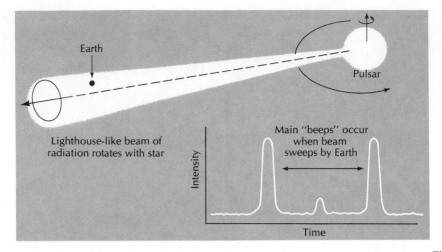

Earth

Pulsar

Lighthouse-like beam of
radiation rotates with star

Intensity

Main "beeps" occur
when beam
sweeps by Earth

Time

FIGURE 14.29 The beep-
ing signal from a pulsar
arises because a pulsar
is a rotating object which
emits radiation, like a
lighthouse, in a narrow
beam.

Thus neutron stars are probably cinders of massive stars, in the same sense that white dwarfs are the cinders of low-mass stars.

The rapid rotation of the neutron stars is a result of the conservation of angular momentum. When the parent star collapses, its rotation rate must increase as its radius shrinks, to maintain constant angular momentum. If the Sun, by some magic stroke, were compressed to a radius of 10 kilometers, it would rotate about 2,000 times per second.

We have now answered the questions: What is a pulsar? Why are its radio signals so regular? The question we have not answered yet is: Why do pulsars beep? If the radio signals were coming uniformly from the star's entire surface, even the rapid rotations would not cause pulses. Instead, the pulsar must emit a lighthouse-like beam that causes a pulse every time the beam sweeps through the Earth (Figure 14.29). Exactly why the signal is beamed is unclear. Scientists believe that neutron stars have immense magnetic fields, with the magnetic poles tipped at large angles to the rotation axis. The radiation is probably produced in this field, perhaps by a process like synchrotron radiation.

Most pulsars are slowing down at measurable rates, as we mentioned above. The slowing down is probably caused by the drag of the magnetic field of the neutron star on the surrounding interstellar matter, which thus acts as a brake. The two pulsars in young supernova remnants (the Crab and Vela) rotate the fastest. The actual rotation rate may be indicative of the time since the supernova explosion which produced the pulsar.

Both the Crab and the Vela pulsars have experienced "glitches." A glitch occurs when the star's rotation rate suddenly increases.

Some people have suggested that a glitch is the result of a starquake, when the crust of the star shrinks a tiny amount, and the rotation rate increases to preserve angular momentum.

Thus we believe that, when a supernova event occurs, a tiny, dense, rapidly rotating neutron star is left behind. A beam of radiation produces pulses detected at the Earth as the star rotates.

Black Holes

If the cinder left after a supernova explosion contains more than about 10^{57} neutrons—that is, more than roughly the mass of the Sun in neutrons—it cannot form a stable neutron star. The cinder will continue to shrink forever, becoming smaller and smaller and denser and denser. At some point, the object will become a **black hole**. If you thought neutron stars strange, read on.

The energy released by fuel as a rocket is fired from a planet goes to overcome the *gravitational potential* at the surface of the planet. On a tiny object like a neutron star, it takes a huge amount of energy to overcome the gravitational potential, but it is possible (at least in theory) to escape from the gravitational pull of the star. Let's look at a massive supernova cinder as it begins to shrink. As it gets smaller and smaller, the gravitational potential at its surface increases. Finally a strange point is reached. When the radius of the cinder is a few kilometers, the amount of energy that an object needs to escape the cinder's gravitational influence is equal to the total energy the object contains: its *rest-mass* energy. The rest-mass energy is the energy that will be released by an object of mass m if it is completely converted to energy. That energy is given by Einstein's $E = mc^2$ formula. For the object to escape from the cinder, it must be completely annihilated. When it gets out, nothing is left. If the cinder shrinks a little more, escape is impossible. The cinder has become a black hole. Anything that gets near its surface is trapped forever. Even light cannot escape from the black hole. An imaginary surface around the black hole, with a radius equal to the radius the cinder had when escape just became impossible, is called its *event horizon*. As the cinder shrinks further, the gravitational potential at the event horizon remains constant. Light cannot escape from inside. We can never know what occurs inside the event horizon of a black hole.

Do black holes really exist? There seems to be some evidence for black holes in a type of object known as X-ray binary stars, a typical example of which is Cygnus X-1 (Figure 14.30). Cygnus X-1 looks like a B-type star. However, its spectrum lines have a variable Doppler shift, indicating that the B star orbits around a

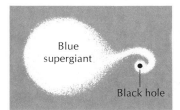

FIGURE 14.30 Possible model of an X-ray binary system like Cygnus X-1.

massive, unseen companion. The strange part of Cygnus X-1 is that it is a strong source of X-rays. Many astronomers believe the unseen companion is a black hole: probably the remains of a very massive star that passed through its life history while the B star remained relatively unchanged. Material may flow from the normal star into the black hole. In the process it will be heated to ten-million-degree temperatures. The X-rays we see are emitted by the hot material before it crosses the event horizon, never to be seen again.

This picture of X-ray binaries is not universally accepted. Many doubt that the companion is a black hole. Some are skeptical about the existence of black holes at all. This is a fascinating area in which research continues at many of the great observatories.

We have devoted a lot of words to the ends of the lives of stars, for two reasons. We believe that we know something about the subject, and the end-product objects—white dwarfs, neutron stars, pulsars, and black holes—are fascinating. To end this chapter, let's look at the other end of the lives of stars: their births. How is a star born?

THE BIRTH OF A STAR

We know of stars of all ages, from the very young O stars, which cannot be more than a million years old, to the ancient globular-cluster stars. For such a wide range of ages to exist within our Galaxy, stars must have been forming throughout its history, at least in the volume of space occupied by the Disk and Arm population of stars. The stars in the globular clusters and in the center of the Galaxy may have been formed in one massive birth process in the early history of the Galaxy.

The observational evidence strongly suggests that stars are created in multiple births. Some of the groups created in this way are stable clusters, like the open or globular clusters. Others, like stellar associations, disrupt soon after formation. The process by which an interstellar cloud fractionates and forms into a group of stars proceeds on many scales, from that of the center of the Galaxy to that of small associations. What happens is not fully understood. All we really know is that it happens. In this section we will talk about two better-understood things. First, we will talk about the theory of how a star contracts to the main sequence after it has been initially formed. Then we will look at some of the objects in the sky that may be embryonic stars.

Evolution Toward the Main Sequence

A star begins as a cloud of cool gas, which slowly contracts under its own gravitation. As the cloud contracts, its interior heats up from very low temperatures. At a certain point, a stage is reached where some of the increased internal energy created by contraction begins to tear molecules apart and ionize atoms. Under these conditions, for a short time, the temperature does not increase as the contraction proceeds, and the internal gas pressure actually decreases. Hence the cloud begins a rapid contraction, or collapse. At this stage the cloud has become a protostar, and we can place it on the H–R diagram.

Figure 14.31 shows in a schematic way the evolutionary track in the H–R diagram for a one solar-mass star. When we first see it, the object has just come from somewhere off the bottom right edge of the diagram, as a dim, cool protostar. It is collapsing rapidly under its own gravitation. That is, each little element of gas in the protostar exerts a gravitational pull on all other elements of gas, causing the whole thing to collapse. In a very short time (a year, perhaps) the protostar is the size of Mercury's orbit, and has a surface temperature of around 4,000° K. At this point, the interior of the protostar has heated up enough that gas pressure can begin to slow the collapse. A slower contraction then occurs, with surface temperature remaining nearly constant. When the core of the

FIGURE 14.31 Evolution of a solar-type star onto the main sequence. At A, the internal pressure in the protostar is high enough to stop rapid collapse. At B, nuclear reactions are just beginning in the protostar's core.

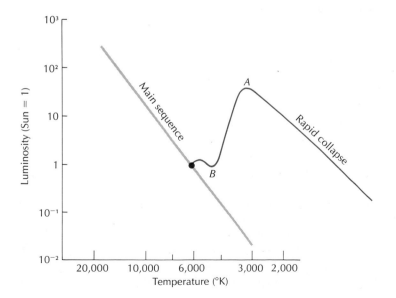

protostar reaches about 10 million °K, nuclear reactions begin. The star then slowly adjusts to its main-sequence configuration.

A planetary system originates at some stage in the development of a star from a cloud of interstellar gas. The beginnings of the system have taken shape by the time the new star has reached its main-sequence phase. The probable steps in the formation of a planetary system were discussed in Chapter 10.

Let's look around the Galaxy for objects that might be protostars which are still evolving toward the main sequence. We may not see any in the collapse phase, since it proceeds so quickly.

Young Stellar Objects

It is clear that the search for young stellar objects must begin in the densest interstellar clouds, for there we are most likely to find knots of material that can begin to contract under their own gravitational pull and ultimately form into stars. Visually, such clouds can be recognized as **globules.** Examples can be seen silhouetted against the bright Rosette nebula in Monoceros (Figure 14.32). Dense interstellar clouds have also been found by means of their infrared or microwave radio radiation. The densest clouds contain water (H_2O) and the radical OH under conditions which cause them to emit copious amounts of radio radiation.

One of the fascinating stellar types that are found in dense interstellar clouds are the T Tauri stars, named after the type-star T Tauri. These objects are variable stars which vary in brightness by roughly two magnitudes and with no hint of periodicity. Fluctuations in as short a time as a few hours have been observed.

The T Tauri stars show an absorption spectrum characteristic of spectral types F, G, and K. An unusual feature of their spectra, however, is extremely intense emission lines, including the Balmer lines of hydrogen, the H and K lines of CaII, and lines of iron and titanium. Apparently the photospheres of these stars are surrounded by a dense, gaseous layer that somewhat resembles the solar chromosphere.

What is the evidence that these unusual stars are pre-main-sequence stars? T Tauri stars are only found in regions of space that contain large amounts of dust and gas, often in association with young, massive O-type stars. In addition, their position in the H–R diagram is clearly above the main sequence, whenever their spectrum and absolute magnitude can be measured. The T Tauri stars emit more infrared radiation than is expected from stars of their spectral types. Astronomers interpret these observations to mean that the stars lie in or near dense knots of dust and gas. Some feel

FIGURE 14.32 Close-up view of the Rosette Nebula in Monoceros (see Figure 13.20). The small dark clouds are known as globules. (Hale Observatories.)

that the infrared radiation comes from material left over after the stars were formed. All these facts added together—location in dense dust clouds; association with other very young stars; location above the main sequence in the H–R diagram—have convinced astronomers that T Tauri stars have not quite reached the main-sequence phase of their existence.

Recently, a source of infrared radiation has been found in the Orion nebula that looks like the infrared from a T Tauri star. No visible light has yet been seen from this source, however. In this infrared source is a protostar, it may be approaching the point in its existence when collapse occurs. Astronomers watch the object expectantly. To witness the birth of a star would be a thrill.

Another group of objects that may be young stars are Herbig-Haro objects, but their identification as protostars is somewhat controversial. Figure 14.33 shows three Herbig-Haro (H–H) objects. You will notice that each appears to be a patch of nebulosity with bright knots included in it. The knots of bright material are the H–H objects. Some astronomers feel that the objects are protostars just before they reach the T Tauri stage. Others, however, believe that H–H objects are compact reflection nebulae (like those

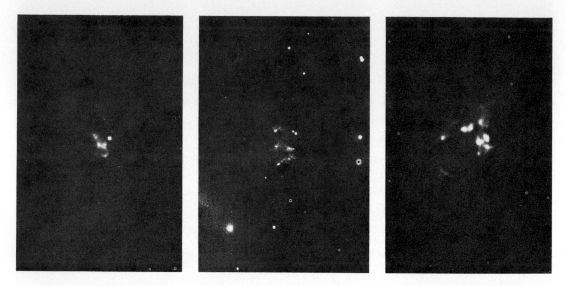

FIGURE 14.33 Three Herbig–Haro objects. (Lick Observatory photograph.)

observed in the Pleiades) surrounding hot, extremely young stars. Whichever hypothesis is correct, the Herbig–Haro objects remain very interesting. They are are among the few objects outside the solar system for which observable changes occur.

The topic of stellar birth is receiving considerable attention by astronomers today, and more is being learned daily. We are rapidly approaching the day when we will understand the life histories of stars from birth to death.

REVIEW QUESTIONS

1. What physical processes occur in stellar interiors?

2. What is a main-sequence star? How do the insides of upper and lower main-sequence stars differ?

3. What happens inside a star when it ends its life on the main sequence? What is the next stage of its life?

4. What evidence do we have for post-main-sequence stellar evolution?

5. Discuss the significance of the following in the life history of stars: (a) white dwarfs; (b) planetary nebulae; (c) T Tauri stars.

6. What is a nova? What is the current theory of the outburst?
7. What is a supernova? What is the current theory of the outburst?
8. Describe the following and discuss their significance in stellar evolution: (a) pulsar; (b) neutron star; (c) black hole.
9. How do stars form? Describe a possible young star.

15

THE UNIVERSE

The universe is filled with a vast array of puzzling and interesting objects. We know the most about ordinary galaxies. But the process by which galaxies are formed, and the stages in their life histories, remain largely a mystery. Other objects, like the quasars, have befuddled astronomers for more than a decade. Finally, the universe itself presents the ultimate puzzle. How big is the universe? Will it end?

In this chapter, we will talk about the amazing objects that inhabit the universe, then proceed to talk about the characteristics of the universe itself. When we have completed the task, the result will be somewhat of a disappointment: the observations that might tell us the size, shape, and lifetime of the whole universe are not yet good enough to permit us to draw any firm conclusions. Thus there remain exciting problems to occupy the creative energies of future generations of astronomers.

GALAXIES

The knowledge that our Galaxy is an "island universe" was established by the pioneering work of astronomers in the early decades of this century, culminating in the studies of E. P. Hubble (Chapter 13). Hubble found Cepheid variable stars in the Andromeda galaxy and established its enormous distance by using the period-luminosity law. Detailed studies of individual galaxies since that time show them to be diverse in form, size, and stellar content.

Types of Galaxies

There are three basic forms of galaxies: **spiral galaxies, elliptical galaxies**, and **irregular galaxies**. Within each form, there is a range of shapes. Several attempts have been made over the years to classify galaxies into easily recognized subtypes. The simplest scheme for doing so was devised by Hubble in 1926.

Spiral Galaxies. Hubble divided spiral galaxies in two parallel groups: the so-called normal spirals and the barred spirals. The barred spirals are easily distinguished by the fact that their spiral arms are attached to a conspicuous bar across the nucleus. Roughly three-fourths of all known galaxies are spirals, and of these one-fourth have bars. Hubble classified subtypes of spirals according to the relative sizes of the nuclei of the galaxies and the tightness of their spiral arms. Figures 15.1 and 15.2 show the sequence of

FIGURE 15.1 Representatives of the types of normal spiral galaxies. (Hale Observatories.)

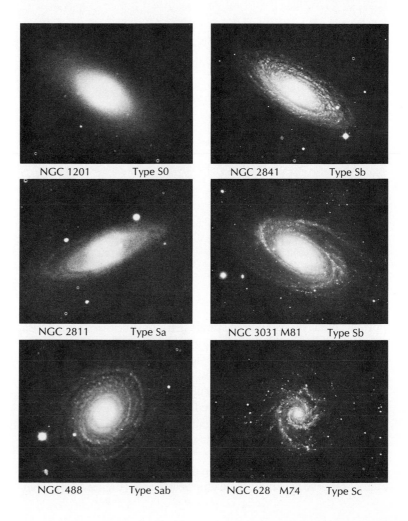

NGC 1201	Type S0	NGC 2841	Type Sb
NGC 2811	Type Sa	NGC 3031 M81	Type Sb
NGC 488	Type Sab	NGC 628 M74	Type Sc

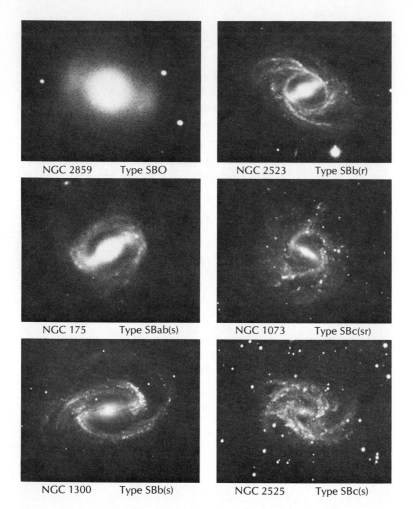

NGC 2859 Type SBO	NGC 2523 Type SBb(r)
NGC 175 Type SBab(s)	NGC 1073 Type SBc(sr)
NGC 1300 Type SBb(s)	NGC 2525 Type SBc(s)

normal spirals (S) and barred spirals (SB). In subclasses Sa and SBa, the nucleus dominates the galaxy, whereas in subclasses Sc and SBc, the nucleus is small and inconspicuous. Classes Sb and SBb are intermediate to the two extremes.

In 1960 Alan Sandage added another detail to Hubble's classification. For both normal and barred spiral classes, he appended (r) if a ring structure surrounded the galactic nucleus, (s) if no ring were present, and (sr) if a poorly defined ring were present.

Elliptical Galaxies. Elliptical galaxies, which make up roughly 20% of observed galaxies, have elliptical shapes (Figure 15.3). They

are grouped into subclasses E0 through E7 according to the shape of the image: E0 galaxies appear to be circular, and E7 galaxies have a long-to-short axis ratio of 10/3, the greatest extreme observed. The Andromeda galaxy has two elliptical companions (Figure 13.25): the one nearer to the spiral is a class E2 elliptical; the one more distant is a class E5 galaxy.

Irregular Galaxies. Irregular galaxies have no clear form that can be easily classified. The **Magellanic clouds** are irregular galaxies (Figures 15.4 and 15.5). They are companions to our Galaxy.

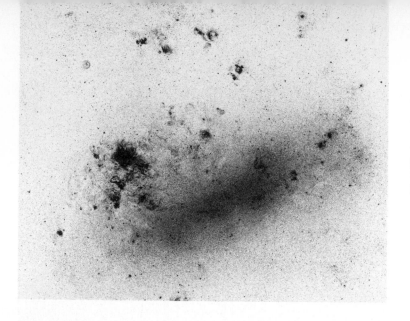

A Closer Look

The classification of galaxies according to their shapes is only one small part of the picture. What are the characteristics of each type? How does each type compare to our Galaxy? Do other galaxies contain dust and gas, like our Galaxy? Our Galaxy is a type Sb spiral (see Figure 15.6). To find the answer to some of these questions, we need to keep the latter fact in mind as we read on.

Spiral Galaxies. Spectra of spiral galaxies help us discover how much fluorescing gas these objects contain. If a galactic spectrum contains emission lines of atoms and ions that are characteristic of the hot, gaseous nebulae in the Milky Way galaxy, then we can be sure the distant galaxy contains hot gas. Type Sa galaxies normally show only slight emission, indicating a low content of emitting gas. The Sb and Sc galaxies, on the other hand, show evidence of a much greater gas content. Gas in the Sa and Sb spirals is concentrated toward the outer parts of their arms, but it is found throughout Sc spirals. The observations of barred spirals lead to similar conclusions. Bright knots seen in the spiral arms of galaxies are HII regions, glowing because of the ultraviolet light emitted by hot O and B stars imbedded in the gas.

Dust in spiral galaxies makes its presence known because it absorbs light radiated by the background of stars and fluorescing gas. In photographs of spiral galaxies seen on edge, for example, the dust is clearly seen silhouetted against the bright nucleus and disk of stars. Figure 15.7 shows a lovely example of an Sc galaxy. Note the dark, spiral-shaped dust lanes near the nucleus. Dark lanes of

FIGURE 15.6 M81, an Sb spiral galaxy. (Lick Observatory photograph.)

FIGURE 15.7 M33, an Sc spiral galaxy. (Lick Observatory photograph.)

FIGURE 15.8 NGC 4565, a spiral galaxy of type Sc seen on edge. (Hale Observatories.)

dust are also visible against the bright bar of the SBb spiral in Figure 15.9. Dust seems to be concentrated more in the nuclei and inner spiral arms of galaxies than in the outer parts of the arms.

Spiral galaxies rotate differentially. The rotation can be shown to occur by measuring the Doppler shifts in the spectra observed at various locations in a galaxy. For a nearby galaxy like the Andromeda spiral, the spectra of individual bright knots of emission can be obtained, and Doppler shifts of their spectrum lines can be measured. Figure 15.10 shows the radial velocity of several knots as a function of distance from the center of the image for a typical nearby galaxy. The average radial velocity of the whole galaxy has been subtracted from the plotted velocities. The radial velocities do not fall along a straight line, because the galaxy, like our own, rotates differentially (Chapter 13).

A major question for many years has been whether spirals rotate with their arms leading or trailing. The answer now appears to be that the arms trail. Recent theoretical work on the nature of spiral arms show them to be spiral-shaped density waves running around the galactic disk (Figure 15.11). Like sound waves propagating through air, the spiral-arm waves move through the material of the disk, and the material is alternately compressed and rarefied as waves pass by. Interestingly, young O and B stars in our Galaxy seem to lie just inside the spiral arms. A probable explanation is

FIGURE 15.9 NGC 1300, a barred spiral galaxy. (Hale Observatories.)

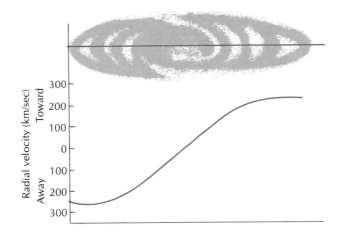

FIGURE 15.10 Radial velocities of knots of emitting material in a typical galaxy show us that the galaxy is rotating differentially.

that the wave compresses the material enough to allow stars to form. When the wave moves on, the new stars remain behind along a curved path that has the shape of the wave. These easily observed, massive stars evolve and die before the wave moves too far. Thus they appear mainly just inside the spiral-arm wave. Less-massive stars, like the Sun, are also formed in the compressed material. However, these stars live so much longer than the massive O and B

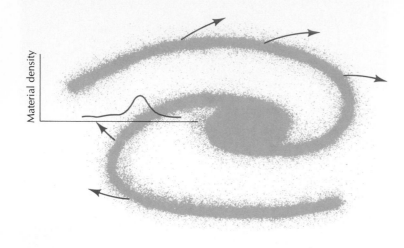

stars that they remain long after the spiral wave passes, forming the galactic disk. The reason why some spiral galaxies show a bar, which, incidentally, rotates rigidly, is unclear.

Elliptical Galaxies. If we assume that elliptical galaxies are rotating objects with various degrees of flattening, then we must think about the relationship between the galaxy's true shape and the shape we observe. Is an E0 galaxy a truly spherical galaxy or a highly flattened galaxy viewed along its long axis? So far, we cannot be sure which is true for an individual elliptical galaxy.

Studies of elliptical galaxies indicate that they are made up entirely of ancient Population II stars, just like the nuclei of spiral galaxies and globular clusters. The E0 galaxy shown in Figure 15.3, by the way, has a huge collection of globular clusters, each of which appears as a small, fuzzy image in the photograph. This entourage of globular clusters is fairly typical of elliptical galaxies. Ellipticals, if we can judge from their spectra, usually do not contain much interstellar gas.

Irregular Galaxies. The Magellanic Clouds are the closest irregular galaxies to us, being about 60,000 parsecs away. They are similar to one another in properties; so we will take a close look at the Large Magellanic Cloud (LMC).

The LMC lies in the direction of the constellation of Dorado, the Swordfish, about 20° from the south celestial pole. The cloud is composed almost entirely of Population I stars, though it does contain a small fraction of Population II stars, including globular clusters. It is also full of gas, both cool, neutral hydrogen and hot, glowing HII regions. The largest of these HII regions, called 30 Doradus, would fill half the sky if it were placed at the Orion nebula's distance from us. Despite the large amount of gas in the LMC, it has a relatively small dust content.

A major question in the study of galaxies is: Why do galaxies look different? Does a galaxy change in appearance as it ages? Very little is known about the answers to these important and fascinating questions. We do believe that the sequence of galaxies, Sa to Sb to Sc, is *not* an age sequence. That is, an Sa galaxy probably will not turn into an Sc galaxy. What does happen, however, is still a mystery.

The Local Group

We live in a collection of galaxies known as the **Local Group**, which contains roughly 20 members (Table 15.1) in a volume about a million parsecs in diameter. Three of the 20 galaxies are relatively recent discoveries, and their status as members remains uncertain. Notice that six of the elliptical galaxies in Table 15.1 are called dwarf ellipticals (Figure 15.12). These tiny elliptical systems contain only a few million stars each.

The two galaxies Maffei 1 and 2 are elliptical galaxies which lie partly hidden by the dust in the plane of our own Galaxy. They

Table 15.1 The Local Group

Galaxy	Type	Distance
The Milky Way galaxy	Sb	—
Large Magellanic Cloud	Irr	60,000 parsecs
Small Magellanic Cloud	Irr	60,000 parsecs
Andromeda galaxy and companions		
M31	Sb	700,000 parsecs
M32	E2 (dwarf)	700,000 parsecs
NCG 205	E5 (dwarf)	700,000 parsecs
M33 in Triangulum	Sc	700,000 parsecs
NGC 6822	Irr	500,000 parsecs
IC 1613	Irr	700,000 parsecs
Ursa Minor System	E4 (dwarf)	70,000 parsecs
Sculptor System	E3 (dwarf)	100,000 parsecs
Draco System	E2 (dwarf)	100,000 parsecs
Fornax System	E3 (dwarf)	250,000 parsecs
Leo I System	E4 (dwarf)	300,000 parsecs
Leo II System	E0 (dwarf)	200,000 parsecs
NGC 147	E6 (dwarf)	700,000 parsecs
NGC 185	E2 (dwarf)	700,000 parsecs
Maffei I	Elliptical	1,000,000 parsecs
Maffei II	Elliptical	

were discovered because of their infrared emission. Recently, a very nearby (nearer than the Magellanic Clouds) suspected dwarf galaxy, close to the plane of the Milky Way, has been found from its 21-centimeter radio emission. Its identification is still being studied. Are there other Local Group members hidden behind the dust clouds of the Milky Way?

The dwarf galaxies in the Local Group could not be observed at all if they were in a distant group of galaxies. If these galaxies are as common elsewhere as in the Local Group, however, they are the most common type of galaxy in the universe.

FIGURE 15.13 A compact group of galaxies known as Stephan's Quintet. Interactions between the galaxies seem to distort their individual structures. (Hale Observatories.)

Clusters of Galaxies

It appears that almost half of all galaxies belong to groups, ranging from small aggregates like Stephan's Quintet (Figure 15.13) to the fascinating large, compact clusters (Figures 15.14 through 15.17). The cluster of galaxies lying in the direction of the constellation Coma Berenices (called the Coma cluster, for short) is a typical example of a large cluster (Figure 15.15). It contains almost 2,000 galaxies.

We have now talked about the form and content of galaxies. We have also seen that galaxies occur in groups of various sizes and richness. Before we can say much more, however, we need to stop and establish a fundamental set of facts: the distance scale in the universe.

FIGURE 15.14 A cluster of galaxies in the constellation Hercules. (Hale Observatories.)

FIGURE 15.15 Part of a cluster of galaxies in Coma Berenices. These galaxies are about 150 million parsecs away. Thus the light that we see left them almost half a billion years ago. (Hale Observatories.)

FIGURE 15.16   Cluster of galaxies in Corona Borealis, at a distance of about 400 million parsecs. (Hale Observatories.)

FIGURE 15.17   A very remote cluster of galaxies over a billion parsecs away. When the light we see left the cluster, the Earth was in its infancy. (Hale Observatories.)

How do we find the distances to remote objects in the universe? We use a chain of arguments which allows us to penetrate farther and farther into the mysterious depths of the universe. In what follows, we will use the megaparsec, abbreviated Mpc, as our unit of distance. One megaparsec is one million parsecs.

We have already discussed one of the first links in the galactic distance chain: the period-luminosity law of the Cepheid variable stars. Cepheids can be recognized to a distance of about 4 Mpc. At that distance, the distance modulus of a star is 28. Thus the apparent magnitude of a star is 28 magnitudes fainter than its absolute magnitude. Even the brightest Cepheids are exceedingly faint at 4 Mpc. Distances to several galaxies in the local group have been calculated from their Cepheids. Studies of novae in some of the galaxies in the Local Group support these distances. Novae have also allowed us to find distances of a few galaxies outside the Local Group, that of the spiral galaxy M81 (Figure 15.6), for example.

Astronomers have made careful studies of the galaxies with well-known distances, including the members of the Local Group and a few bright galaxies just outside the Local Group. Two facts stand out. First, the absolute magnitudes of the very brightest supergiant stars in these galaxies average around -9. Second, the largest H$\textsc{ii}$ regions in each galaxy appear to have roughly the same linear sizes.

The Virgo cluster of galaxies lies roughly 11 Mpc from our Galaxy. It is a large cluster, containing more than 1,000 members. The distance to the cluster was estimated from the H$\textsc{ii}$ regions in its spiral galaxies. The apparent sizes of the H$\textsc{ii}$ regions depend on their distance from us. The smaller the H$\textsc{ii}$ region appears to be, the more distant the galaxy. The Virgo cluster, with its known distance, becomes an important link in the extragalactic distance chain. It contains one of the brightest, most easily recognized objects in the universe, a giant elliptical galaxy. Knowing the distance of the cluster by means of its H$\textsc{ii}$ regions, we can find the distance and the absolute magnitude of the giant elliptical galaxy.

Studies of giant elliptical galaxies by Alan Sandage of the Hale Observatories shows that they are all more or less of the same brightness, with absolute magnitude around -23. Thus a giant elliptical galaxy could be seen more than a billion parsecs away. Of course, other types of bright galaxies can also be used as distance indicators, since their total brightnesses are very similar.

Several thousand clusters of galaxies exist throughout the universe, and many of them are observed to contain giant elliptical galaxies. Their distances can be estimated by using the absolute magnitude calculated for the Virgo cluster.

Once a way to find the distances to remote galaxies was devised, the door was opened to one of the most incredible discoveries ever made about our universe.

THE EXPANDING UNIVERSE

In the early decades of this century, V. M. Slipher of the Lowell Observatory devised a spectrograph by which he could obtain the spectra of faint, distant galaxies. From these spectra, he calculated their radial velocities. A puzzling result was noted immediately. Except for the Andromeda Spiral and one or two other members of the Local Group, all galaxies show red-shifted spectrum lines, which indicates that they are moving away from us.

Hubble announced the great discovery in 1929: the galaxies are all apparently moving away from our Galaxy, and the speed of recession increases linearly with increasing distance (Figure 15.18).

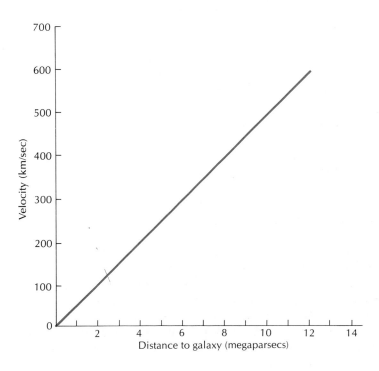

FIGURE 15.18 The velocity-distance relationship. The average radial velocities of galaxies depend on their distance from us.

That is, if galaxy B is twice as far away as galaxy A, it moves away from us twice as fast as galaxy A. This can be expressed by a simple formula:

$$V = H \times r;$$

that is, the velocity V of a galaxy is a constant H—called the Hubble constant—times its distance r. The measured velocities and distances discussed earlier let us estimate the value of the constant H.

Astronomers have been arguing over the value of the Hubble constant for years, primarily because of the uncertainties in the distances calculated for the more remote galaxies. A safe value for H seems to be 50 kilometers/second/Mpc; that is, for each megaparsec of its distance, an object's velocity increases by 50 kilometers/second. An object at a distance of a billion parsecs (1,000 Mpc) moves away from us at a speed of 50,000 kilometers/second, or one-sixth the speed of light.

Does the fact that all galaxies move away from us mean we are the center of the universe? The answer is no: it means that the universe is expanding at a uniform rate. We can understand this statement by studying Figure 15.19. At time 1, the separations between adjacent galaxies are all the same. Between time 1 and time 2, the distances between galaxies grow uniformly, so that, at time 2, the separations between adjacent galaxies are again all the same. However, the separations at time 2 are all larger than the separations at time 1. To achieve the uniform separation, the following occurred. As seen from galaxy 1, galaxy 2 has moved a distance d, galaxy 3 a distance $2 \times d$, galaxy 4 a distance $3 \times d$, and so on. Since the motions all took place in the same time, galaxy 3 appeared to move twice as fast as galaxy 2, galaxy 4 appeared to move three times as fast as galaxy 2, and so on. That is, the speed with which each galaxy moved away from galaxy 1 depended on its distance from galaxy 1. If we were to view the expansion from galaxy 3, we would observe the same thing. Between time 1 and time 2, each galaxy appeared to move away from galaxy 3 at a rate depending on its distance. The same thing is observed from galaxy 6, or any other galaxy.

From a study of the figure we conclude two things: (1) A uniform expansion causes objects to have a velocity of recession which increases as a function of distance from the observer. (2) The same result is observed no matter where the observer's vantage point happens to be. If we extend this simple line model to the three-dimensional universe, the same conclusions are valid. The fact that all galaxies appear to move away from us with a velocity that

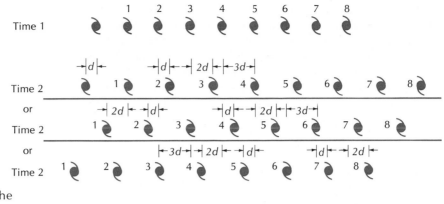

FIGURE 15.19 The expanding universe (see text).

increases with distance means the universe is uniformly expanding. Furthermore, no matter what galaxy we observe from, all galaxies will appear to move away from us. We cannot assume that our Milky Way galaxy is the center of the expansion.

Galaxies also have random velocities, just like the stars. The random velocity of the Andromeda spiral, because it is close to us, is greater than its velocity of recession as seen from the Milky Way Galaxy. That is undoubtedly why this nearby galaxy is moving toward us rather than away. The random velocities of galaxies can also be observed in a cluster of galaxies. Despite the fact that all cluster members are more or less the same distance from us, they show measurable differences in radial velocities.

The expansion of the universe with its velocity-distance relationship can be used to calculate distances to remote objects once the Hubble constant is known. If we see an object receding from us with a velocity of 100,000 kilometers/second, we infer that it is two billion parsecs away. Implicit in this distance calculation is the assumption that the object's velocity is due to the expansion of the universe (that is, that it is a *cosmological* velocity) and not due to some other cause.

RADIO GALAXIES

The existence of small, intense sources of radio radiation with fixed position on the sky was established early in the development of the science of radio astronomy. In those early days, the sources were identified by the name of the constellation in which they were situated and by a letter indicating the order of discovery. Thus the point sources included Cygnus A, Cassiopeia A, Virgo A, Centaurus A, Taurus A, and, of course, many others. Later, detailed radio surveys of the sky revealed so many discrete sources that the old naming method had to be supplanted by several new ones. One of the most famous surveys of the radio sky was the *Third Cambridge Catalogue* (1960) of radio sources, in which the sources were arbitrarily designated 3C1, 3C2, and so on.

When radio astronomy was in its infancy, it was difficult to measure the positions of the discrete sources of radio radiation with accuracy. However, that situation began to change by 1950, because of the development of radio interferometers. As we pointed out earlier (Chapter 7), if one properly combines the signals from two dish antennas separated by a distance D, the system can achieve the same resolution as a solid dish antenna with a diameter D. The size of the smallest object an antenna will resolve is related to the accuracy with which it can pinpoint an object's location. If the image of a radio source is a big smudge on the sky, we know the source is somewhere inside the smudge, probably near its center. The smaller the smudge, the better we know the position.

Around 1950, the positions of the radio sources were known well-enough that optical astronomers could hope to find them with their giant reflectors. Some of the sources, like Taurus A and Casseopeia A, turned out to be supernova remnants. Taurus A was identified with the Crab nebula, for instance. Other radio sources were associated with peculiar galaxies.

Virgo A was found to be the giant elliptical galaxy M87 in the Virgo cluster of galaxies. A short-exposure photograph of this galaxy reveals a jet of material, apparently ejected from inside the galaxy (Figure 15.20). The radio emission arises from the jet. The emission is clearly synchrotron radiation, since it is highly polarized and its spectrum has the tell-tale shape of a synchrotron source.

Apparently, giant explosions sometimes occur in the centers of galaxies. We have already spoken of evidence for an explosion in the center of our own Galaxy. The galaxy M82 was also wracked by a violent explosion several million years ago. Its irregular shape, seen in Figure 15.21, is due to a mass of hydrogen ejected from the nucleus of the galaxy. Centaurus A (Figure 15.22) is another peculiar object that seems to have resulted from an explosion or explosions during the last few million years. Perhaps such an explosion ejected the jet which produces the radio emission from M87. In galaxies of this type, the energy of the radio emission is small compared to the energy emitted in the visible part of the spectrum.

The source Cygnus A is another matter. In 1951, Walter Baade discovered a very peculiar object at the radio position (Figure

FIGURE 15.20 The radio galaxy M87, which was identified as the Virgo A radio source. Radio emission comes from the jet of material seemingly ejected from the galaxy. (Lick Observatory photograph.)

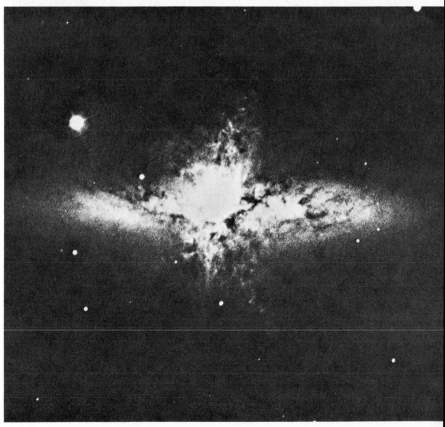

FIGURE 15.21 The galaxy M82. Faint filaments of material above and below the plane of the galaxy are rapidly moving away from it, indicating a violent explosion a few million years ago. (Hale Observatories.)

FIGURE 15.22   NGC 5128, the galaxy identified with the radio source Centaurus A. (Hale Observatories.)

FIGURE 15.23   The radio source Cygnus A. (Lick Observatory photograph.)

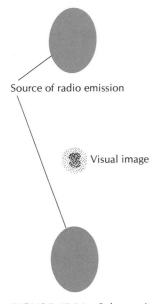

Source of radio emission

Visual image

FIGURE 15.24 Schematic picture of the radio source Cygnus A, compared to the visual image of the galaxy.

15.23). Its spectrum revealed red-shifted lines that indicated a distance of well over 100 Mpc. This object, nevertheless, was an exceptionally intense source of radio radiation, emitting as much energy at radio wavelengths as it does at visual wavelengths. The first hypothesis suggested by astronomers was that Cygnus A is two galaxies in collision. Certainly the double shape of the visual image suggests such an hypothesis. Ultimately they concluded that a collision would not generate the huge quantities of energy the object radiates. Furthermore, high-resolution studies show that the radio radiation comes from two sources, one on each side of the visual image, as shown schematically in Figure 15.24. The feeling today is that the intense energy radiated by Cygnus A resulted from an explosion of unimaginable magnitude in the distant galaxy.

The objects we have described above are a few examples of a whole series of types of radio galaxies. Without doubt, the most amazing and puzzling of these objects are the **quasars**.

Quasars

When telescopes like the great Hale Reflector were first turned to the measured positions of many radio galaxies, a clearly unusual object was evident. However, often all that could be seen were point images that seemed to be stars. Astronomers asked themselves: could there be a class of stars that emit radio energy on a grand scale? We observe radio waves from the Sun only because it is a mere astronomical unit away. If it were a few parsecs away, we could not detect the signals. Thus any radio star must emit huge amounts of radio energy compared to the Sun.

By the early 1960s, radio interferometers had pinned down the locations of several of these objects with enough precision that astronomers were able to identify the specific starlike object that emitted the radio noise. One of these, 3C273, seemed to be ejecting a jet of luminous material. Perhaps we were seeing an event like that in the radio galaxy M87, but on a smaller, stellar scale. An interesting new observation of the object came in 1962. Radio astronomers pointed their telescopes at 3C273 and measured the changes in its radio signals very carefully as the Moon occulted it. The limb of the Moon is a sharp edge. As it moved over the source, the manner in which the signal changed with time allowed astronomers to infer the structure of the source. They discovered that 3C273 contains two sources: one source fell right at the position of the starlike visual image, and the second source fell at the tip of the jet. The radiation did not come from the entire jet: 3C273 was not like the radio galaxy M87. (See Figure 15.25.)

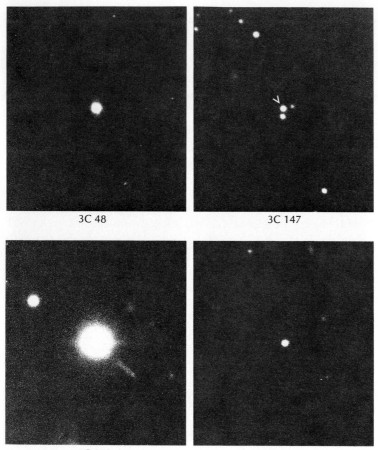

FIGURE 15.25 Quasi-stellar radio sources. (Hale Observatories.)

3C 48

3C 147

3C 273

3C 196

These objects were finally shown to be something other than stars by Maarten Schmidt at Palomar Observatory. He obtained several spectra of the starlike objects. At first he was perplexed: no stars ever seen before had such spectra. What elements produced the lines? Schmidt soon found the key to the interpretation of the spectra. The unusual lines were in fact lines that ordinarily fall in the ultraviolet region of the spectrum. However, because of enormous redshifts, they were shifted from the ultraviolet into the visible region. If those redshifts indicate velocities which are cosmological; that is, which are due to the expansion of the universe, the objects are at distances that place them near the edge of the observable universe. The distance for 3C273 is around 1.5 billion

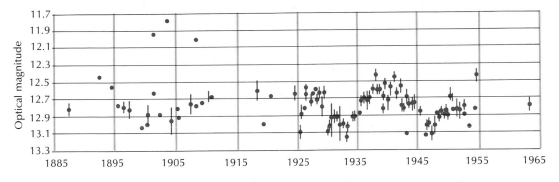

FIGURE 15.26 Observed fluctuations in the brightness of the quasar 3C273. The older observations were obtained by examining plates in the collection of the Harvard College Observatory. (From *The Problem of the Quasistellar Objects* by G. Burbidge and F. Hoyle. Copyright © 1966 by Scientific American, Inc. All rights reserved.)

parsecs, and the light we see today left it 5 billion years ago, probably before the Earth was formed. The name quasistellar radio source, soon shortened to quasar, was given to these fascinating objects.

Today we can use radio telescopes on different continents in conjunction to form a radio interferometer that has dimensions nearly as large as the Earth. With these intercontinental baseline interferometers, the angular sizes of a number of quasars have been measured. Typical sizes turn out to be considerably less than 0″.1. Some may be as small as 0″.001. Even at the enormous distances suspected for these objects, their linear diameters, as inferred from the tiny angular sizes, turn out to be a few parsecs, miniscule compared to a typical galaxy.

The small diameters inferred from the measured angular sizes are supported by another observation. Quasars vary in brightness, with fluctuations occurring on time-scales of weeks or months (Figure 15.26). It is not too hard to see that an object that fluctuates with a time-period of, for example, a year must be smaller than the distance light travels in a year. Otherwise, light from different parts of the source will arrive at the observer at different times and mask the fluctuations (Figure 15.27). Since some quasars fluctuate in fractions of a year, they must be fractions of a light year across.

The quasar 3C273 emits around 10^{48} ergs of energy each second. For comparison, the Sun emits 10^{33} ergs/second. Thus that quasar emits the same energy as 10^{15} Suns. This is thousands of times more energy than any entire galaxy emits.

We have now had a brief look at a few of the facts about quasars. They are probably at great distances; they are small; and they emit an unbelievably large amount of energy. How can this be? What are these objects?

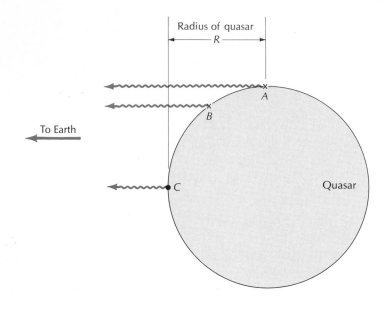

Radius of quasar — R —

To Earth

A
B
C

Quasar

FIGURE 15.27 How brightness fluctuations tell us about the sizes of quasars. Pretend the quasar is one light year in radius. Then, among the photons arriving at Earth now, those from point A left one year before those from point C and half a year before those from point B. If the brightness of the whole surface fluctuated with a one-year period, then photons we see now might have left A when it was brightest, C when it was faintest, and B when it was at an intermediate brightness. The combined light would not show the fluctuations.

The Nature of Quasars. Some astronomers have tried to avoid the difficulties by assuming that quasars are nearby objects. One idea is that the violent explosion in the nucleus of our Galaxy spewed out the quasars at great speeds. Thus the redshifts we see would not be due to the expansion of the universe. Another idea is that they are nearby, very massive objects. Light must do work to overcome the gravitational field of any object. As a result the energy of each photon is decreased, effectively redshifting its wavelength. Even for as massive a body as the Sun, the gravitational redshift is miniscule. But if the photon were emitted from the surface of a neutron star or from near the event horizon of a black hole, the gravitational redshift might be enormous. Maybe the observed spectrum of a quasar comes from emitting material near the event horizon of a black hole. That would explain the small size and the large redshift. Furthermore, if the quasar were nearby, say, in our own Galaxy, the energy it radiated could be as small as 10^{38} ergs/second. Of course, that is still 10^5 times the solar energy output. The problem with the black-hole idea is that the emitting material would be compressed and very hot. The emission would then be expected to be similar to that observed from celestial X-ray sources, not like that observed from quasars.

Some quasars have been observed in clusters of galaxies. This in itself is not significant, since the quasar could be nearby and merely

FIGURE 15.28 A Seyfert galaxy, NGC 4151. (Hale Observatories.)

lie in the same direction as the cluster. What is significant, however, is that the quasar has the same redshift as the galaxies in the cluster. This suggests the quasar is truly within the cluster.

Another bit of evidence that suggests that quasars may be galaxies is provided by the **Seyfert galaxies** (Figure 15.28). A Seyfert galaxy is a spiral galaxy that emits large amounts of energy, perhaps as much as 100 times the energy emitted by our Galaxy. The enormous energy outflow comes largely from a bright, starlike nucleus. Studies of these nuclei indicate that they are a few parsecs in diameter. In almost all respects, the nuclei of Seyfert galaxies look like a cross between the nucleus of an ordinary galaxy and a quasar. Perhaps the energetic processes that may have produced explosions in the center of our Galaxy come in all sizes, from those of our Galaxy, through those of Seyfert galaxies, to those of quasars.

Recently, a quasar has been found that is very near another galaxy, with the quasar seemingly connected to the galaxy by a faint bridge of material. The redshifts of the two objects are significantly different. Does this suggest that at least part of the redshift from quasars is gravitational? Some think that the faint bridge connecting the objects is a photographic effect and does not really exist.

The question of the nature of quasars remains open. The majority of astronomers favor cosmological distances for them, but that hypothesis cannot be proven conclusively. The question of where the enormous energy output of quasars comes from is still an enigma. Today astronomers continue to study these exciting objects in the hopes that a key discovery is just around the corner. It is fascinating objects like this that keep the field vigorous.

We have now looked at some of the beasts that populate the universe. We have studied galaxies of all types, including those that emit radio radiation. We have seen evidence of incredible explosions in the nuclei of galaxies, and we have studied the mysterious quasars. Let's now turn from the inhabitants and look at the universe itself.

COSMOLOGY

Olbers' Paradox

Sometimes a vital fact can stare us in the face for years without being recognized. Then someone finally points out the importance of the fact, and people think, "That's so simple, why didn't I think of it?" In 1827, the German astronomer Wilhelm Olbers made such a discovery: he noticed that the sky is dark at night. Why is this seemingly simple fact so important?

Let's assume that the universe is infinite, static, and uniformly filled with galaxies of stars. In imagination, we divide the universe into a large number of thin spherical shells, each the same thickness, and each with its center at the Earth, as indicated in Figure 15.29. The volume of each successive thin shell increases as the square of its radius. The total number of galaxies in any shell is the density of galaxies in the universe times the volume of the shell, and we assume the density of galaxies to be constant everywhere in the universe. Thus the number of galaxies in successive shells increases as R^2. However, the apparent brightness of each galaxy depends on its distance according to the inverse-square law; that is, the apparent brightness of a galaxy depends on $1/R^2$. Since the number of galaxies in a shell depends on R^2 and the brightness of each galaxy depends on $1/R^2$, the total brightness of all the galaxies in any shell is a constant, independent of the size of the shell. That is, each shell adds the same brightness to the night sky. If the universe is infinite in extent, there are an infinite number of shells, each adding something to the sky brightness. Thus the sky should be very, very

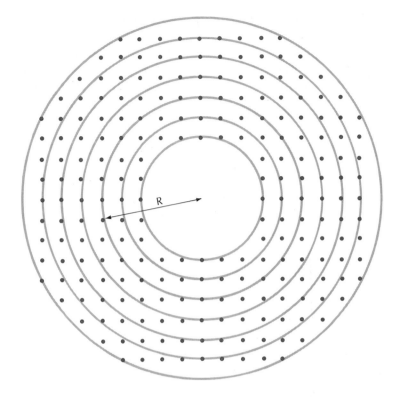

FIGURE 15.29 Olbers' paradox. The total brightness as seen at Earth of all galaxies in a given shell is the same for each shell (see text).

bright. If we assume that parts of stars in very distant galaxies are blocked by stars in less distant galaxies, we can conclude that the sky should be completely covered with stars and should therefore be as bright as the surface of a star. Roughly, the whole sky should be as bright as the Sun. By this argument, Olbers concluded that, in an infinite universe uniformly filled with stars, the sky should be very bright, not dark. His argument is known as Olbers' paradox.

Astronomers tried to resolve the paradox by arguing that dust in galaxies would dim the light of remote stars. This argument does not work, however. If the universe were as filled with radiation as Olbers' paradox claims, the dust would be heated up until it was as hot and bright as the stars. The dust would not obscure the brightness, it would add to it. No, something else must be wrong.

Olbers' made three assumptions: the universe is infinite; the universe is static; and the universe is uniformly dense. One or more of these assumptions must be incorrect. If the universe were finite in size, for instance, the number of shells contributing to the sky brightness could be small.

We know that the universe is not static. Hubble and others found that it is expanding uniformly, as discussed earlier. The expansion will reduce the brightness of the sky. The radiation from very distant objects is highly redshifted. The redshift is equivalent to a decrease in energy of each photon. Photons from very distant objects lose a very large fraction of their energy to the redshift. Furthermore, in the time the photons from a very distant source travel to the Earth, the universe will have expanded considerably. Thus the number of photons per volume of space will decrease because the volume of space increases. The combination of decrease of the energy per photon and of the number of photons per volume of space causes distant objects to seem much fainter than they would in a static universe. Even in an infinite universe, expansion provides a way out of Olbers' paradox.

Cosmological Models

There is reasonable agreement on the big picture of the universe today. Some of the details remain unsettled, however. The universe is expanding, and it is possibly finite in extent, though it really does not have an "edge." The universe seems to obey the **cosmological principle**, which says that, no matter where an observer's vantage point lies, the universe looks the same. Uniform expansion clearly obeys this principle, as we explained earlier. No matter where we are, all galaxies appear to move away from us with a speed that increases with increasing distance. The principle also requires that the same kinds of galaxies occur everywhere. If we were to view the universe from a galaxy a billion parsecs from the Milky Way galaxy, we would see spiral galaxies, elliptical galaxies, Seyfert galaxies, quasars, and so on, just as we do now. Observational evidence certainly suggests that this is true. However, the universe does not seem to obey the **perfect cosmological principle**, which says that the universe looks the same at all locations and at all times. The universe expands, and its density decreases; so it will look somewhat different a billion years from now than it does today.

Big Bang. Why is the universe expanding? A simple explanation says that all the matter in the universe was clumped together at one spot a long time ago, then a gigantic explosion (the **big bang**) occurred and started the expansion. When did the big bang occur? The answer is hidden in the Hubble constant. The Hubble expansion law says that a galaxy recedes from us at a rate of 50 kilometers/second per megaparsec of its distance. We need only ask, How long would it take a hypothetical galaxy to move 1 Mpc at a speed of 50 km/sec? When the universe began, the material in our Galaxy and in that distant galaxy was all together in the **primeval egg**. During the lifetime of the universe, that galaxy has moved 1 Mpc away from us at its speed of 50 km/sec. One megaparsec is roughly 3×10^{19} kilometers. An object moving at 50 km/sec will cover that distance in 6×10^{17} seconds or 20 billion years. Thus we conclude that the big bang that started the expansion occurred 20 billion years ago. If this is correct, then the oldest stars in our Galaxy have been around for at least half the life of the universe, and our Sun has been around for one fourth the life of the universe.

The big bang was, almost by definition, the most energetic event ever to occur in the universe. During the first few seconds of the event, hydrogen and helium, and perhaps some of the heavier elements we see today, were created. In this great primordial fireball, the energy density of radiation was huge. The small universe was filled with photons. Some of these photons remain in the universe today. In 1965, two scientists at the Bell Laboratories (A. Penzias and R. Wilson) discovered radio-wavelength radiation arriving at earth uniformly from all directions in space. It corresponded to radiation of a black body at only 3° K. It seems clear today that this 3° K radiation is the remains of the primordial fireball. How can this be true? A hypothetical experiment will show how.

We will build a box with perfectly reflecting, mirrored walls, and fill it with photons from a very hot black body. The photons will bounce around inside the box, but will not be absorbed by the walls. Now we will slowly increase the volume inside the box, so that the photons occupy more space. Interestingly, as the volume inside the box increases, the radiation continues to look like the emission from a black body, but the apparent temperature of the black body decreases. The radiation from the primordial fireball was very high-temperature black-body radiation (10 billion degrees Kelvin at one second after the bang). As its container (the universe) expanded, it continued to be black-body radiation. Today it looks like radiation from a 3° K black body.

Dynamics of the Universe. As the universe expands, it does so against the gravitational attraction of the matter. Each galaxy, for instance, pulls on each other galaxy with a gravitational force that tends to slow down the expansion. The big question is, will the expansion stop and be replaced by a collapse back to the primeval-egg state? The answer to the question depends on the average density of the universe. If the density is above a critical value,

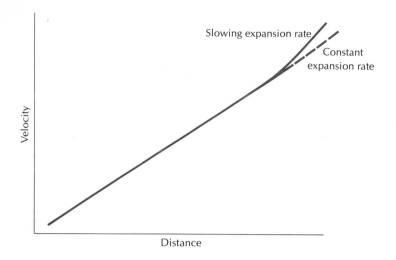

FIGURE 15.30 Velocity-distance relation for a slowing universe. The linear relationship bends upward for very distant galaxies, since light left them billions of years ago, when the expansion was faster.

gravitational forces will stop the expansion. If the average density is below that value, the expansion will continue forever. If the Hubble constant is 50 kilometers/second/megaparsec, then the critical density is 5×10^{-30} grams/cubic centimeter.

The calculation of the average density of the universe is difficult, indeed. The value depends on how much dark matter exists outside galaxies, for instance. Fortunately, there is another observational test that can be carried out.

If the expansion of the universe is slowing down, the rate must have been greater in the distant past. Objects very far from us, say, at a billion parsecs distance, are seen as they were billions of years ago. Thus they should appear to have the expansion rate the universe had billions of years ago. Thus if the universal expansion is slowing, we should see the observed velocity-distance relationship curve upward, above the linear velocity-distance relationship, at great distances (Figure 15.30). One form of the velocity-distance curve is the redshift-magnitude relation shown in Figure 15.31, where the apparent magnitudes of galaxies of a previously chosen type are taken as a measure of their distance. The problem is that the most distant galaxies are seen as they were billions of years ago, and their brightnesses may have been significantly different in the past. The brightness difference could offset the expansion-rate difference.

The result of all the observations is uncertain. The evidence leans slightly toward an ever-expanding universe.

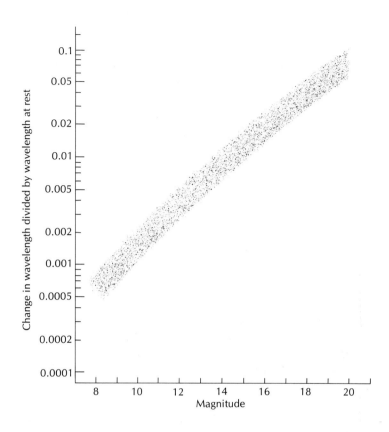

Change in wavelength divided by wavelength at rest

0.1
0.05
0.02
0.01
0.005
0.002
0.001
0.0005
0.0002
0.0001

8   10   12   14   16   18   20

Magnitude

The Shape of the Universe. Is the universe finite or infinite in extent? If it is finite, what is beyond the edge? What is its shape? To answer these questions, we must call upon the general theory of relativity devised by Einstein—and the concepts become very complex, indeed.

If the universe has greater density than the critical density, then it is *closed* (Figure 15.32). It has a finite volume. However, it does not have an edge. Think of ourselves as two-dimensional creatures living on a sphere. We only know motion on the sphere's surface. We cannot even conceive of digging into the sphere or flying off its surface. We live in a finite universe (the area of the surface); yet it has no edge, no beginning point or end point. If we travel far enough, we will merely return to our starting point. The closed universe is the three-dimensional analog of this surface.

If the density of the universe is just equal to the critical density, it is infinite in extent and will just expand forever. The two-dimensional analog is a flat plane. If the density is less than the critical density, the universe is also infinite in extent and will expand forever. The analog in two dimensions is a saddle-shaped surface.

These analogs suggest a way to discover the shape of the universe by means of observations. First, we construct a sphere, a plane, and a saddle-shaped surface, and draw dots on the surface to represent galaxies. Next we take a bow compass and draw circles around a fixed point on the surface. The radius of each circle is measured from the compass pin to drawing point. On the plane the areas of the circles and the number of galaxy dots inside the circles increases as the square of the radius of the circle. However, on the sphere, the area inside each circle increases more slowly than the radius squared, and on the saddle-shaped surface it increases more rapidly than the radius squared; likewise, the number of galaxy dots inside the circles increases more slowly or more rapidly than r^2, accordingly. In the three-dimensional universe, the analogous argument holds. Volumes inside successive spheres centered at the Earth should increase as the cube of the radius of the sphere, in a "flat" universe. As a more concrete example, let's draw two spheres around the Earth, one with a radius of 1 billion parsecs, one with a radius of 2 billion parsecs. In a flat universe, the larger sphere should contain $2^3 = 8$ times more volume and therefore 8 times more galaxies than the smaller. However, in a closed universe (like the surface of a sphere), the larger volume should be less than 8 times the smaller, and in an open universe (like the saddle-shaped surface), the larger volume should be more than 8 times the smaller. If we count the numbers of galaxies as a function of distance from us and compare the rate of increase in the numbers with r^3, we should find the curvature of the universe. Once again, the extreme difficulty of the observations has prevented a definite conclusion from being reached.

We have talked about three means for discovering the characteristics of the universe. The variation of the velocity-distance relationship at large distances from us tells us whether the rate of expansion of the universe changes with time. The average density of the universe tells us whether it will expand forever or will stop and begin to contract. The number of galaxies as a function of distance tells us the shape of the universe. As our observational instruments and techniques improve in the future, we may find the key that will tell us what the universe is like.

The Steady-State Universe. A school of thought that developed in the late 1950s said that the universe obeys the perfect cosmological principle; that is, it does not change with time. Thus as expan-

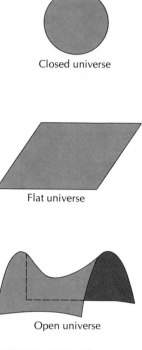

Closed universe

Flat universe

Open universe

FIGURE 15.32 Three-dimensional shapes that are the analogs of possible shapes of the universe.

sion occurred, new matter has to be created to keep the density of the universe constant. This theory then permitted the universe to have no beginning: it has always been. Observational evidence, particularly the 3° K background radiation, appears to rule out this theory today.

Fred Hoyle, who was one of the authors of the steady-state theory, agrees that the original version of the theory is not consistent with modern observations. However, Hoyle is not prepared to abandon the steady state completely. The concept of a universe that has always been and always will be, remaining unchanged on the large scale, has a certain philosophical appeal.*

We have now finished our discussion of the astronomical universe. Before we close the book, however, it is interesting to think about one more problem. Are we alone? Does life exist elsewhere? Let's look briefly at the meagre information that we have on the subject.

---

*Hoyle's latest thoughts on the subject are discussed in his book *Astronomy and Cosmology: A Modern Course* (San Francisco: W. H. Freeman and Company, 1975).

REVIEW QUESTIONS

1. Describe the events that led to the recognition of galaxies for what they are.
2. Describe each of the three types of galaxies.
3. How do we determine the presence of dust in distant galaxies?
4. What kinds of stars occur in elliptical galaxies?
5. What are the nearest irregular galaxies?
6. Describe several links in the galaxian distance chain.
7. Why do astronomers believe the universe is expanding? Are we the center of expansion?
8. Describe a typical quasar. What do astronomers believe quasars to be?
9. What is Olbers' paradox? What does it tell us about the universe?
10. Describe the *big bang* theory of the origin of the universe.
11. What will the universe look like in the future? Describe one observation that will allow us to test various possible answers to the question.

16

LIFE IN THE UNIVERSE

For most of the history of civilization, mankind has believed that Earth stands at the center of the universe: that all of creation revolves about it. In the five centuries since Copernicus was heretical enough to doubt this orthodoxy, we have learned step by step that our position is far from central. We live on one of many planets that orbit the Sun. For us, the Sun is all-important: life can exist on Earth only so long as the Sun continues to provide its warmth. Yet, in the over-all scheme of things, the Sun is insignificant. It is an average star: average in mass, average in diameter, average in temperature. Furthermore, the Sun is located near the edge of the Milky Way galaxy, merely one of 100 billion stars making up that giant system. Today we believe the universe has no center. It is like the surface of the Earth: it has no beginning and no end. Thus, our Galaxy, by definition, cannot be at the center of the universe.

The realization of our puny place in the scheme of things has prompted the question: if its location is so unimportant, why should Earth be home for the only intelligent life in the universe? This question is not a scientific question; yet it makes us think. It makes us pose scientific questions that we can try to answer logically. Let's look briefly at what science can tell us about life in the universe.

HOW MANY PLANETS ARE THERE?

There are 100 billion stars in our Galaxy. That's a lot of stars. If we tried to count them, one a second, without rest, it would take us 300 years. The first important question we must answer is: how many of these stars have planets? If planetary systems are exceedingly rare, then perhaps we live on the only one. If the nebular hypothesis discussed in Chapter 10 is correct, then planetary systems form as a natural by-product of star formation. Planetary systems would then be common. The current opinion among astronomers is that planets probably are common.

Solid planets form closest to the star, where it is warm enough that the volatile, light gases, hydrogen and helium, are driven away, leaving behind a solid residue of silicon, magnesium, carbon, and some of the heavier gases such as nitrogen and oxygen. The old Population II stars in the nucleus and globular clusters of our Galaxy were formed in its early history. The material from which they were created probably had the composition of the primordial matter of the universe just after the big bang, that is, mostly hydrogen and helium. Thus these old stars may not have had enough of the heavier elements to create solid planets. Perhaps only the more recent stars in the galactic disk, which are made from metal-enriched material, have solid planets.

Observational evidence for planets is almost nonexistent. Even mighty Jupiter could not be detected from the distance of the nearest star. Perhaps someday giant telescopes orbiting the Earth above its turbulent atmosphere will help find other planetary systems. Astronomers keep this possibility in mind as they plan the space ventures of the closing decades of this century.

Planets are probably common. But what of conditions necessary for the development of life? This is a complex question that we cannot fully answer. In fact, you will encounter this latter statement again and again as you read on. We do not know the answers. We can only give our best guesses. We know the conditions under which life developed on Earth. However, we are not at all sure just how different conditions could be and still permit any kind of life to develop. It is difficult, and dangerous, to infer general rules from a single example. In what follows, we will take a rather narrow view, and ask: how frequently will we encounter conditions on other planets that would allow life more or less as we know it to form? Thus we will assume that life needs water to form, and that it is based on hydrocarbon organic chemistry.

THE ORIGIN OF LIFE

Chemistry

The organic molecules that make up all life on Earth are hydrocarbons; that is, they consist primarily of hydrogen and carbon with small amounts of other atoms, notably oxygen, nitrogen, sulfur, phosphorus, iron, and copper. We have seen earlier that simple

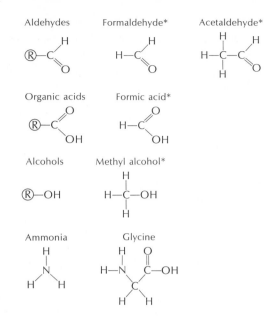

Aldehydes   Formaldehyde*   Acetaldehyde*

Organic acids   Formic acid*

Alcohols   Methyl alcohol*

Ammonia   Glycine

FIGURE 16.1 Some important organic molecules. Those marked with an asterisk have been seen in interstellar space by radio telescopes.

organic molecules have been discovered between the stars in our Galaxy: formaldehyde (HCHO), methyl alcohol (CH_3OH), and cyanoacetylene (C_2HCN), to mention a few. Other substances, such as ammonia (NH_3), water (H_2O), and formic acid (HCOOH), have been detected.

It is interesting to speculate whether more complex molecules also exist in space. For instance, if ammonia exists, then the amino radical ($—NH_2$) may exist. If an amino group and a carboxyl group (organic acid, —COOH) became attached to a methyl group in the appropriate manner, the molecule glycine might be formed (Figure 16.1). Glycine is the simplest of the amino acids, and one of the basic building blocks of proteins. Astronomers will be very excited if they discover glycine in interstellar space. It will prove once again that, given a cosmic "soup" with the proper ingredients and the proper cooking conditions, complex organic molecules will form spontaneously.

An experiment carried out in the 1950s is interesting, in this context. Scientists filled a bottle with a broth of methane, hydrogen, ammonia, and water, and caused an electrical spark to discharge in the mixture. Analysis of the contents of the bottle showed that amino acids were produced by the discharge. Of course, it is a long way from glycine, and the 21 other biologically important amino acids, to even the simplest proteins, which contain literally thousands of atoms. Biochemists feel, however, that the evidence is becoming more and more compelling that complex organic molecules predated "life" on Earth, that they were formed in *prebiotic* conditions.

The step from biologically important molecules to unicellular life is immense. Although a single cell seems simple compared to the human organism, the cell is a dazzlingly complex structure. Scientists believe that the first single-celled organisms—bacteria and blue-green algae—existed about 3 billion years ago. Thus it took almost two billion years of the Earth's existence for the miracle of life formation to occur.

The Cambrian period began 600 million years ago. Fossils from that geological period consist of marine plants and invertebrates, chiefly the trilobites. All life now on Earth has evolved from these simple invertebrates and marine plants during the last 600 million years. We must assume that some multicellular predecessors of the trilobites existed before the Cambrian period, and we might guess the Earth to have been three billion years old when they first formed.

It is probably safe to assume that biological evolution is exceedingly slow anywhere in the universe. Thus we can rule out certain types of stars as sites for life-bearing planetary systems. When a star swells to the red-giant stage of its existence, it will engulf the solid planets near it and destroy any life that formed. No star having a main-sequence lifetime of less than 3 billion years will ever have planets with other than simple multicellular life. This restriction rules out all stars with masses greater than 1.5 times the mass of the Sun (all O, B, and A stars) as abodes for higher forms of life. This is not a severe restriction, however, since O, B, and A stars make up a very small fraction (less than 1%) of all main-sequence stars.

The massive, hot O and B stars are vitally important for life, however. Within their interiors, the heavy elements are synthesized by nuclear reactions, then spewed out into space by supernova explosions. These heavy elements are the substance from which we are made.

We have, so far, come to the following conclusions. Most stars have planets, and the majority of these exist long enough for complex life forms to come into being. Furthermore, the chemical processes that lead to biologically important, simple organic molecules seem to occur spontaneously throughout our Galaxy. So far, things seem to be favorable for life to exist elsewhere.

The Proper Temperature for Life

Yellowstone National Park contains one of the greatest assemblages of hot springs and geysers on Earth. In some of these hot springs, the water temperature reaches as high as 85° C (185° F). Incredibly, algae grows in the almost-boiling water. At the other end of the temperature scale, the lichen that provide the main food for reindeer thrives in the intense cold of the Arctic. Of course, we are speaking of two different organisms with special adaptations. No single organism is known to thrive over this broad range of temperature.

Humans have existed for centuries in the burning deserts of Africa. Timbuktu experiences several months in the summer when the average daily high temperature exceeds 40° C (105° F). In the frozen arctic home of the Eskimo, several months of winter weather pass with temperatures never above −29° C (−20° F). However, if we average the temperatures, day and night, and for all seasons, of human habitats, we find that humans live in locales with mean temperatures between 0° C (32° F) and 32° C (90° F). Since life as we know it requires liquid water for its existence (see next section), we conclude that the proper temperature range for life is the same as the proper temperature range for liquid water. At sea level on Earth, this temperature range is 0° C to 100° C. If atmospheric pressure is decreased, the boiling point of water decreases. That is why cooking can become complicated in the high mountains. It is difficult to hardboil an egg in water that boils at 88° C (190° F), say.

We will assume in what follows that the temperature needed for life to exist is between 0° C and 100° C. These temperature figures translate into a range of possible distances of a planet from its central star, which we will call the **ecosphere**. If a planet is closer to its star than the ecosphere, it will be too hot; if it is further away, it will be too cold. As we shall see, other factors also affect the planet's temperature: for instance, rotation rate and atmospheric density.

Atmosphere and Water

Life as we know it requires an atmosphere and water for its existence. Our atmosphere serves many purposes. It provides us with the oxygen that our body needs to function; it screens out the deadly ultraviolet radiation from the Sun; it maintains a balance between daytime and nighttime temperatures; it provides pressure to prevent our bodily fluids from boiling away, as they would in a vacuum. An atmosphere is absolutely essential for our existence; and no doubt is essential for the existence of any life form. The same is true for water.

An atmosphere will escape from a planet if many molecules near the top of the atmosphere have velocities which exceed the planet's escape velocity. This condition is met if the planet has a low surface gravity, or is near its star and is very warm. The Moon has a low surface gravity and could not retain an atmosphere. Mercury has a low surface gravity and is warm: it, too, has no atmosphere. Mars, with 38% of the Earth's surface gravity, has an atmosphere that is too thin for us to breathe comfortably. Scientists believe a planet with the same density as the Earth, and about 0.75 its radius (making it 50% larger than Mars), would have an atmosphere with a sea-level pressure about half that on the Earth. This is roughly equivalent to the Earth's atmosphere at the elevation of the Altiplano of Bolivia, where people do live. Such a small planet would not hold its atmosphere with as tight a pull as the Earth does, and the rate of decrease of pressure with height would be much less.

Both the oceans and the atmosphere of the Earth were formed in part by volcanic activity in our planet's early history. If a planet has a low mass, it will probably not develop high internal temperatures. There is clear-cut evidence for volcanic activity on Mars, for instance, in the giant Olympus Mons (Figure 9.23). The Moon experienced great lava flows in the past, probably when its crust was broken by the impacts of giant meteoroids. However, there are no great volcanic cones on the small body, indicating low volcanic activity.

Retention of an atmosphere requires a planet to be at least three-quarters the size of the Earth, if it has the same density. Formation of an atmosphere and of life-giving oceans requires volcanism, which in turn argues for a planet not much smaller than Mars (half the size of the Earth). Mars seems to have had liquid water on its surface in the past, but it is hidden today, perhaps in the ice caps and in permafrost. Let's be brave and guess that no planet smaller than half the size of the Earth could support significant life forms. As this is being written, Viking continues its search for signs of life on Mars. So far the results seem to be negative.

Is there an upper limit on a planet's mass? If we look at the frail human body, the answer is yes. We could not stand the bone-crushing gravity of Jupiter—if that planet had a solid surface to live on. Yet there is no reason to suppose that creatures could not have supporting structures (skeletons?) that are massive enough to endure Jupiter's gravity. However, there is the problem of whether a planet has a solid surface. If a planet is too massive, it can retain its primordial hydrogen and helium, and look like Jupiter, which may have no solid surface at all.

Other Conditions

What other conditions can we place on a planet for it to be habitable? The planet's orbit should not be too eccentric; otherwise its distance from its star will vary dramatically, and its temperature will show large variations. The inclination of its axis of rotation to the plane of its orbit should not be too great; otherwise its seasons will be too extreme. Uranus, with its rotation axis in the plane of its orbit, for instance, has the most extreme possible seasons. At times one hemisphere is pointed toward the Sun, receiving heat and light, while the other hemisphere points away from the Sun, and is in perpetual darkness for almost 20 years (or a fourth of the planet's 84-year period of revolution).

Rotation Rate.   If a planet rotates very slowly, and its atmosphere is more or less like the Earth's, the difference between daytime and nighttime temperatures may be too great for habitability. Basically, the planet's surface and atmosphere would become too cold during the long night. Venus rotates very slowly—once roughly every 240 days—but it does not show a large day-night temperature difference. This can be accounted for if mass motions in the planet's incredibly dense atmosphere carry heat from the dayside to the nightside. If Venus had an atmosphere like Earth's, it would show a large day-night temperature variation.

The rotation rates of the inner planets of the solar system seem to have been slowed by tidal effects. The Sun raises tides in the bodies of the planets; these tides move through the material as the planet rotates, causing friction and slowing the rotation. Evidence compiled by astronomers shows that Mercury, Venus, and Earth have been slowed by the effects of tides caused by the Sun. A tiny M0 star emits less than 1% of the energy the Sun emits. Thus a planet must be quite near the star to receive enough radiation to maintain a habitable temperature. On the other hand, the M0 star has 50% of the Sun's mass. Thus, if a planet is close enough to the M star to maintain a habitable temperature, it experiences a much larger tidal force than the Earth does, and its rotation will be slowed considerably. Scientists believe the planet's rotation will be slowed to a point where it will not be habitable. Thus we conclude that M stars do not have habitable planets. Since we ruled out O, B, and A stars earlier, we are left with F, G, and K type stars as possible sites for life-bearing planets. These stars make up 25% of all main-sequence stars.

Double Stars. Let's think of a planet orbiting in a double-star system. The science of celestial mechanics tells us that either the planet must orbit very close to one of the two stars, in an orbit that is small compared to the separation between the stars; or it must orbit both stars, in an orbit that is large compared to the separation of the two stars. The restriction on orbit sizes makes it less likely, but not impossible, that a particular double-star system will have a planet within the ecosphere. It has been estimated that half of all F, G, and K type stars are double stars. Let's be conservative, and say that only the single stars are likely to have planets within their ecosphere.

Number of Life-Bearing Planets

From the previous discussion, we conclude that all disk population, single, main-sequence stars of spectral types F, G, and K can have planets within the habitable temperature range. How many stars is this? To answer this question, we must multiply the mass of the Galaxy by a number that is close to 1, to weed out stars that are not on the main sequence, and to account for the mass of the gas and dust; the number is close to 1 because most of the mass of the Galaxy is in main-sequence stars. We then assume that the Sun is an average star, and divide the mass of the Galaxy by the mass of the Sun to see how many main-sequence stars make up the mass. We then multiply the number of main-sequence stars by the following fractions: first, by $\frac{1}{2}$, to arrive at the total number of all stars that have enough heavy metals to produce solid planets (this is a very rough guess); then by $\frac{1}{4}$, to arrive at the number of F, G, and K stars; then by $\frac{1}{2}$, to arrive at the number of single stars (stars not in double star systems). The answer to this calculation turns out to be 10^{10} stars. Thus there seem to be 10 billion stars with planets of some kind that could be habitable.

Now, we must estimate how many of these planets really are habitable, according to the factors we have discussed briefly in this chapter. In his book *Habitable Planets for Man*, Stephen Dole concludes that roughly 5% of the F, G and K stars have planets that could be habitable by mankind.* In our discussion we have tried to

---

*Dole's complete discussion of the number of habitable planets is too long to be summarized here; and I think his book is important reading. If you have read this entire book, you can understand most of what Dole presents.

be less restrictive than Dole, and have talked about planets that are habitable by anything more advanced than a single-celled creature. Thus we must find more than 5% of the F, G, and K stars habitable. However, let's accept Dole's number as the lower limit to the number of habitable planets. That is, there must be at least as many planets habitable by any multicellular creatures as there are habitable by humans. If we accept this argument, then there are 500 million habitable planets in the galaxy (5% $\times$ 10 $\times$ 10^9 stars).

HOW MANY CIVILIZATIONS ARE THERE?

If a planet is habitable, it will be inhabited. That is to be our basic tenet in what follows. But how advanced will the inhabitants be? That question is exceedingly difficult to answer, because we do not know all the ways biological evolution can proceed. So we must base our answer on what we know about the Earth. Even then, we have many unanswered questions to cope with.

We will assume that one or another of the 500 million planets has life at each stage of evolution represented by the last 2 billion years on Earth: simple marine creatures; creatures just emerging from the sea; the great reptiles; mammals; intelligent life and civilization. This is not to assume that all planets go through a stage when they are dominated by dinosaurs. There is no reason at all to suppose that life on another planet evolved just like life on Earth. However, we will assume that each planet has one or more long epochs when it is dominated by some life form other than intelligent beings.

Human beings have been on Earth for 2 or 3 million years, so far. How much longer will we exist? Will our warfare, thoughtless squandering of our resources, and pollution of our environment soon cause us to become extinct? Or will we prove that we are an intelligent life form and come through the current crisis to exist for another million years? If we do come through and exist until biological evolution changes us into something else, perhaps something even more intelligent, then the time span of intelligent life on Earth could be 5 million years, 10 million years, or perhaps much more. If intelligent life is around for 5 million years, it represents a fraction

$$\frac{5,000,000}{2,000,000,000} = 0.0025$$

of the time-span of life on Earth. If all planets follow the same pattern, then there are 0.0025 $\times$ 500 $\times$ 10^6, or more than a million planets with intelligent life in the Galaxy. Actually, it is not likely that other planets follow the same pattern as Earth, but since we don't know how they might differ, we will have to use Earth as our pattern.

If mankind continues to exist for several million years into the future, and if we assert that we have become an advanced technological civilization in the twentieth century, then we will be in the advanced technological state for roughly half our time on Earth. If this fraction is true for all intelligent life in our Galaxy, then there are half a million technological civilizations. To be more specific, we will define a civilization to be in the advanced technological state when it develops radio transmitters and receivers that are powerful enough to communicate with civilizations in other planetary systems. The great radio telescope at Aricebo, Puerto Rico, for example, is equipped with powerful transmitters and receivers. It could communicate with any similarly equipped instrument in the Galaxy.

Incidentally, if we accept the optimistic view that advanced civilizations last for millions of years, then we are just at the beginning. Most of the civilizations out there are far more advanced than we are. Just think what we can learn from them. There is an incentive to try to communicate. We hope that once communication is established, they will view us as willing pupils rather than as food.

Let us now accept that there are hundreds of thousands of advanced technological civilizations in the Galaxy. How do we communicate with them?

INTERSTELLAR COMMUNICATION

If there are 500,000 advanced civilizations in our Galaxy, then the average distance between such civilizations is about 140 parsecs. It takes light about 460 years to travel this distance. A two-way conversation would not be very lively. We send a greeting: "Hey out there, how are you?" More than nine centuries later, we receive the reply: "Not bad, and you?" By the time the reply came, the record that the message was sent would have been lost in antiquity. How much do we recall of what went on in A.D. 1000? Of course, the comparison is not quite fair. The mass media records more of our daily life (good or bad) for posterity today. In the following paragraphs, we will take a closer look at the problems presented by interstellar communication.

What Message Do We Send?

If we start sending "Hey out there, how are you?" messages into space, we have two problems. How do we send the message? What strategy do we use in order to choose a channel to which someone might be listening? What channel do we monitor, awaiting a reply?

One opinion says that we should choose a radio-wavelength band where there is little natural radiation from the multitude of interstellar gas clouds in our Galaxy. Such bands do exist at microwave wavelengths. Then, what signal do we send? A standard A.M. broadcast of the English-language phrase, "Hey, how are you out there?" would not give the listener a clue that signal was from intelligent beings. Some have suggested sending a picture made up of black or white elements. We would send a series of "beeps" which represent, say, dark spots. If we sent beeps on the average one a second, for instance, then leaving out a beep would be a white spot. How would we tell the listener that the beeps should be arranged in a square to get a picture? First, we would repeat the sequence of "beeps" several times to let the listener know it was not a natural signal. The picture could be a square, or a rectangle. If it is a rectangle, its dimensions should be the product of two prime numbers (numbers, like 17, that are divisible only by themselves and 1). If the picture had 391 beeps and spaces, it would have to be a rectangle 17 by 23 elements on a side; whereas, if it had 392 elements, its dimensions could be 14 by 28, 8 by 49, or one of several other combinations. What do we depict in the picture we send? Perhaps stick figures of humans, with other information about ourselves. A message that has been beamed toward the great globular cluster in Hercules from the Aricebo telescope is shown in Figure 16.2.

Another attempt at communication was made on the Pioneer 10 and 11 spacecraft, which passed Jupiter in 1973 and 1974. When the spacecraft passed close to Jupiter, the giant planet's tremendous gravitational pull whipped the spacecraft to sufficiently great speeds that they will leave the solar system. Both craft carry plaques which convey a message about their origin (Figure 16.3). They show a man and a woman. A line drawing of the spacecraft behind them gives their height. Also on the plaque is a drawing of our solar system, and a coded "starburst" figure showing the direction to 14 pulsars as seen from our solar system. Each direction line has a binary code of the period of the pulsar in terms of the frequency of the 21-centimeter line of hydrogen. The use of pulsars as natural ways to indicate directions seemed logical, since their constant-period beeps are such outstanding features. Pulsars are known to

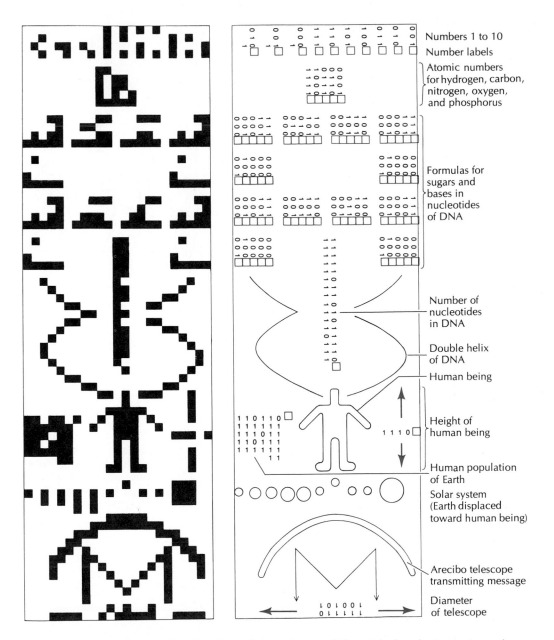

FIGURE 16.2 A message that has been beamed toward the globular cluster in Hercules from the Aricebo Observatory. (From *The Search for Extraterrestrial Intelligence* by C. Sagan and F. Drake. Copyright © 1975 by Scientific American, Inc. All rights reserved.)

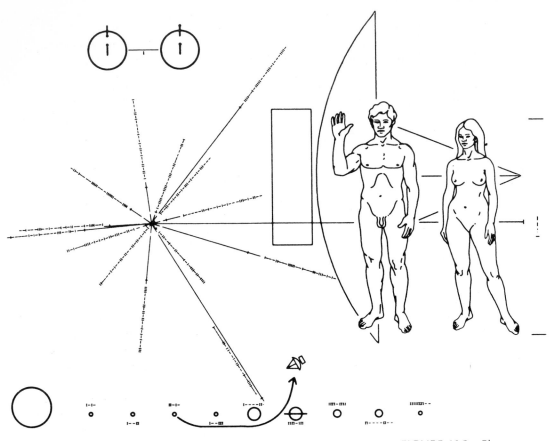

FIGURE 16.3 Plaque that was attached to the Pioneer 10 and 11 spacecraft, which will wander about our Galaxy for eons. (Cornell University.)

slow down at a fixed rate. The finder of the plaque can deduce the travel time of the spacecraft by comparing the periods of the pulsars as depicted, with their periods when the spacecraft was found.

How Do We Listen?

Scientists advocate, with a voice that becomes stronger each year, that we should make an all-out effort to listen for signals that might come from extraterrestrial intelligence. One group advocates a project known as "Cyclops" (Figure 16.4). Cyclops is a circular field of 1,500 radio antennas, each 100 meters in diameter. The antennas together can detect a very faint signal. The signal is sent through a sophisticated computer system which is programmed to detect intelligible messages.

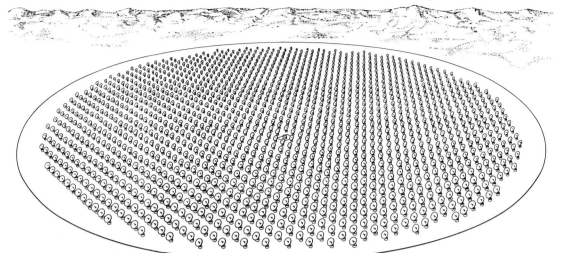

FIGURE 16.4 Cyclops antennas proposed to search for signals from extraterrestrial beings. (From *The Search for Extraterrestrial Intelligence* by C. Sagan and F. Drake. Copyright © 1975 by Scientific American, Inc. All rights reserved.

What If We Hear Someone?

What will happen if we institute something like the Cyclops system and hear a "Hey out there, how are you?" message arriving from another star? We will certainly know we are not alone in the universe. Then we will begin the communication process. "Who are you?" "What are you like?" "Here is what we look like." I suspect the philosophical impact will be small. When Copernicus toppled us from the center of the universe, the biggest change in our philosophical outlook occurred. I think we are ready to learn that we are not alone. But when we discover our celestial neighbors, can we pay them a visit?

INTERSTELLAR TRAVEL

Einstein based his special theory of relativity on one basic premise: that the speed of light, as seen by an observer moving at a constant speed relative to the light source, is always the same no matter what the relative speed of the source and observer happens to be. One of the deductions from this theory is that matter cannot be accelerated to the speed of light. It can get very close to the speed of light, but cannot actually reach that ultimate speed. Physicists have devised numerous tests of Einstein's theory, and in each case the theory passed the test. Thus physics as we know it today says that matter cannot exceed the speed of light.

Physics therefore seems to put a fundamental limitation on interstellar travel. If the average distance between civilizations in the galaxy is 140 parsecs, even at the speed of light it would require 960 years for a round-trip visit.

Have We Been Visited?

There has been much speculation that we have been visited by intelligent beings at various times. Although the thought is interesting, there is no real evidence to support the contention that such visits have occurred. It is my own personal conviction that we have not been visited by extraterrestrial beings at any time. The evidence for such visits is all circumstantial and, in my opinion, can be explained without involving intelligent creatures from another world.

The subject of extraterrestrial life is an important area of research that is attracting the attention of more and more serious scientists. Many experts believe we are not alone in the universe, and many advocate that we make a continuing effort to contact and communicate with other worlds. In this chapter we have discussed briefly the evidence in favor of the belief in extraterrestrial life. In these few pages, we could only scratch the surface of the subject. We have left many, many questions either unanswered or inadequately answered. To adequately review the knowledge in the field would fill another book.

REVIEW QUESTIONS

1. Is there life elsewhere in the universe? How would you convince a friend of your viewpoint?

2. If you heard a strange radio signal from outer space, how might you differentiate between natural and artificial origins for the signal?

APPENDIX A

The Skinny Triangle

FIGURE A.1 Illustration
of the concept leading
to the small angle for-
mula. As the angle a gets
smaller and smaller, the
arc of the circle and the
chord cut by the angle
become closer and closer
to the same length.

Astronomers often deal with skinny triangles, where one angle is
very small. Such a triangle is particularly easy to solve, using the
small-angle formula. In Figure A.1 we cut a sector out of a circle
with an angle a, and draw a chord across the sector. When a is
large, the arc cut out of the circumference of the circle is much
longer than the chord. However, when a is small, the arc and the
chord are close to the same length. The smaller the angle a, the

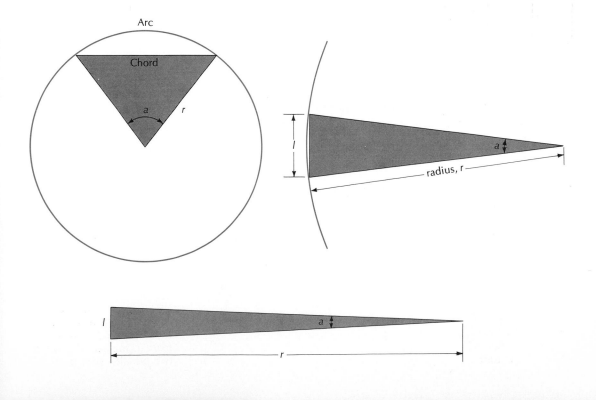

FIGURE A.1 Illustration of the concept leading to the small angle formula. As the angle a gets smaller and smaller, the arc of the circle and the chord cut by the angle become closer and closer to the same length.

closer the two lengths are to one another. When the angle a is very small, we can assume that the chord and the arc are the same length.

We can easily find the length of the arc by the proportion

$$\frac{\text{length of arc}}{\text{circumference of circle}} = \frac{a}{360°},$$

or, using symbols,

$$\frac{l}{2\pi r} = \frac{a}{360°}.$$

If a is so small that it is measured in seconds of arc, the proportion can be written

$$\frac{l}{r} = \frac{a''}{206{,}265},$$

where $206{,}265 = 60 \times 60 \times 360 \div 2\pi$. This is the small-angle formula. Normally in astronomical problems, one measures the angle a, and knows either l or r. The unknown r or l can be found from the small-angle formula.

Let us look at three examples of how the small-angle formula is used in situations that are discussed in the text. For each, the radius of the big circle of Figure A.1 is the distance D, and the chord is the small side of the triangle R.

THE DISTANCE TO MARS

Suppose we want to calculate the distance to Mars from its measured geocentric parallax (see Figure A.2). We know the Earth's radius, R, and the parallax, a, because we have measured them; and we want to find the distance to Mars, D. The formula we use here is

$$D = \frac{206{,}265}{a''} \times R,$$

where D and R are in kilometers, and a is in seconds of arc.

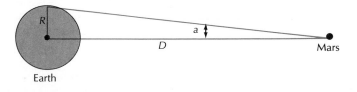

FIGURE A.2 The skinny triangle used to calculate the distance to Mars from its measured geocentric parallax (see p. 97).

At Mars' opposition to Earth in 1956, a was measured to be 23.3 seconds of arc. We can now insert the numbers into the formula, work through the calculations, and find that

$$D = \frac{206,265}{23.3} \times 6,371 = 56,000,000 \text{ km}.$$

THE RADIUS OF MARS

Having calculated the distance to Mars, we can use this new information to find the radius of Mars (or of any other planet whose distance we know). Looking at Figure A.3, we see that we know the distance to Mars, D, and can measure the angular radius of the planet's image, a. We can then calculate the actual radius of the planet by means of the formula

$$R = \frac{a''}{206,265} \times D,$$

where again R and D are in kilometers, and a is in seconds of arc.

FIGURE A.3 The skinny triangle used to calculate the radius of a planet from its measured angular radius (see p. 224).

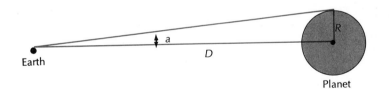

When Mars was 56,000,000 kilometers distant from us in 1956, its disk had an angular radius of 12.2 seconds of arc; so again we plug the numbers into the formula, and find the radius,

$$R = \frac{12.2}{206,265} \times 56,000,000 = 3,300 \text{ kilometers}.$$

THE DISTANCE TO A STAR

In calculating the distance to a star, we work with the skinny triangle shown in Figure A.4. Here we know the radius of the Earth's orbit, R, and the star's parallax, a, because we can measure

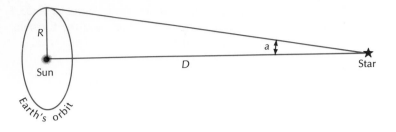

FIGURE A.4 The skinny
triangle used to calculate
the distance to a star
from its measured helio-
centric parallax (see
p. 326).

them. We can calculate the distance to the star, D, by means of the same formula we used for planets:

$$D = \frac{206{,}265}{a''} \times R.$$

Here a is again in seconds of arc; however, D and R are not in kilometers, but in astronomical units, the distance from the Earth to the Sun. Since that last distance is simply the radius of the Earth's orbit, in this formula $R = 1$; so it does not affect the calculation.

If a star were at a distance from us where its parallax was exactly one second of arc, then we can see from the formula that its distance D would be exactly 206,265 astronomical units. It is this distance that is *defined* as a **parallax second** or **parsec**. Using this definition makes life a little simpler for astronomers, because, if we want to find the distance D to a star in parsecs, the formula becomes simply

$$D = \frac{1}{a''}.$$

There are, in fact, no stars as close to us as one parsec. The nearest star is about 1.3 parsecs away; so its parallax is about three quarters of a second of arc.

APPENDIX B

Powers of Ten

Throughout this book, we use what is called scientific notation, or powers-of-ten notation, to write very large or very small numbers. The mass of the Earth, for example, is roughly 6,000,000,000,000,000,000,000,000 kilograms. It is inconvenient to write such long numbers, and difficult to read them. Powers-of-ten notation is a shorthand way to write such numbers. The mass of the Earth can be written as 6×10^{24} kilograms.

Any number written in the form 10^n is to be understood as a 1 followed by n zeros. The number n is called the **exponent** of the number. The number 10^n is, in fact, 10 multiplied by itself n times. It should be read "ten to the nth power." The mass of the Earth is "six times ten to the twenty-fourth power."

The pattern for large numbers is as follows.

$$1 = 10^0;$$
$$10 = 10^1;$$
$$100 = 10^2 = 10 \times 10;$$
$$1,000 = 10^3 = 10 \times 10 \times 10;$$
$$1,000,000 = 10^6 = 10 \times 10 \times 10 \times 10 \times 10 \times 10;$$
$$1,000,000,000 = 10^9; \text{ and so on.}$$

Small numbers are written in a similar way, with negative exponents. The number 10^{-n} represents the number that is written as a decimal point followed by $n - 1$ zeros and a 1. For example, $10^{-6} = .000001$. The mass of a hydrogen atom is 1.7×10^{-27} kilograms. The number 10^{-n} is the reciprocal of 10^n, as follows.

$0.1 = 10^{-1} = 1/10;$
$0.01 = 10^{-2} = 1/10^2 = 1/100;$
$0.001 = 10^{-3} = 1/10^3 = 1/1,000;$
$0.000001 = 10^{-6} = 1/10^6 = 1/1,000,000;$
$0.000000001 = 10^{-9} = 1/10^9 = 1/1,000,000,000;$ and so on.

How do we do arithmetic using powers of ten? For addition and subtraction, both numbers must have the same power of ten:

$$
\begin{array}{r}
3 \times 10^4 \\
+2 \times 10^4 \\
\hline
5 \times 10^4
\end{array}
$$

$$
\begin{array}{r}
6 \times 10^{-3} \\
+2 \times 10^{-2} \\
\hline
\end{array}
\longrightarrow
\begin{array}{r}
6 \times 10^{-3} \\
+20 \times 10^{-3} \\
\hline
26 \times 10^{-3}
\end{array}
$$

or

$$
\begin{array}{r}
0.6 \times 10^{-2} \\
+2 \times 10^{-2} \\
\hline
2.6 \times 10^{-2}
\end{array}
$$

To multiply two numbers in powers-of-ten notation, (1) *multiply* the quantities before the times sign in each number together, and (2) *add* exponents:

$$(6 \times 10^8) \times (3 \times 10^7) = 18 \times 10^{8+7} = 18 \times 10^{15},$$

which can be written 1.8×10^{16}.

To divide two numbers in powers-of-ten notation, (1) *divide* the quantities before the times signs in each number, and (2) *subtract* exponents.

Example: How many times more massive is the Earth than the hydrogen atom?

$$\frac{6 \times 10^{24}}{1.7 \times 10^{-27}} = 3.5 \times 10^{24-(-27)} = 3.5 \times 10^{51}.$$

APPENDIX C

Temperature Conversion

The formula to remember if you wish to convert between Centigrade and Fahrenheit temperatures is

$$\frac{T_f - 32°}{T_c} = \frac{9}{5},$$

where T_f is the temperature in Fahrenheit degrees and T_c is the temperature in Centigrade degrees. Let's look at a couple of examples.

Sulfur boils at 445° C. What is the corresponding Fahrenheit temperature?

$$T_f - 32° = 455 \times \frac{9}{5} = 801° \text{ F},$$

or
$$T_f = 833° \text{ F}.$$

If the outside temperature is 28° F, what is the temperature in degrees Centigrade?

$$T_c = \frac{5}{9} \times (T_f - 32°) = \frac{5}{9} \times (-4°) = -2° \text{ C}.$$

APPENDIX D

The Elements

Name	Symbol	Atomic Number*	Atomic Weight**
Actinium	Ac	89	(227)
Aluminum	Al	13	27.0
Americium	Am	95	(243)
Antimony	Sb	51	121.8
Argon	Ar	18	39.9
Arsenic	As	33	74.91
Astatine	At	85	(210)
Barium	Ba	56	137.4
Berkelium	Bk	97	(245)
Beryllium	Be	4	9.0
Bismuth	Bi	83	209.0
Boron	B	5	10.8
Bromine	Br	35	79.9
Cadmium	Cd	48	112.4
Calcium	Ca	20	40.1
Californium	Cf	98	(248)
Carbon	C	6	12.0
Cerium	Ce	58	140.1
Cesium	Cs	55	132.9
Chlorine	Cl	17	35.4
Chromium	Cr	24	52.0
Cobalt	Co	27	58.9
Copper	Cu	29	63.5
Curium	Cm	96	(245)
Dysprosium	Dy	66	162.5
Einsteinium	Es	99	(254)
Erbium	Er	68	167.2
Europium	Eu	63	152.0
Fermium	Fm	100	(253)
Fluorine	F	9	19.0
Francium	Fr	87	(223)
Gadolinium	Gd	64	156.9
Gallium	Ga	31	69.7
Germanium	Ge	32	72.6
Gold	Au	79	197.0
Hafnium	Hf	72	178.6
Helium	He	2	4.003
Holmium	Ho	67	164.9
Hydrogen	H	1	1.008
Indium	In	49	114.8
Iodine	I	53	126.9
Iridium	Ir	77	192.2
Iron	Fe	26	55.8
Krypton	Kr	36	83.8
Lanthanum	La	57	138.9
Lawrencium	Lw	103	(257)
Lead	Pb	82	207.2
Lithium	Li	3	6.9

Name	Symbol	Atomic Number*	Atomic Weight**
Lutetium	Lu	71	175.0
Magnesium	Mg	12	24.3
Manganese	Mn	25	54.9
Mendelevium	Md	101	(256)
Mercury	Hg	80	200.6
Molybdenum	Mo	42	95.9
Neodymium	Nd	60	144.2
Neptunium	Np	93	(237)
Neon	Ne	10	20.2
Nickel	Ni	28	58.7
Niobium	Nb	41	92.9
Nitrogen	N	7	14.0
Osmium	Os	76	190.2
Oxygen	O	8	16.0000
Palladium	Pd	46	106.7
Phosphorus	P	15	31.0
Platinum	Pt	78	195.2
Plutonium	Pu	94	(242)
Polonium	Po	84	(209)
Potassium	K	19	39.1
Praseodymium	Pr	59	140.9
Promethium	Pm	61	(145)
Protactinium	Pa	91	(231)
Radium	Ra	88	226.0
Radon	Rn	86	(222)
Rhenium	Re	75	186.3
Rhodium	Rh	45	102.9
Rubidium	Rb	37	85.5
Ruthenium	Ru	44	101.1
Samarium	Sm	62	150.4
Scandium	Sc	21	45.0
Selenium	Se	34	79.0
Silicon	Si	14	28.1
Silver	Ag	47	107.9
Sodium	Na	11	23.0
Strontium	Sr	38	87.6
Sulfur	S	16	32.1
Tantalum	Ta	73	180.9
Technecium	Tc	43	(99)
Tellurium	Te	52	127.6
Terbium	Tb	65	158.9
Thallium	Tl	81	204.4
Thorium	Th	90	232.0
Thulium	Tm	69	168.9
Tin	Sn	50	118.7
Titanium	Ti	22	47.9
Tungsten	W	74	183.9
Uranium	U	92	238.1
Vanadium	V	23	50.9
Xenon	Xe	54	131.3
Ytterbium	Yb	70	173.0
Yttrium	Y	39	88.9
Zinc	Zn	30	65.4
Zirconium	Zr	40	91.2

*Atomic number is the number of protons in the nucleus or the number of electrons orbiting the nucleus of a neutral atom.

**Atomic weight is the weight of the atom on a scale in which the atomic weight of naturally occurring oxygen is exactly 16. In general, naturally occurring elements are a mixture of isotopes in a more or less fixed ratio. Atomic weights given in parentheses are for the isotope of the element that has the longest half-life; the elements represented in this way are the rare, radioactive species.

Periodic table of the elements:

1 H																	2 He
3 Li	4 Be											5 B	6 C	7 N	8 O	9 F	10 Ne
11 Na	12 Mg											13 Al	14 Si	15 P	16 S	17 Cl	18 Ar
19 K	20 Ca	21 Sc	22 Ti	23 V	24 Cr	25 Mn	26 Fe	27 Co	28 Ni	29 Cu	30 Zn	31 Ga	32 Ge	33 As	34 Se	35 Br	36 Kr
37 Rb	38 Sr	39 Y	40 Zr	41 Nb	42 Mo	43 Tc	44 Ru	45 Rh	46 Pd	47 Ag	48 Cd	49 In	50 Sn	51 Sb	52 Te	53 I	54 Xe
55 Cs	56 Ba	57 La	72 Hf	73 Ta	74 W	75 Re	76 Os	77 Ir	78 Pt	79 Au	80 Hg	81 Tl	82 Pb	83 Bi	84 Po	85 At	86 Rn
87 Fr	88 Ra	89 Ac															

Rare earths

58 Ce	59 Pr	60 Nd	61 Pm	62 Sm	63 Eu	64 Gd	65 Tb	66 Dy	67 Ho	68 Er	69 Tm	70 Yb	71 Lu

Actinide series

90 Th	91 Pa	92 U	93 Np	94 Pu	95 Am	96 Cm	97 Bk	98 Cf	99 Es	100 Fm	101 Md	102	103 Lw

FIGURE D.1 Periodic table of the elements.

APPENDIX E

The Greek Alphabet

α	alpha		ν	nu
β	beta		ξ	xi
γ	gamma		o	omicron
δ	delta		π	pi
ε	epsilon		ρ	rho
ζ	zeta		σ	sigma
η	eta		τ	tau
θ	theta		υ	upsilon
ι	iota		ϕ	phi
κ	kappa		χ	chi
λ	lambda		ψ	psi
μ	mu		ω	omega

APPENDIX F

Star Charts

The following star charts can be used to identify the constellations visible in various parts of the sky. (See Chapter 2 for an explanation of the terms used here.)

If you are in the Northern Hemisphere, facing North, use Chart 1. To find the proper orientation at 10:00 P.M. Standard Time (or 11:00 P.M. Daylight Time), turn the chart until the name of the current month is up. For each two hours later, rotate the chart one month later; for each two hours earlier, rotate the chart one month earlier. For example, if it is midnight Standard Time in February, the chart should be used with March up. If you are a Southern Hemisphere observer facing South, use Chart 2 in exactly the same manner. The altitude of the celestial pole (the center of the charts) is equal to your latitude.

Charts 3, 4, 5, and 6 are designed to be used by a Northern Hemisphere observer facing South. To find the proper chart and its proper orientation at 10:00 P.M. Standard Time (or 11:00 P.M. Daylight Time), find the name of the current month along the bottom of the chart. Constellations above the month name are due South in the sky. For each two hours later, use the name of the month one month later; for each two hours earlier, use the name of the month one month earlier. Remember, the altitude of the point where the Celestial Equator (declination 0°) crosses your meridian is 90° *minus* your latitude.

Numbers along the bottom of Charts 3, 4, 5, and 6 are right ascension, and numbers up and down the middle of the charts are declination. On the circumpolar charts, the numbers on the concentric circles are declination.

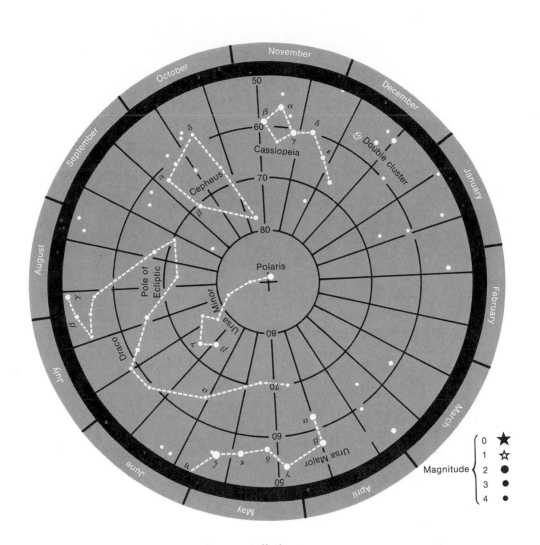

CHART 1. Northern circumpolar constellations.

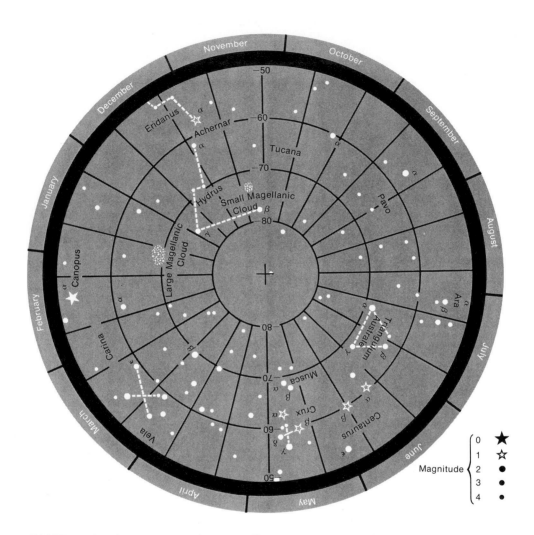

CHART 2. Southern circumpolar constellations.

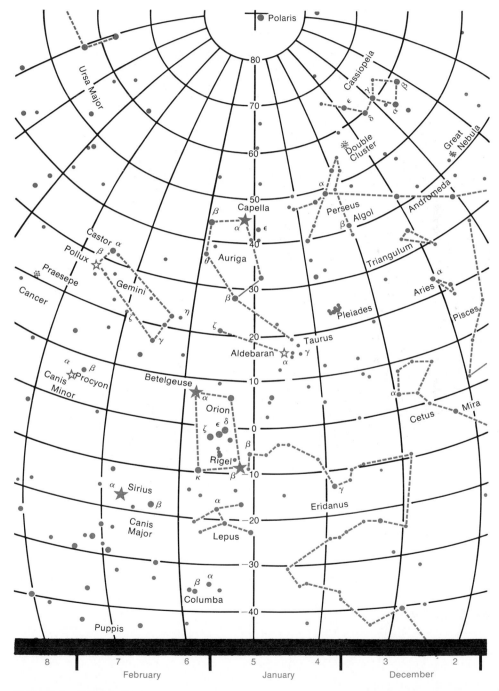

CHART 3. Winter sky.

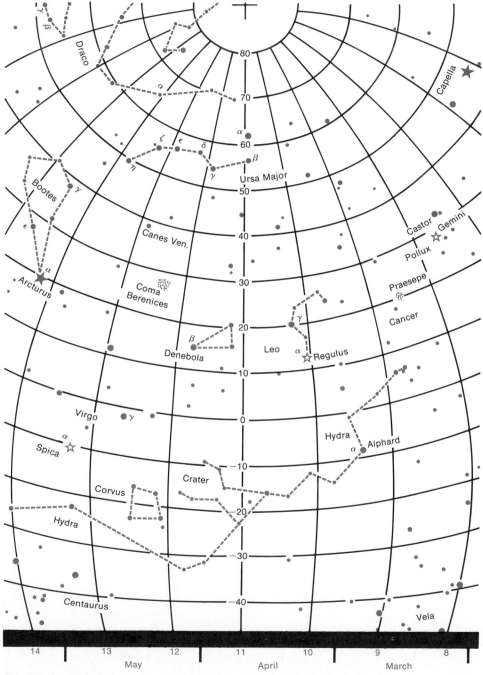

CHART 4. Spring sky.

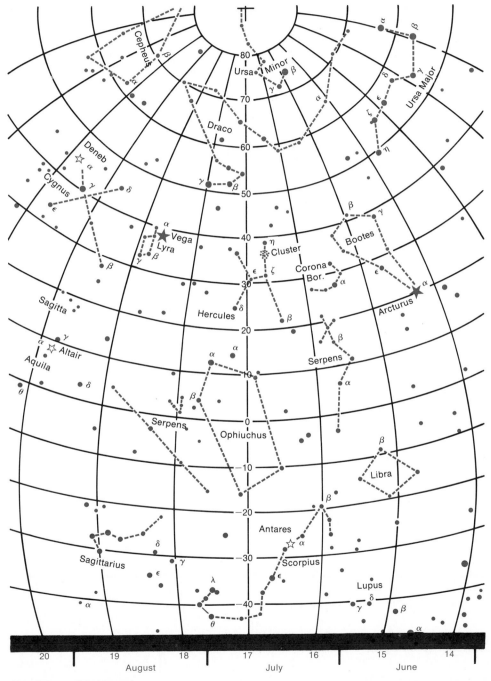

CHART 5. Summer sky.

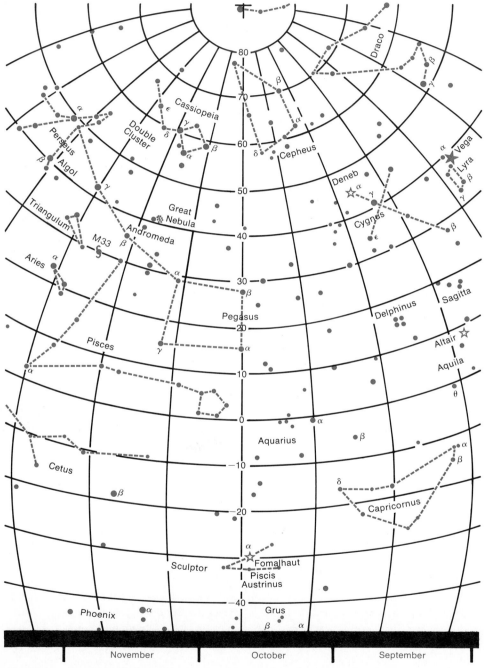

CHART 6. Autumn sky.

GLOSSARY

absolute magnitude. The magnitude that a star would appear to have if it were 10 parsecs away from us.

absolute zero. The lowest temperature that can be achieved.

absorption line. A narrow line of missing color in the continuous spectrum, indicating that the source of the continuous spectrum is viewed through a cooler gas. The wavelengths of all the absorption lines indicate the types of atoms, ions, or molecules in the cooler gas.

acceleration. The rate of change of velocity.

airglow. A faint, steady glowing of the Earth's atmosphere.

albedo. The reflectivity of a planet, satellite, asteroid, or other celestial body. It is the fraction of incident light that the body reflects back to space.

amplitude. The amount by which a physical quantity varies.

Ångstrom. A unit of length. There are 10^8 Ångstroms in a centimeter.

angular diameter. The angle subtended by the diameter of an object. The angular diameter of the Moon is half a degree.

angular momentum. A measure of the rotation or revolution of an object about an axis or a point. If a ball of mass m is twirled by a string of length l at a speed v, its angular momentum is $m \cdot l \cdot v$.

annual motion. The motion of the Sun through space in one year.

aphelion. The point in the path of an object orbiting the Sun where it is farthest from the Sun.

apparent magnitude. The magnitude that a star appears to have as viewed in the sky.

apparent motion. A motion that a celestial object appears to have because of a motion of the observer. The sky appears to rotate, because of the Earth's rotation.

association. A group of stars with a common origin. Associations differ from star clusters because of their low star densities and because they are not stable.

asteroid. One of the many small bodies which orbit the Sun, primarily between the orbits of Mars and Jupiter.

astronomical unit. Roughly, the average distance from the Earth to the Sun. Actually, a more complex modern definition makes the Earth's mean distance from the Sun 1.0000002 a.u.

atom. The smallest recognizable unit of a chemical element. An atom is composed of a nucleus and orbiting electrons.

atomic number. The number of protons in the nucleus of an atom.

atomic weight. The combined number of protons and neutrons in an atomic nucleus.

aurora. Light emitted by atoms and ions in the upper atmosphere, after they are excited by collisions with energetic particles from the magnetosphere.

Balmer series. A series of spectrum lines in hydrogen, due to atomic transitions which begin or end in the second energy level.

barred spiral galaxy. A spiral galaxy with a bar running across its nucleus. Spiral arms are attached to the ends of the bar.

barycenter. The center of mass of two bodies which are orbiting one another.

basalt. A type of rock found on both the Earth and the Moon, composed of crystallized lava.

big bang. The explosive event that began the expansion of the universe.

binary star. A system of two or more stars orbiting one another.

black body. A body which absorbs and reemits all the radiation which falls on it.

black hole. A body with such an intense gravitational field that nothing, including light, can escape from it.

Cassegrain focus. The focus of a reflecting telescope in which a convex secondary mirror reflects light to a focus behind the primary mirror through a hole in the center of the primary.

celestial equator. The circle on the sky midway between the north and the south celestial poles.

celestial poles. The points on the sky where the Earth's rotation axis, if extended in both directions, intersects the celestial sphere.

celestial sphere. The immense imaginary sphere surrounding the Earth; the sky.

center of mass. The point at which a two-body system would balance if connected by a rod.

Cepheid. One of several types of pulsating variable stars, which include δ Cephei-like classical Cepheids, RR Lyrae stars, and W Virginis stars.

chromosphere. The layer of the solar atmosphere above the photosphere, where the temperature increases from about 4,500° K to about 100,000° K.

circumpolar stars. Stars near the celestial poles which neither rise nor set as seen by a particular observer. The fraction of stars that are circumpolar increases with increasing latitude of the observer.

cluster of galaxies. A gravitationally bound group of galaxies.

color index. An index which indicates the color of a star. It is the difference between a star's magnitudes as measured in two different wavelength bands, using either photoelectric detectors and filters or different kinds of photographic emulsions.

color-magnitude diagram. A form of the Hertzsprung-Russell diagram with color as the horizontal scale and magnitude as the vertical scale. Points representing individual stars are plotted on the body of the diagram.

comet. A solar-system body that consists of a tiny, icy nucleus in which is imbedded meteoroid debris. As the ices of the nucleus sublime, the comet forms a coma and eventually a tail.

constellation. A group of stars in an area of the sky, supposedly in the form of an animal, person, or thing.

continuous spectrum. A spectrum which contains light of all colors.

convection. Bulk motion of matter which carries energy from one place to another.

corona. The very hot outer layer of the Sun's atmosphere.

cosmology. The study of the universe.

crescent. The phase of the Moon or a planet when less than half the visible disk is illuminated.

crust. The outermost, solid layer of the Earth.

dark nebula. A cloud of interstellar dust which obscures the light of distant stars.

deferent. A circle around which the center of an epicycle moves.

degenerate gas. A superdense gas that does not behave like an ideal gas when it is compressed.

density. The mass per unit volume of a body.

differential rotation. Rotation in which different parts of a body move at different angular rates.

diffuse nebula. A bright cloud of interstellar gas.

dispersion. The name of the phenomenon where refraction of light by a substance changes with wavelength.

distance modulus. The difference between apparent and absolute magnitude of an object such as a star or galaxy. It is a measure of the object's distance.

diurnal. Daily.

dome. A lunar surface feature, shaped like a dome.

Doppler effect. The shift in wavelength and frequency of light or sound due to a relative motion of the source and observer.

eclipse. The obscuration of all or part of one celestial body by another.

ecliptic. The apparent annual path of the Sun around the sky; the intersection of the plane of the Earth's orbit and the celestial sphere.

ecliptic plane. The plane of the Earth's orbit.

ecosphere. The volume of space around a star in which a hypothetical planet would have the conditions needed for life as we know it to exist.

electromagnetic radiation. Radiation in the radio, infrared, visible, ultraviolet, X-ray, and gamma-ray regions of the spectrum.

electron. A negatively charged elementary particle found orbiting the nuclei of atoms.

element. A chemical substance made up of atoms all of which have the same atomic number.

ellipse. A closed conic section with two axes of differing length.

elliptical galaxy. A galaxy consisting of old stars and little or no interstellar matter. The galaxy's image is circular or elliptical.

emission line. A bright spectrum line, caused by downward atomic transitions between two specific energy levels.

energy. The capacity to do something, such as moving a body or raising its temperature. Kinetic energy, for instance, exists in moving bodies.

energy level. Atoms exist in a series of states, each of which has a specific amount of energy associated with it. Each of these states is an energy level.

epicycle. The circular orbit of a planet, according to some old theories. The center of the epicycle travels around the circumference of another circle, called the deferent.

equation of state. The relation between temperature, pressure and density for a gas.

equinox. The points in the sky where the Sun crosses the equator. Day and night are of equal length at an equinox.

escape velocity. The smallest velocity an object needs in order to escape from the gravitational field of another body.

evening star. Venus, when it appears in the evening sky.

excitation. The process of raising an atom from one energy level to a higher energy level by adding energy.

exciting collision. A collision between an atom and another particle which excites the atom to a higher energy level.

extragalactic. Beyond our Galaxy.

eyepiece. A lens used to view the image produced by the objective of a telescope.

filament. A dark, linear feature on the Sun, seen in Hα.

flare. A sudden outburst on the Sun which brightens a portion of the chromosphere, and produces X-radiation, as well as ultraviolet and radio emission.

focal length. The distance from a lens or mirror to its focus.

focal point. The point where light rays that were initially parallel to the axis of a lens or mirror cross the optical axis after being refracted or reflected.

force. A push on a body that produces an acceleration.

frequency. The number of oscillations per second in a periodic phenomenon such as a light wave.

galactic cluster. *See* open cluster.

galactic rotation. Rotation of the Milky Way galaxy.

galaxy. A very large object in the universe, consisting of billions of stars, usually accompanied by gas and dust.

gibbous. The phase of the Moon or a planet when more than half but less than all the visible disk is illuminated.

globular cluster. A spherically shaped, rich cluster of stars.

globule. A small, dark cloud of interstellar dust, often seen silhouetted against bright nebulae.

granulation. The pattern of convective cells on the solar photosphere.

gravitation. The force by which masses attract one another.

greenhouse effect. The effect in planetary atmospheres that permits sunlight to warm the surface and lower layers of the atmosphere, but does not permit the heat to escape readily.

Gregorian calendar. A calendar introduced by Pope Gregory XIII. It differs from the Julian calendar (q.v.) in that century years are not leap years unless divisible by 400.

H⁻ ion. An ion consisting of a hydrogen atom to which is attached a second electron.

HI region. A cloud of neutral interstellar hydrogen.

HII region. A cloud of ionized interstellar hydrogen. Also called a bright diffuse nebula.

half-life. The time required for half of a mass of radioactive atoms to decay.

head of a comet. The nucleus and coma of the comet, combined.

heavy elements. All elements except hydrogen and helium.

heliocentric. Anything centered on the Sun.

helium flash. The explosive initial phase of helium burning in the degenerate core of a red-giant star.

Hertzsprung-Russell diagram. A plot of absolute magnitude versus spectral type for a group of stars. A variant is the color-magnitude diagram.

H-R diagram. The Hertzsprung-Russell diagram.

Hubble constant. The constant that relates the recession velocity of extragalactic objects to their distance from us. The constant is currently thought to be around 50 kilometers/second/Megaparsec.

hyperbola. An open curve formed by the intersection of a cone and a plane. The plane is parallel to the axis of the cone.

ice point. The temperature at which ice and water exist in equilibrium. The freezing point.

induction (or inductive reasoning). A form of reasoning that draws general conclusions from specific observations.

infrared. That part of the electromagnetic spectrum between visible and radio wavelengths. It covers roughly 7500 Ångstroms to 1 millimeter, although the division between radio and infrared at the long wavelengths is somewhat fuzzy.

integral (number). A whole number: 0, 1, 2, and so on.

interference. The process by which two light waves combine to produce a new wave whose amplitude is the sum **(constructive interference)** or difference **(destructive interference)** of the original amplitudes.

interstellar. Existing in the space between the stars of the Milky Way galaxy.

ion. An atom that has either fewer or more electrons than the number it would have if it were neutral. Ions can be either positively or negatively charged.

ionic bond. The name of the bond where a molecule is held together by an electrostatic attraction between two oppositely charged ions.

ionization. The removal of electrons from an atom to produce a positive ion.

irregular galaxy. A galaxy that has no particular shape or form.

isotope. Atoms with the same atomic number but different atomic masses are called isotopes. The difference is due to differing numbers of neutrons in the nuclei.

Jovian planet. The Jupiter-like planets: Jupiter, Saturn, Uranus, and Neptune.

Julian calendar. A calendar introduced by Julius Caesar. The 365-day normal year is made up of twelve months. Every fourth year is a leap (366-day) year.

kinetic energy. The energy (q.v.) contained in the motion of an object.

light curve. A graph on which the variations in magnitude of a variable star are plotted as a function of time.

limb. The edge of the observed disk of a celestial body.

limb-darkening. The decrease in brightness of the solar disk from its center to its limb. Other stars show evidence of limb-darkening, too.

line of nodes. A line across an orbit, connecting the two nodes (q.v.).

Local Group. A group of roughly 20 galaxies which includes the Milky Way galaxy and its neighbors.

luminosity. The total energy radiated by an object.

Magellanic Clouds. Two irregular companion galaxies of the Milky Way galaxy, seen in the southern-hemisphere sky.

magnetosphere. The magnetic field of a planet, along with the neutral and ionized gas it encloses.

magnitude. A measure of the brightness of a star.

main sequence. A diagonal line across the H–R diagram on which the majority of stars lie. The chief phase of the life of a star.

mantle. The part of the Earth's interior below the crust and outside the core.

mare. A flat plain on the lunar surface. Plural is **maria**.

mascon. An apparent concentration of mass under several lunar maria. They appear as irregularities in the Moon's gravitational field.

mass. A measure of the amount of matter in an object.

mass-luminosity relation. An empirical relation between the masses and the luminosities of main-sequence stars.

mean sun. An imaginary Sun that moves at a constant rate of about 1° per day around the celestial equator.

meridian. A circle around the sky which goes through the zenith and the north and south points on the horizon. Each observer has his own meridian.

mesosphere. The layer of the atmosphere just above the stratosphere.

Messier. An eighteenth-century astronomer who compiled a catalog of faint, fuzzy objects on the sky, to avoid confusing them with comets at future sightings. The objects, which turn out to be star clusters, nebulae, and galaxies, are referred to by the names M1, M2, . . . , M101.

meteor. A shooting star. The flash of light produced by a meteoroid burning up in the atmosphere.

meteor shower. A group of meteors, seen on one or more successive nights, that radiate from the same spot on the sky.

meteorite. A meteoroid that has passed through the atmosphere and reached the Earth's surface.

meteoroid. A small body in interplanetary space which produces a meteor as it streaks through the atmosphere, and becomes a meteorite if any of it reaches the ground.

mid-Atlantic ridge. A line of mountains along the floor of the Atlantic Ocean, where material wells up from the Earth's interior to form new crust.

Milky Way. The name of our Galaxy. The diffuse band of light around the celestial sphere.

molecule. A structure composed of two or more atoms bound together in a specific way.

momentum. A measure of the motion of a body. It is the product of the mass and velocity of the body.

morning star. Venus, when it appears in the morning sky.

nebula. A cloud of interstellar matter.

neutrino. A fundamental particle produced in certain nuclear reactions. The neutrino has no charge and no mass, but it does have momentum.

neutron. The neutral particle, with roughly the same mass as the proton, which occurs in most atomic nuclei.

neutron star. A star whose interior is an unbelievably dense neutron gas.

Newtonian focus. The focus in a reflecting telescope in which a flat mirror set at 45° to the light beam, at the upper end of the telescope tube, diverts the beam to the side of the tube for viewing.

NGC. The New General Catalog of star clusters, nebulae, and galaxies, compiled at the end of the nineteenth century. The objects listed in the catalog are referred to by the designations NGC 1, NGC 2, and so on to roughly NGC 7000.

node. One of two points on the orbit of the Moon, a planet, an asteroid, or other celestial body, where the plane of the orbit crosses the plane of the ecliptic.

normal. The line that is perpendicular to a surface at a specific point.

nova. A star that suddenly brightens by up to 10 magnitudes, then subsides to its previous brightness in the course of roughly a year.

nuclear fusion. The process by which two atomic nuclei fuse together to form a heavier nucleus, and release energy.

nucleus. The central part of an atom, a galaxy, or other object is frequently called its nucleus.

objective. The main light-gathering and image-forming element of a telescope. It may be a lens or a mirror.

open cluster. One of many irregularly shaped clusters of stars which lie in or near the plane of the Milky Way.

opposition. The name of the configuration in which a planet is opposite the Sun on the sky.

orbit. The closed path on which one body moves around another.

parabola. A curve formed by the intersection of a cone and a plane, when the plane is parallel to a line on the surface of the cone that is drawn through its apex.

parallax, geocentric. The apparent shift of an object (for example, Mars) relative to background objects (for example, stars) caused by the observer's motion on the rotating Earth.

parallax, heliocentric. The apparent shift of nearby stars relative to distant stars and galaxies caused by the observer's motion as the Earth revolves around its orbit.

parsec. The distance a star would be at if its heliocentric parallax were one second of arc. A parsec is 206,265 a.u., or 3×10^{13} kilometers.

partial eclipse. An eclipse (q.v.) in which the Sun, Moon, or other celestial body is not fully blocked from view.

penumbra. The brighter portion of a sunspot surrounding the umbra.

perihelion. The point in the path of an object orbiting the Sun where it is nearest to the Sun.

period-luminosity relation. An empirical relation between the brightness and the pulsation period of a Cepheid variable star.

perturbation. The gravitational effect of one small body on another that slightly changes its motion around the Sun. One body is said to **perturb** the motion of the other.

photoionization. The removal of an electron from an atom by its absorption of a sufficiently energetic photon.

photosphere. The visible surface of the Sun. The corresponding layer of other stars.

photon. A unit of electromagnetic radiation.

plage. A region of increased chromospheric emission on the Sun. An active region.

planetary nebula. A bright gas cloud, usually circular in outline, ejected from a star at a late stage in its evolution.

plate tectonics. The motion of large plates of the Earth's crust that is responsible for continental drift.

polarization. The process by which electromagnetic waves with planes of oscillation in all but one particular direction are removed from a light beam.

precession. A slow conical motion of the axis of a rotating body.

prime focus. The focus inside the tube of a large telescope where light from the primary mirror is focused without encountering a secondary mirror.

prominence. A cloud of glowing gas in the solar corona, seen extending beyond the limb of the Sun.

proper motion. The apparent change in the position of a star on the sky due to its real motion through space.

proton. The positively charged fundamental particle found in the nuclei of atoms.

protostar. A mass of gas on its way to becoming a main-sequence star.

pulsar. A celestial object that emits radiation in highly regular pulses; thought to be a rotating neutron star.

quasar. Also known as a **quasi-stellar radio source**. A radio-emitting object which looks like a star on the sky, but is probably very distant, judging from the redshifts of its spectrum lines.

radial velocity. The part of an object's velocity that is directly toward or away from an observer.

radiant. The spot on the sky from which meteors in a shower appear to radiate.

radioactivity. The breakup of an atomic nucleus with the emission of an energetic particle that could be dangerous to living tissue.

red giant. A large red star that has completed its main-sequence lifetime and has expanded to hundreds of times its initial size, and whose surface has cooled.

redshift. The Doppler shift of lines to longer wavelengths in the spectrum of an object that is moving away from the observer. Seen in the spectra of all but a few of the nearest galaxies.

reflection nebula. A cloud of interstellar dust illuminated by a nearby star.

reflector. A **reflecting telescope;** a telescope whose objective is a mirror.

refraction. The bending of a light beam when it passes from one medium to another.

refractor. A **refracting telescope;** a telescope whose objective is a lens.

relativity. A theory of the phenomena of time, space, motion, and gravitation developed by A. Einstein. The Special Theory of Relativity deals with the physics of steady motions; the General Theory of Relativity deals with accelerated motions and gravitation.

resolving power. The ability of an optical system to distinguish fine details in an image.

retrograde motion. The motion of a planet that is in a direction opposite to its normal motion.

rille. A lunar feature that appears to be a long, narrow, shallow channel.

satellite. Any body that orbits a larger body is a satellite of the larger body.

scalar. A quantity that is characterized by a single magnitude or size.

seeing. Blurring of an astronomical image due to turbulent motions in the Earth's atmosphere.

Seyfert galaxy. One of the many galaxies with a bright star-like nucleus, which is the site of some type of explosive event.

sidereal day. The time between two crossings of the upper branch of the meridian by the vernal-equinox point.

sidereal period. The rotation or revolution period of an object relative to the stars.

sidereal year. The Earth's period of revolution around the Sun relative to the stars.

solar motion. The motion of the Sun relative to nearby stars.

solar system. The Sun and its retinue of planets, satellites, asteroids, comets, meteoroids, and interplanetary material.

solar wind. The flow of gas from the Sun out through the solar system.

solstice. One of two points where the Sun is its maximum distance above or below the celestial equator.

space velocity. The total velocity of a star relative to the Sun.

spectral type. A pigeonhole in which a star is placed based on the observed characteristics of its spectrum. The spectral type is designated by a letter and number; for example, the spectral type of the Sun is G2.

spectrograph. An instrument for making a photographic record of a spectrum. The photograph is a **spectrogram**.

spectroscope. An instrument for direct visual viewing of a spectrum.

spectroscopic binary. A binary-star system in which two stars cannot be resolved, but whose orbital motions can be inferred from the Doppler shifts of their spectrum lines.

spiral galaxy. A galaxy with a spiral-like structure outside its nucleus.

steam point. The temperature at which water and steam exist in equilibrium. The boiling point.

stratosphere. The layer of the Earth's atmosphere directly above the tropopause.

substorm. An outburst in the magnetosphere, producing such effects as the rapid fluctuation of compass needles.

sunspot. A dark spot on the solar photosphere.

sunspot cycle. The 11-year cycle in which the number of sunspots increases and decreases.

supernova. A rare stellar explosion in which the entire outer envelope of the star is blown away in a violent outburst, leaving a dense core behind.

synchrotron radiation. Radiation emitted by very energetic electrons circling in a magnetic field.

synodic period. The time interval in which a planet moves from some initial Sun-Earth-planet configuration back to that same configuration.

tangential velocity. The velocity of a star across the line of sight.

temperature. The measure of the amount of heat energy per atom or molecule that a body contains.

terminator. The sunrise or sunset line on the Moon or on an airless planet.

terrestrial planet. The earth-like planets: Mercury, Venus, Earth, Mars, and perhaps Pluto.

ultraviolet. The region of the electromagnetic spectrum between the X-ray and visible wavelengths.

umbra. The dark, inner part of a sunspot.

universe. Everything; the cosmos.

variable star. A star whose brightness changes.

vector. A quantity that is characterized by a magnitude and a direction.

velocity of escape. *See* escape velocity.

vernal equinox. The intersection point of the celestial equator and the ecliptic which the Sun crosses from south to north around March 21.

visual binary. A binary-star system in which both components can be seen.

wavelength. The distance between successive waves in a wave train.

weight. The force of gravity on an object.

white dwarf. A star, below the main sequence in the H-R diagram, which is small and dense. The energy it radiates comes from the cooling of its interior.

X-ray binary A binary-star system that emits X-rays.

zenith. The point on the celestial sphere that is directly overhead.

zero-age main sequence. The line on the H-R diagram occupied by stars that have just evolved to the main sequence.

zodiac. The twelve constellations around the ecliptic.

INDEX

Absolute magnitude. *See* Magnitudes
Absolute zero, 166
Absorption spectrum, 162, 174
Acceleration, 49, 50
Achromatic lens, 75, 76
Active regions. *See* Sun, active regions
Adams, John Couch, 55
Airglow, 86
Albedo, 232
Algol, 349, 351
Altitude, 14
Amplitude, 61
Andromeda galaxy, 86, 380–381, 432, 447
Ångstrom, 64
Angular momentum, 53
Aperture synthesis, 90
Aristotle, 15
Associations, 365–366
Asteroids, 268
    discovery, 269
    orbits, 269–273
    physical makeup, 273
    Trojan, 271–272
Astronomical unit, 97, 98
Astronomy
    definition, 3
    infrared, 92
    radio, 89
    short-wavelength, 93
Atmosphere, escape, 306
Atomic nucleus, 112
    decay of, 112–113
Atoms, structure, 169–174
Aurora borealis, 312
Autumnal equinox, 101

Baade, Walter, 388, 450
Babcock, H. W., 220

Barnard's star, 328
Barycenter, 98
Basalts, 136
Bayer, Johann, 321
Betelgeuse, 344
Big bang, 460–461
Big Dipper, on four different dates, 19
Binary stars, 345
    center of mass, 347
    eclipsing, 349–351
    physical, 346
    spectroscopic, 348–349
    visual, 345–347
Black body, 163–166, 167–168
Black dwarfs. *See* Stars, black dwarfs
Black holes, 423–424, 456
Bode's law. *See* Titius-Bode Law
Bohr, N., 170
Brahe, Tycho. *See* Tycho Brahe
Bremsstrahlung, 377
Brownian motion, 177
Bruno, Giordano, 41

Caesar, Julius, 28
Calendar, 28–29
Cannon, A. J., 335
Cassegrain focus, 81, 82
Catadioptric telescopes, 83–84
Celestial equator, 99, 323
Celestial meridian, 25–26
Celestial pole, 14
    change of altitude with position, 13–14
Celestial sphere, 11, 15. *See also* Sky
Centaurus A radio source, 450, 452
Center of mass, 98, 99, 347
Cepheid variable stars, 353–356, 380, 388, 432
    pulsation, 354
Chromatic aberration, 75, 76

Chromosphere. *See* Sun, chromosphere
Chromospheric network. *See* Sun, chromospheric network
Civilizations, number in Galaxy, 476–477
Clark, Alvin, 405
Clusters, star. *See* Star clusters
Clusters of galaxies, 443–445
Color, relationship between color, wavelength, and frequency, 69
Color index, 340
Color-magnitude diagram, 361
Comets, 281–291
  coma, 283
  discovery, 281–282
  long-period, 283
  names, 284–285
  nature of, 285–289
  nucleus, 283
  orbits, 283–284
  origin, 290–291
  short-period, 283, 289
  tail, 283, 288
Conic section, 43
Constellations, 10, 13, 15–20
  Cygnus, 16
  myths, 15–17
Continental drift. *See* Plate tectonics
Convection, 206, 395, 396
Convection zone, 206
Copernican universe. *See* Cosmology, Copernican
Copernicus, Nicholas, 36–41
Corona. *See* Sun, corona
Cosmological models, 460–465
Cosmological principle, 460
Cosmology, 32, 458–465
  ancient, 32–34
  Copernican, 36–41
  Galileo's views, 47
  Newtonian, 54
  Ptolemaic, 34–36
Coudé focus, 83
Crab Nebula, 415, 416, 417, 420, 450
Cygnus A radio source, 450, 452

Deceleration, 50
Declination, 323
Deferent, 34
Degenerate gas, 400
Deimos. *See* Mars, moons
Descartes, René, 41
Deslandres, H., 205
Diffraction, 73
Dispersion, 69
Distance modulus, 354
Doppler effect, 177–179, 329, 447
Dwarf novae, 413–414
Dynodes, 88

Earth, 95–120
  age, 112–113
  albedo, 232
  asthenosphere, 114
  atmosphere, 301–304
  core, 115
  crust, 114; chemistry, 115–116
  density, 109, 111
  interior, 114–116
  lithosphere, 114
  magnetic field, 309
  mantle, 114
  mass, 110–112
  mesosphere, 114
  motion, 37, 48
  orbit, 96–98
  path of north celestial pole on sky, 107
  precession, 106–108
  revolution, 96–102
  rotation, 47, 102
  rotation axis, 99
  shape, 25, 103–104
  size, 108–110
Earthquake, 114
Eclipse seasons, 134
Eclipsing binary stars. *See* Binary stars, eclipsing
Ecliptic, 21
  plane of, 99
  relationship to celestial equator, 101
Einstein, Albert, 56, 87, 193
Electromagnetic spectrum, 69–70. *See also* Light
Electron, 87
Ellipse, 43
  focus, 43
  how to draw, 43
Emission lines, 159
Energy transport
  convective, 395
  radiative, 395
Epicycle, 34
Equation of state, 176
Equator, terrestrial, 13
Equatorial mounts, 85
Eratosthenes, 109
Escape velocity, 306
Evening star. *See* Venus, evening star
Event horizon, 423
Excitation, 174–177
  collisional, 175
Eyepiece, 72

Faculae. *See* Sun, faculae
Falling stars. *See* Meteors
Filaments. *See* Sun, filaments
Filtergrams, 206
Flamsteed, John, 321
Flamsteed's numbers, 321

Flares. *See* Sun, flares
Focal length, 70
Focal point, 68, 70
Force, 49, 50
   gravitational, 52
Fossil magnetism, 119
Foucault, Jean, 102
Foucault pendulum, 102–103

Galactic clusters. *See* Open clusters
Galaxies, 432–447
   classification, 433–434, 435
   clusters. *See* Clusters of galaxies
   distances, 446–447
   dust in, 436–438
   dwarf, 441, 442
   elliptical, 432, 434–435, 440
   gas in, 436, 441
   globular clusters, 440
   irregular, 432, 435, 440–441
   radio, 449–453
   rotation, 438, 439–440
   spiral, 432, 433–434, 436, 440
   spiral arms, 436, 438–439
   stellar populations, 440
Galaxy (Milky Way), 320, 378–390
   center, 389–390, 457
   mass, 385
   position of Sun, 379
   rotation, 383–384
   size, 379
   spiral structure, 385–387
Galileo, 46
Gamma rays, 93
Gaposchkin, Cecilia Payne, 385
Gauss, Carl Friedrich, 269
Geomagnetic substorms, 313
Globular clusters, 364–365, 379, 402
   distribution, 379
Globules, 426
Gravitation, universal law of, 53
Gravitational constant, 111
Gravitational differentiation, 280
Gravitational potential, 423
Gravity, 52
Greenhouse effect, 237–238, 243
Greenwich mean time, 27
Gregorian calendar, 29
Gregory XIII, Pope, 29
Gyroscope, 106

H⁻ ion, 198, 199
HI regions, 371–374, 375
   temperature, 377
HII regions, 375–378, 385
   density, 375
   temperature, 375

Hale, G. E., 201–202, 205
Half-life, 112
Hall, Asaph, 250
Halley, Edmund, 49, 54, 328
Halley's comet, 54, 282
Helium burning, 401
Helium flash, 401
Helmholtz contraction, 192
Henry Draper Catalogue, 335
Herbig-Haro objects, 427–428
Herschel, William, 54, 55, 378
Hertzsprung-Russell diagram, 341–343, 363
Hipparchus, 34, 324
Horizon, 11, 13
Hoyle, Fred, 465
H-R diagram. *See* Hertzsprung-Russell diagram
Hubble, Edwin P., 380, 432
Hubble constant, 448, 462
Hyades open cluster, 362
Hydrogen
   Balmer series, 171
   energy-level diagram, 172
   Lyman series, 172
   spectrum, 159
   structure, 169–172
Hydrogen convection zone, 198
Hydrostatic equilibrium, 393–394
Hyperbola, 43
Hyperfine structure, 373

Ice point, 166
Image, 70
Inductive reasoning, 43
Interference, 63
   constructive, 62
   destructive, 62
   fringes, 64
Interstellar absorption lines, 371, 372, 373
Interstellar communication, 477–481
Interstellar dust, 362, 366–370
Interstellar gas, 370–378
   composition, 378
Interstellar matter, 366–378
Interstellar molecules, 373, 374, 377
Interstellar reddening, 368
Interstellar travel, 481–482
Ionic bond, 174
Ionization, 174–177
Ionosphere, 303
Ions, 176
Isotope, 112

Jansky, Karl, 89
Jolly, P. Von, 110
Julian calendar, 28
Jupiter, 252–257
   appearance, 252

Jupiter (*continued*)
  atmosphere, 254–256
  belts, 252
  interior, 256–257
  magnetosphere, 311
  mass, 252
  moons, 47, 253
  red spot, 254
  shape, 253–254
  size, 254
  temperature, 252
  zones, 252

Kepler, Johannes, 42
  laws of planetary motion, 43–46
Kilogram, 51
Kinetic energy, 176
Kirchoff, G., 161
Kirkwood, D., 269
Kirkwood's gaps, 269, 271

Leap year, 28, 29
Leighton, R. B., 220
Lens, 70–72
Leverrier, Urbain, 55, 56
Life, 2
  chemistry, 469–471
  origin, 469–476
  requirements, 472–475
  in the universe, 467–482
Life-bearing planets, number of, 475–476
Light
  brightness, 87
  detection, 86–88
  frequency, 66, 69
  interference, 62
  inverse-square law, 93–94
  nature of, 60–69
  photoelectric detectors, 87–88
  velocity, 64–66
Limb-darkening. *See* Sun, limb-darkening
Lindblad, Bertil, 383
Lippershey, Jan, 46
Little Dipper, 11
Local apparent solar time, 26
Local Group, 441, 447
Loop prominences. *See* Sun, loop prominences

Maffei 1 and 2, 441
Magellanic clouds, 354, 435, 436, 440, 441
Magnetic field, 308. *See also* Earth, magnetic field
  line of force, 308
Magnetopause, 310
Magnetosphere, 308–311
Magnitudes, 324–325
  absolute, 333–334
  apparent, 334
  photographic, 339

  photovisual, 339
  UBV, 340
Main sequence, 341. *See also* Stars, main sequence
Mars, 238–251
  atmosphere, 239, 242
  canals, 239
  moons, 250–251
  polar caps, 239
  retrograde motion, 24, 39
  surface features, 244–249
  tilt of axis, 239
  water on, 242–244, 248
Mascons. *See* Moon, mascons
Mass, 50, 51
Mass-luminosity relation, 351–352
Maunder minimum, 315
Mean sun, 26
Mercury, 226–230
  apparent motion, 23
  interior, 230
  magnetosphere, 311
  phases, 226
  precession of orbit, 57
  maximum elongation, 39
  revolution period, 226
  rotation, 226–227
  surface, 227–230
  temperature, 227
  transit, 232
Meridian, celestial. *See* Celestial meridian
Meteor showers, 274–275
  radiant, 274
Meteorites, 273, 277–279
  impacts on Earth, 276–277
Meteoroid streams, 275
Meteoroids, 273
  origin, 280–281
Meteors, 273
  sporadic, 275–276
Mid-Atlantic ridge, 116
Midocean ridges, 119
Milky Way, 10, 47. *See also* Galaxy (Milky Way)
Minor planets. *See* Asteroids
Monochromators, 206
Moon, 22, 121–155
  Alpine Valley, 142
  binocular view, 146
  craters, 135, 138–141; origin of, 150–153
  distance, 128
  domes, 141
  eclipses, 132–135
  far side, 147–148
  first quarter, 124
  full, 124
  gibbous, 124
  hourly motion, 124
  lunar highlands, 135
  "man in the Moon," 136

map, 136
maria, 135; origin of, 154–155
mascons, 130
mass, 129–130
mountains, 142
new, 124
"Old Moon in the new Moon's arms," 123
orbit, 130–131
origin of surface features, 149–155
phases, 123–128
rilles, 144
ringed basins, 135, 147
rising of, 124–127
rotation, 131–132
shape, 129
sidereal period, 131
size, 129
surface, 135–148
synodic period, 131
terminator, 146
"woman in the Moon," 136
wrinkle ridges, 135
Morgan, W. W., 385
Morning star. See Venus, morning star
Mythology, 2

Nebula
dark, 370
reflection, 369
Neptune, 261, 262
atmosphere, 261
discovery, 55–56
size, 261
Neutrinos, 196
Neutron stars, 420–421, 423
Neutrons, 112
Newton, Isaac, 49
laws of motion, 49, 52
Newtonian focus, 81, 82
Nodes, 134
line of, 134
North celestial pole, 11, 13
North Star, 11
Northern lights. See Aurora borealis
Nova, 410–414
light curve, 412
maximum brightness, 413
origin, 413
Nuclear fusion, 193
Nucleosynthesis, explosive, 414

Objective lens, 72
Oceanic trenches, 118
Olbers' Paradox, 458–460
Oort, Jan H., 383
Open clusters, 360–363, 402
distribution, 379
sizes, 367

Oppenheimer, J. Robert, 421
Opposition, 23
Ozone, 302, 303, 304

P-wave earthquake waves, 114, 115
Pangaea, 116
Parabola, 43
Parallax, 326–327, 329
geocentric, 97
statistical, 333
Parsec, 327
Perfect cosmological principle, 460
Period-luminosity relation, 354, 379, 380
Periodic table, 173, 490–492
Perturbation, 54
Peru-Chile trench, 118
Phobos. See Mars, moons
Photocathode, 88
Photoelectric effect, 87
Photoexcitation, 175
Photographic plates, 86
Photography, 86
Photomultiplier tube, 88
Photons, 60, 87
Piazzi, G., 269
Plages. See Sun, plages
Planck, Max, 167
Planck's law, 167
Planetary nebulae, 407–410, 414
central stars, 410
Planets, 22–25, 224–264. See also specific planets
inferior, 231
Jovian, 225
motion, 43
orbits, 43
sizes, 224–225
terrestrial, 225
Plasma, 308
Plate tectonics, 116–119
Pleiades, 360, 368
Pluto, 262–264
Polar axis, 85
Polaris. See North Star
Polarization, 66–67
Polaroid filter, 66
Population I and II stars. See Stellar populations
Power-of-ten notation, 487–488
Precession of the equinoxes. See Earth, precession
Prime focus, 81, 82
Prime meridian, 27
Primeval egg, 461
Prominences. See Sun, prominences
Proper motion, 328, 329
Protons, 112
Protostars, 425, 426, 427
Ptolemaic system. See Cosmology, Ptolemaic
Ptolemy, 34
Pulsars, 420–423

Pulsars (*continued*)
  "glitches," 422–423
  theory of, 421–422

Quantum mechanics, 169–170
Quasars, 453–458
  distances, 454
  energy output, 455
  fluctuations, 455
  nature, 456–458
  sizes, 455, 456
  structure, 453

Radial velocity, 329–331
Radiation
  extreme ultraviolet, 88
  infrared, 88, 92
  radio, 88
Radiation laws, 167–168
Radicals, 287
Radio astronomy, 89
Radio interferometers, 90, 91, 92, 450, 453, 455
Radio telescopes. *See* Telescopes, radio
Radioactivity, 112
Red-giant stars. *See* Stars, red giants
Redshift, 447. *See also* Universe, expansion
Redshift-magnitude relation, 462
Reflecting telescopes. *See* Telescopes, reflecting
Refracting telescopes. *See* Telescopes, refracting
Refraction, 68, 69
Relativity, general theory of, 57
Retrograde motion, 23–24, 35, 38, 41
Right ascension, 324
Ritchey-Chrétien system, 81
Römer, Ole, 65
RR Lyrae stars, 356, 364, 379, 402

S waves, 114
Saturn, 257–261
  atmosphere, 261
  density, 257
  mass, 257
  rings, 258–261
  size, 257
  temperature, 258
Scalar, 50
Schiaparelli, G., 239
Schmidt, Maarten, 454
Schmidt camera, 84
Schwabe, H., 202
Scientific method, 4
Sea-floor spreading, 118
Seasons, 99–102
  lag of, 101
Seeing, 74
Seyfert galaxies, 457
Shapley, Harlow, 379
Shooting stars. *See* Meteors

Sidereal revolution period, 40
Sidereal time, 28
Sirius, companion of, 404–405
Sky, 10–29
  annual motion, 18
  apparent motion, 11, 13, 14
  as calendar, 10
  diurnal motion, 11
  maps, 494–500
  night, 10
  positions on, 323–324
  why it is blue, 305
Slipher, V. M., 447
Sodium, spectrum of, 160
Solar activity, 205
  theory of, 220–221
Solar disk, 187
Solar flares, 314–315
Solar motion, 332–333
  apex, 332
Solar power output, 191–194
Solar radiation, 299–301
Solar system
  origin, 292–294
  regularities, 267–268
  scale, 266–267
Solar-terrestrial relations, 316–317
Solar variability, 301
Solar wind, 214–215, 307–308
Solar XUV, 299–300. *See also* Sun
Sound, 62
Space velocity, 331
Spacecraft, lunar landing, 122
Speckle interferometry, 344
Spectra, 334–339
  classes, 335–339
Spectral line, 159
Spectrograph, 334
Spectroheliograms, 206
  calcium, 207, 209, 210
  hydrogen, 207, 208, 209, 210
Spectrophotometer, 163
Spectroscope, 158
Spectrum, continuous, 158, 163–166
Spectrum, electromagnetic, 68–70
Speed, 49
Spicules, *See* Sun, spicules
Star clusters, 360–366
  ages, 401–403
  color-magnitude diagram, 402
Stars. *See also* Binary stars and specific star names
  birth, 424–428
  black dwarfs, 403
  brightness, 324–326
  colors, 339–340
  distances, 326–327
  energy generation, 396
  evolution, 398–428; to red giant, 399–401

giants, 342, 352
internal structure, 393–398
late stages of evolution, 403
main sequence, 342, 352, 396–397
main-sequence lifetime, 398–399
masses, 345–347
motions, 328–331
names, 321–323
red giants, 401
sizes, 343–345
supergiants, 342, 352
white dwarfs, 342, 352, 403, 404–407, 414
Steam point, 166
Stellar populations, 388–389, 440, 441
Stephan's Quintet, 443
Subduction zones, 119
Sublime, 286
Substorms. *See* Geomagnetic substorms
Sudden ionospheric disturbance, 314
Summer solstice, 101
Sun, 22, 181–221, 336
 active regions, 209
 age, 398
 annual motion, 20–22
 atmosphere, 199
 chromosphere, 206–211
 chromospheric network, 209
 convection zone, 198
 corona, 188, 211–217
 differential rotation, 191
 energy transport in, 197
 eruptive prominences, 214
 faculae, 187
 filaments, 209
 flares, 218–220
 granulation, 186
 how to view, 184
 limb-darkening, 187
 loop prominences, 215
 model of, 196, 398
 photosphere, 186, 206
 plages, 209
 prominences, 209, 212
 rotation, 189–191
 spectrum, 160, 200
 spicules, 209, 210
 structure, 194
 supergranulation, 209
 transient, coronal, 212
 transition region, 211, 216
Sunspot cycle, 201–205
Sunspots, 183, 185, 201–205
 magnetic polarity, 204
 penumbra, 183
 umbra, 183
Supergiants. *See* Stars, supergiants
Supergranulation. *See* Sun, supergranulation
Supernova, 414–419

Kepler's 417
maximum brightness, 417
origin, 419
Tycho's 417
Synchrotron radiation, 389, 418–419
Synodic revolution period, 40

T Tauri stars, 294, 426, 427
Tangential velocity, 329
Telescopes, 46, 70–85
 catadioptric, 83–84
 light-gathering power, 74–75
 parameters of, 72
 radio, 89–92
 reflecting, 75, 77–84
 refracting, 75
 resolving power, 73–74
 Schmidt camera, 84
 types of, 75–84
 Wolter-type, 93
Telescopes, famous
 Hale 5-meter reflector, 74, 81, 82, 83, 85
 Hooker 2.5-meter reflector, 380
 Lick 3-meter reflector, 77, 78, 79
 Mayall 4-meter reflector, 79, 80
 Palomar Schmidt, 84
 Soviet 6-meter reflector, 79
 Yerkes 1-meter refractor, 76, 77
Temperature, 166–167
Thermal equilibrium, 394–395
3° K background radiation, 461
Tides, 104–106
 interval between, 106
Time, change with locale, 26–27
Titius-Bode Law, 267–268, 294
Transition region. *See* Sun, transition region
Transverse wave, 60, 61
Triangulation, 97
Tropical year, 28
Trumpler, Robert J., 366
Turbulent convection, 197, 206
21-cm line, 373–374, 375, 377, 386
Tycho Brahe, 41–42

Universal time, 27
Universe
 closed, 463, 464
 critical density, 462
 distance scale, 446–447
 dynamics, 461–462
 Egyptian, 2
 expansion, 447–449
 flat, 464
 open, 464
 shape, 463–464
 steady-state, 464–465
Uranium, 112
Uranometria, 321

Uranus, 261–262
  atmosphere, 261
  discovery, 54
  rings, 262
  size, 261

Van Allen belts, 310
Vector, 50
Velocity, 49
Velocity of escape. *See* Escape velocity
Venus, 231–238
  albedo, 232
  apparent motion, 23
  atmosphere, 233, 235–237
  evening star, 231
  maximum elongation, 39
  morning star, 231
  phases, 47, 48, 233
  rotation, 233
  surface, 233–234
  synodic period, 231
  temperature, 233, 237
  transit, 232
Vernal equinox, 101, 324
Virgo A radio source, 450
Visitors from space, 482

W Virginis stars, 356, 388
Wave
  electromagnetic, 61
  longitudinal, 62
  ocean, 60
Wavelength, 61, 69
Wegener, Alfred, 116
Weight, 52
White dwarfs. *See* Stars, white dwarfs
Winter solstice, 101

X-ray binary stars, 423
X-ray sources, 389, 456
X-rays, 93

Young stellar objects, 426–428
Young, Thomas, double-slit experiment, 62, 87

Zeeman effect, 202
Zenith, 11, 13, 15
  changing position among stars, 25
Zero-Age Main Sequence, 397
Zodiac, 21
Zone time, 27